Space and Society

Series editor

Douglas A. Vakoch, SETI Institute, Mountain View, CA, USA
and California Institute of Integral Studies, San Francisco, CA, USA

More information about this series at http://www.springer.com/series/11929

Michael Ashkenazi

What We Know About Extraterrestrial Intelligence

Foundations of Xenology

Springer

Michael Ashkenazi
Villalonga
Spain

ISSN 2199-3882 ISSN 2199-3890 (electronic)
Space and Society
ISBN 978-3-319-44455-0 ISBN 978-3-319-44456-7 (eBook)
DOI 10.1007/978-3-319-44456-7

Library of Congress Control Number: 2016948280

Cover design: Paul Duffield

Printed on acid-free paper

This Springer imprint is published by Springer Nature
The registered company is Springer International Publishing AG
The registered company address is: Gewerbestrasse 11, 6330 Cham, Switzerland

This book is dedicated to three people whose personality and support are the real foundations of this book:

My father,

who would have loved to read this book had he lived, and in the hope his grandchildren will have the chance of testing the ideas laid down here in practice.

Stanley R. Jacob,

my brother-in-law, in whose home in Palo Alto this book was conceived and partly gestated, and whose mischievous sense of humor helped keep me going.

and

John "JB" Billingham,

formerly of the NASA SETI project, who encouraged me to proceed with this book, and whose incisive point of view is reflected to some degree here.

Preface

In 1961, a young radio astronomer, Frank Drake, proposed a simple formula—now known by his name—for estimating the number of detectable extraterrestrial civilizations in the Milky Way galaxy. The formula is based on what was then a large number of unknowns: the average rate of star formation per year in our galaxy, the fraction of stars with planets, the fraction of planets in the life zone (where water is neither permanently frozen nor permanently a vapor), and so on. In the five decades since the formula was first published, the knowledge base has grown: Astronomers have answers to the first few questions, and the question of the existence of life-supporting planets is only a matter of time.

The title of this book is deliberately presumptive and provocative. Over the years of working on SETI, I have been bothered by the fact that most of the speculation about ETI have been done by physical scientists, with a handful of social scientists providing input. There are two problems with this setup. First, physical scientists live in an intellectual world where fairly rigid rules hold sway. These are demonstrable by experiment and formed on mathematical bases. Social scientists, in contrast, have not (yet) been able to prove their contentions mathematically (for the large part), have few theories bolstered by experimentation (less true of some such as psychologists), and are beset by the subjects of their inquiries questioning their methods, assumptions, and conclusions. Yet, after many years of researching and teaching social sciences, I know, and am prepared to state, that certain social phenomena are, in a sense, inevitable. That is, they represent what physical scientists accept as a given: fixed scientific laws.

I argue throughout this book that certain of these 'laws' make it possible to assemble a bare, perhaps weak, set of general knowledge about ETI. Some of these are based on human experiences, which *may* make them invalid for ETI, unless it is possible to peel back the human genotypical elements and look beneath them to realities based, as physical sciences are, on the generalized laws of parsimony, evolution, communication, and other features. Consider, for example, one thing we all know about ETI, which is absolutely fundamental: No researcher has ever denied, nor, to my mind, is it possible to deny that ETI are social beings. It is, I would argue, impossible to have an intelligent species consisting of one member, let alone an empire or civilization. As such, we know (until and unless proven or demonstrated otherwise, as in all scientific thinking) that any ETI we come into contact with will be members of a society. How that society works, what its beliefs, activities, and interests are, we cannot know. Yet we can know, with a great deal of certainty certain things about these ETI from the very fact that they are social beings. This is the intent of this book: To identify what these realties are, and what their implications might be. As a consequence, the reader will be exposed to much discussion of the nuts and bolts of *human* behaviors, as we tease out the universals of behavior and society.

The term 'xenology' used throughout this book is an example of convergent evolution. I originally thought of it (and thought it was my own invention, to quote the White Knight) while on sabbatical at the University of Calgary in Canada. Subsequently, during research, I found that the term had been proposed by Freitas (Freitas, ND) and, indeed, defended by him as a scientific term for the study of non-human intelligent species. I make no apologies, though I do acknowledge Freitas's precedence. This book is about convergent evolution: about the convergence of species into intelligence, and what it means for the human species, and perhaps about the convergence of ideas.

Villalonga, Spain Michael Ashkenazi

Acknowledgments

Many people encouraged me to write this book. Two are foremost. My wife Jeanne, who is used to my many strange ideas, thought nothing of encouraging this one. She also bore more or less patiently my apparently odd ideas, flights of fancy, and frequent absences from home as I attended conferences and seminars where the ideas were examined and tested. No less importantly, she was a frequent fount of well-thought, original ideas which stimulated my thinking, notably when I was in an intellectual chasm, trying to bridge two apparently opposite concepts. All my love, and many, many thanks.

John Billingham, JB to some, was the first one to seriously encourage my interests in the realms of SETI. He was also the first to encourage me to set down my speculations in serious scientific terms. As chief of the now defunct NASA SETI project, he was my host at NASA, as well as at numerous other conferences and workshops. He has always been at one and the same time a good friend and an excellent critic, bringing the cold light of reason—scientific, technical, and political—to flights of fancy. We, the SETI community, myself included, miss him greatly.

Many others are owed thanks. Frank Drake, one of the greatest SETI pioneers encouraged me by example, and by helping me to attend fora where some of the ideas in this book were presented. Frank and his wife Amahl hosted an evening at their home in which ideas were bounced about, and some caught my fancy. Thank you both. Laurance Doyle at NASA first got me interested in planetology and its implications for xenology. My brother- and sister in-law, Stanley and Linda Jacob of Palo Alto hosted me often during periods of research with NASA and with the SETI Institute. They were intrigued by the ideas, and always welcoming and helpful. *Salamat* to both of you. Poul Anderson and his wife Karen led a fascinating informal seminar during one of the CASETI meetings that remains in my memory, and for which I am grateful. The members of NASA's CASETI seminars were a source of inspiration and ideas. Carolyn, my research assistant at BGU ran down obscure references and, notwithstanding healthy skepticism about the entire topic, provided an excellent brake on my imagination. Dani Censor generously gave me access to his paper on truly exotic sensoria and spent time explaining those elements obscure to a layman such as myself. A number of people read parts of the manuscript during its preparation or discussed elements of it and critiqued specific ideas: Michael Green, Dan Birenbaum, Yaakov Haimovic, Ayal Green, Chris Koegler, and Andreas Heinemann-Grueder. Thank you all.

As always, I want to thank Keith Blount and his team for *Scrivener*, Jon Ashwell for *Bookends*, and the team of *DEVONThink* for their products: Without those constantly used tools, this book would have taken far longer and may never have come into being.

I was encouraged, perversely, by the incredulity and stuffiness of the staff at Ben Gurion University in Beersheva. The reaction to their lack of imagination and intolerance for strange ideas sustained me during writing.

Finally, I want to thank Springer for accepting this book for publication.

All errors of fact and speculation are of course my own and no one can be blamed for them but myself.

Contents

Chapter 1
Introduction and Objectives

As the human species progresses in knowledge, new puzzles and questions occur. Yet some very old questions persist. What, for example are the relative influences of nature and nurture on humans and their society? How do we define the moral good? How can we manage our relationships with other people?

One question that is very old, and which is rapidly becoming more interesting and less speculative is whether we are alone in the universe. The human species, to date, has only had one standard—the human species itself—to measure itself against. We do not *know* (except inductively) that our moral, ethical, social, physical or intellectual standards are universal, since we have never had an opportunity to measure them against those of another intelligence species.

In this book I am going to assume that ETI do exist. As a consequence, I am impelled to ask "How do we deal with the neighbors?" which constitutes the major issue that this book addresses.

Beyond the issue's intellectual interest, the possibility of other intelligent species has a number of benefits and dangers. In the positive column we should list learning new sciences, arts and ways of doing things; the possibility of testing our assumptions against another perspective; greater knowledge of the universe; and even the possibility of economic benefits as we trade with new partners. In the negative column we ought to put the risk of being overwhelmed—intellectually, socially, economically, even militarily—by 'people' who have no empathy nor understanding of our condition. As a species we have been at both ends of the giving and receiving of the cruelty of indifference, as those from 'superior' civilizations oppressed, murdered, and wiped out those they considered their inferior. Many, including intellectual leaders, fear the same could happen to us if we meet an ETI civilization.

We need to add two issues to the mix. First, so far, there has been no evidence of the existence of ETI. Flying saucers, ecstatic visions, and imagined sightings to the contrary, there is no evidence. Zip. Nothing. As a purely scientific endeavor, does this mean there is nothing out there, or that out instruments and intellectuals tools are not up to scratch? Time, and the development of scientific instruments and

M. Ashkenazi, *What We Know About Extraterrestrial Intelligence*,
Space and Society, DOI 10.1007/978-3-319-44456-7_1

reasoning will tell. It should be pointed out that for those of us who hope there is someone out there, there is progress. The elements of the Green Bank/Drake equation which forms the basis of SETI (Search for Extra-Terrestrial Intelligence) about which more anon, have gradually proven to be very real and evidence is accumulating for, for instance, habitable worlds elsewhere in the Galaxy.

Second, the physics of our universe make the study of ETI very complex in many ways. Given what we know of the real universe, for example, it is highly unlikely we (or they) could ever visit us in person. Even electronic communication is likely to be punctuated by years, centuries, even millennia of silence as each side waits for the electronic message to crawl its way through Einsteinian space. So, really, we need not fear a fleet of interstellar warships appearing suddenly in our skies.

So, as a confirmed opsimist, I prefer to believe that (a) the answer to the question asked above is probably, no, we are unlikely to be the only ones in the universe; and that (b) notwithstanding the nature of humans, the poverty of our instruments, the physical nature of the universe, *and* the lack of any evidence of ETI, we need to think very carefully about what happens when we and a non-human intelligent species manage to, somehow, communicate.

Methodologically, this book is based upon a controversial premise. I strongly believe that there are certain social laws, which, while unproven mathematically so far, are as certain and to some degree predictable as natural laws we are familiar with. This is a very strong claim, and many social scientists will dispute it vigorously. Nevertheless, as shall be shown subsequently, certain social features that human society displays are very likely to be true for any social intelligent species: not the least that 'intelligent" and 'social' are absolutely contingent.

1.1 A Social Scientist's Perspective

My scientific background is in the social sciences, and though I have a layman's knowledge of the 'hard' sciences—physics, chemistry, biology, astronomy—my perspective throughout this book is that of someone interested in social behavior and its consequences. Social scientists have labored for generations, under a problem that has not affected their physical science colleagues. Except in certain cases, the data on which they work, has not easily been amenable to quantification. In fact, as more than one physical scientist has told me, it hardly qualifies as data at all. This has led many behavioral/social scientists to despair of matching the accuracy and elegance of physical science creations, and has led physical scientists, quite often, to be suspicious, sometimes even contemptuous, of the fumbling attempts of psychologists, anthropologists, and economists to be 'scientific', though the degree to which such accusations are justified varies from case to case. In the behavioral scientist's defense, it must be said that no physical scientist of my acquaintance has ever been faced with a subject of an experiment offering comments on the procedure, quoting theoretical literature, or wondering, as happened to me, whether I "… intend to use a Malinowskian theoretical frame for analysis of what we were doing?".

The behaviors of people, consistent, pattern-laden behaviors that are learned by interaction with other human beings, are what make up a *culture*. Much ink, and some academic blood, has been spent over the importance of the various definitions for this phenomenon. As an anthropologist, it is useful to accept, if only for the moment, the generally accepted (though not without various assorted caveats) very broad definition that culture is what is learned socially. Implicitly, this definition has referred to human beings, though there have been some attempts to extend it, in some form, to non human species on Earth.

I cannot but help wondering if the concept of culture is universally valid. If so, it would mean that some of the rules we have identified for cultures *should* be applicable to non-human intelligent species, because they could be as universal as physical laws. Perhaps this may be so because social rules follow from physical and biological laws. Some assumptions and discussion in this book are based upon that premise. We might, if we accept that assertion, be able to make some assumptions (that would need to be demonstrated empirically) about how ETIs might behave. If we knew how ETIs might behave, we could presumably calculate our steps while communicating with them, in an appropriate manner. Perhaps we might fear the unknown a bit less, and certainly we would be a good deal better prepared should we meet with ETI. If, in some future time, we should meet with ETI, some of the rules of analyzing cultures might turn out, indeed, to be universal. In such a case, we would have something of a head start on the problem of communicating. This argument, insofar as it is one, is the justification for this book.

1.2 Objectives and Rationale

In this book I shall leapfrog the question surrounding the existence of ETI. I accept as a hypothetical the existence of extraterrestrial intelligence, and on that basis, I shall go on to suggest what can be known about their social and cultural aspects. Some of what I deal with here has been addressed by my colleagues. Some has been addressed by science fiction writers (many of whom are also scientists themselves in 'real life'). The possibility of communicating with ETI has been addressed from a linguistic perspective (Freudenthal 1960) and from the perspective of the effect of discovering ETI on human society (Billingham et al. 1994). This book attempts to address the various questions relating to contact with ETI systematically. Specifically, I would like to provoke my readers to challenge *and examine* the assumptions and conclusions I, and themselves have reached concerning themselves as 'human beings'.

Whether intelligences other than human exist in the universe (known as the $N > 1$ theory) or whether humans are the only intelligent species (the $N = 1$ theory) go beyond the scope of this book. Those interested in a full discussion are referred to Billingham (1981), Sagan and Drake (1975), Hart and Zuckerman (1982), Regis

(1985), and especially Brin (1983) for a summary of the main arguments. The issue in this book is not the physical, chemical, or biological question of *whether* ETI exist, but what the consequences are if they do. The foundation issues regarding physics, chemistry, and biology are necessary for understanding how ETI might vary (or, perhaps more importantly, how they might not) from humanity, and are presented here in skeletal form only. The core of the discussion tries to lay the foundations for communicating with ETI at a level more advanced than merely announcing our (or their) presence, or developing a language or diplomatic protocols for what we 'should' do when contact is achieved. I argue here that there are certain necessary and sufficient universalities which can be exploited to make communication easier. This in my view is a direct extension of the anthropological enterprise: the attempt to understand another culture in its own terms.

The problems of communication between human cultures are onerous enough. Few of us can claim, even less demonstrate, that they are able to understand any culture other than their own. Communication between intelligent species is likely to be far more difficult, even when both species are aware of the pitfalls. The reason for attempting to predict the behaviors of hypothetical ETI is an exercise in being able to simulate and foresee the problems and stumbling blocks, even dangers, that might occur if the human species is to communicate with others. Perhaps some lessons might be learned about communicating with ourselves as well.

Hidden within the discussion about ETI is an argument about humanity, and specifically about human culture. Wilson (1975) has argued persuasively that human culture is not separable from human genetic patterns. If this is the case, then much anthropological effort has been wasted in the wrong directions: rather than uncovering the roots of culture, we should, as Wilson recommends, be busy uncovering the genetic components in human behavior. The opposite has been claimed by White (1977): culture is independent of human biological behavior, he says. It is the patterns that emerge, and their relationship to the one thing humans, as opposed to animals, do best: they think. Whether we adopt Wilson's or White's viewpoint, it should be possible to construct a science of culture *if* we had more than one example to generalize from.

1.3 Basic Terminology

In the discussion that follows I distinguish between several terms. ETI are Extra-Terrestrial Intelligence(s). This refers to intelligent beings outside the Earth, in whatever form. Since the discussion is fairly wide ranging, I also use the term ILF (borrowed in modified form from Ursula LeGuin) to refer to any Intelligent Life Form, including humans, non-human terrestrial intelligences (if any), and extraterrestrials.

An important term derives from the Drake/Green Bank equation, about which more later. Proposed by Frank Drake, the equation attempts to evaluate the size of the number N: the number of intelligent, communicating species in the Galaxy.

I will follow this convention, with one modification: here the term N is the number of intelligent species whether they communicate or not.

Following convention, too, 'galaxy' shall refer to any galaxy—a massive cluster of millions, or billions of stars and greater mass of dust—in the universe, whereas the term 'Galaxy' shall refer to our own galaxy, otherwise known as the Milky Way. Similarly, 'solar system' refers to any star and its accompanying planets, if any. 'Solar System' is reserved to the system of Sol, of which the Earth is one planetary member.

Throughout the book I will make a tripartite distinction between society, culture, and civilization. I admit, for the benefit of my social scientist readers that I am not altogether comfortable professionally with this distinction. The arguments concerning what difference there is, if any, between these three terms, and the validity of the terms themselves, have been going on for decades. However, for heuristic purposes, it is useful to use the term 'society' to describe the organizational arrangements and practices of a species or part of it. Culture then refers to society, plus that society's ideas and its artifacts at any point in time. Civilization refers to a species' culture, where that species achieves a technological and organizational level allowing it to subsume many cultures and societies, and to its relations at any given point in time, with other species, with its environment, and with the total cumulative results of its history (cf. Umesao 1986). A civilization in this view, may include many societies and cultures. The need for such a distinction will become apparent, as issues relating to ETI are unfolded. Later in the book it will be possible and necessary to explore these terms and to refine them in a useful way.

One of the major problems of dealing with a subject about which there is little empirical evidence, is that willy nilly one is forced to use analogies and extensions of one's own experience. That is a problem which has bedeviled xenology—the science of studying ETI—from the start. Often, such analogies offer the only way to attack a problem under the circumstances. However, this book is intended to show that one can use such analogies in a sensible and restricted way (Denning 2013). This is likely to cause a problem. On the one hand, the evidence we *do* have about the Galaxy and universe around us, point to the fact that basic physical and chemical processes are the same everywhere. On the other hand, there are some teasing, tantalizing hints, some based on little more than intuition, that other possibilities will occur. I would rather err on the side of caution, and agree with J.B.S. Haldane that the universe is not only queerer than we suppose, it is queerer than we *can* suppose. The term "exotic" used throughout this book refers to such possibilities, to forms of life, and of activity which, on the face of them, go against what is concretely known, if only in small measure. Exotic possibilities include life on neutron stars (Forward 1988), non-Darwinian ecologies, and several other possibilities which, though unlikely, cannot be totally discounted. Nevertheless, principally, I adopt for this book what is called the "assumption of mediocrity" which claims that Earth and humanity are not exceptional, but normal (in the statistical sense) for planets and intelligent species. This is a useful heuristic device for discussing ETI, and stands until disproved by facts.

Two terms used here derive from SETI, the search for extraterrestrial intelligence. Enrico Fermi, the great physicist, is reported to have asked in a conversation about ETI, if they exist "Where are they?" (Jones 1982). This question, which has aroused intense interest in the SETI community, is the basis for what is generally known as "Fact One" (Hart and Zuckerman 1982): to date, there is no evidence of any sort for the existence of ETI. Fact One is important because it casts doubt on the SETI enterprise, and, to a lesser degree, on the subject of this book. To our knowledge, indeed, no ETI have ever been met by humanity. As this book is being written, however, efforts are being put into effect, most notably by the SETI Institute's Project Phoenix, the META program, and the Columbus Optical SETI observatory, to nullify Fact One.

A second important concept derives from SETI as well. This is the final variable L: how long are civilizations able to maintain themselves? This question may well be the determining factor of discussions about SETI because of its physical reality: the universe has existed, and will exist for a very long time. Species (on Earth, at least) survive for perhaps tens of millions of years. The only civilization we know of—the human one—has been around for an eye blink (if one is generous), or has not yet come into being (if one is more pedantic). So if an ETI civilization exists for ten million years, and started developing one hundred million years ago, we will have missed it, by a very small amount of time, measuring cosmically, but a very large period, measuring by humans standards. The SETI community further narrows L, for good and material reasons, to the period of time during which an ETI civilization builds and operates large radio telescopes: a period which may be very short, either because the civilization fails, or because they stop using radio telescopes for some reason. Most of the following discussion ignores the issue of L. Whether a species is intelligent for a short, or relatively long period of time affects the discussion of *communicating* with that species, but not the question of that species intelligence or sentience. However, the reader should keep in mind one important fact. Species have a finite span. Whether this is also true of intelligent species is unknown. Given biological and physical facts as known to us, however, the significant likelihood is that L is a finite, not an infinite quantity.

1.4 Method

The most basic fact about ETI is that none have been identified, nor have we, the human species, any empirical data about their existence. This, 'Fact One' (Hart and Zuckerman 1982), is crucial to the debate on SETI, about which more in the following chapter. It is significant for the discussion to follow but only because there is no empirical way *at present* to prove or falsify any of the assertions made here. This is not an insurmountable problem. The behavioral sciences have operated for decade without any *direct* evidence of many of the factors we discuss. Who among social scientists has ever seen a 'culture'? A 'society'? A 'group'? Or an 'unconscious'? What we have seen are *behaviors* from which we have deduced the existence of those phenomena. For most behavioral scientists, the very meat of their activities are second-order generalizations, not observable phenomena. The method, generally speaking has been to observe those phenomena that are observable, then to deduce from the empirical fact the existence of greater connectivity that relate the phenomena. This system has proven itself, if not without difficulty. Different schemes have been proposed to overcome the difficulty, some more, some less successful.

A similar method is employed here. The process of analysis here is borrowed, in modified form, from Julian Steward's proposals about cultural ecology (Steward 1968, 1977). Steward assumed, as had others before him, that there is some form of relationship between a society's form and its environmental surroundings. He posited that societies, given a technology and an environment, are likely to exploit such an environment intelligently for their benefit. The nature of that society's social activities would be limited by the nature of the environment, modified by their culture. He provides a crucial example in the lifestyle of the Basin Shoshone, a native American society, relying mainly on hunting and gathering for survival. In their natural ecology, the Shoshone were reliant on certain natural foodstuffs, most of which were present in limited quantities. This had two effects. First, the Shoshone could never aggregate into larger units than families, for any length of time, since *local* food scarcity would soon break up such an aggregation. Second, because of the nature of their resources—widespread over the Shoshone range, but limited in any given place—the Shoshone had to adopt a nomadic lifestyle. This in turn placed limitations on their technology. It had to be lightweight, human-portable, made from local materials. One consequence of that was that the Shoshone could not devote much attention (by way of experimentation, trading, or exchange) to changing their cultural ecology. Seasonal opportunities—the ripening of piñon nuts—allowed the Shoshone, for a brief period of time each year, to assemble into larger collectives, but these could not and did not last long, since eventually the food gave out. It took disturbances from the outside, including the introduction of new technology—the horse and gun—and of other humans, from nearby tribes and the European settlers, for Shoshone culture to change.

Cultural forms that are similar to the Shoshone in essence (e.g.!Kung bushmen (Lee and Devore 1976; Marshall 1976), and Australian aborigines (Sharp 1973)), but different in detail have evolved elsewhere in desert environments. The similarity does *not* derive from some ineffable "primitiveness" of those groups, but rather from the application of a species' intelligence to the solution of problems posed by existence. In each case, severe limitations of resources, not imagination, forethought, or tradition, limited the form of the culture concerned. Where these limitations were transcended, by ecological, technological, or social change, the society concerned changed as well. Moreover, not all desert-dwellers necessarily evolved the same culture. Others, in the Middle East, East Asia, and the Americas, organized themselves in other ways, largely because their ecological limitations were different, partly because they organized themselves *initially* in different ways, which led to different consequences.

Two important methodological consequences arise from the discussion in the previous paragraphs. First, we can view the complex interaction between a culture (roughly, for our purposes here, a system of thinking about things, and of doing them) and an ecology (for our purposes here, the availability of material resources) in terms of limitations that those two factors impose. In the absence of surpluses, for example, little can be devoted to experimentation in methods of preservation or exploration (at the simpler end of the scale) or to communicating with far off cultures (at the more complex end of the scale). Second, seen as *limiters*, cultural-ecological factors do not determine in what ways a culture *will* evolve. The ranges of intelligent thinking, restricted psychologically as they may be, are effectively limitless. The conceptualization, transmission, and execution of such ideas, however, is *limited* by cultural-ecological factors. It is thus easier (though still complex) to state what a society *cannot* do, or *cannot* produce (even intellectually) than what it can. This is the view adopted here. We cannot state, or perhaps even imagine, what an ETI society will think or do. But we can state, with apparent certainty, what, given knowledge of some factors, they *cannot* do. To do so, of course, we need some understanding of the limiting factors.

Rather than considering the cultural ecology of simple human tribes, we intend here to consider the cultural ecology of intelligent species and complex societies. This will require, as all such analysis does, some understanding of the physical background. Some of the data is known, but it emerges from disciplines such as physics, astronomy, and chemistry, disciplines that are normally of little salience to the behavioral scientist. From this background it should be possible to come to some conclusions regarding the parameters under which life forms must operate, and therefore, possibly, to come to some conclusions regarding their likely behaviors.

A further word is required about the term 'likelihood' which will be employed often. *Likelihood* is not probability. A probability is measured based on assigning mathematical weights to phenomena and then calculating the future occurrence of such phenomena under conditions as specified. "Likelihood" on the other hand is a subjective measure in which weights are *deducted* from all possible occurrences based on how logical they seem to the observer from his current knowledge and logical processes: a highly speculative prognostication system.

For the purposes of this analysis, the physical rules of the universe are weak propositions. That is, they are commonly accepted, have been verified many times by observation and theory, and need little defense. I take the analysis one step further, however. For the purposes of the exercise of discussing ETI, it will be necessary to assume that *some* rules behavioral scientists have deduced about society and behavior are *universal*. This is a strong assumption indeed. It has no observable direct evidence since we are generalizing from a single case: *Homo sapiens*. But these rules may well derive from the background of physical laws of the universe, which we know (or at least, so the physicists and cosmologists claim) are universal. I will attempt to defend this proposition in general, and some specific rules that derive from it as the discussion progresses.

I have accepted as given several meta-rules for discussion of natural processes. The most important is the acceptance of the principle of parsimony, or Occam's Razor. Briefly stated, of two possible processes that could bring about a given event, the simpler one is the more likely to be the correct explanation for the event. That is, even when speculating upon the possible forms of exotic life, simpler explanations have been preferred to more complicated ones. Occam's Razor has been challenged a number of times, but, insofar as empirical evidence and logical inference can bring us, it is still the most powerful tool in the armory of any speculative scientist.

1.4.1 Rules and Laws

Two terms—"rules" and "laws" —are used throughout this book in very specific senses. We are all familiar with the "Law of Gravity" . This "law" is, in fact, a statement of distilled, observed knowledge. It is a "law" in the sense that no violations of it are known to occur empirically or theoretically. There is no physically observable case in which the law is violated (which does not mean that it cannot be opposed, only that other forces account for apparent inconsistencies such as airplanes staying in the air), and it is part of a set of complex mathematical relationships, each hinging on the other: gravity relates to light, to the structure of the universe, in many complex, but mathematically definable ways.

Physical laws are easy to define in a sense. They are quantifiable, usually inviolate, and obviously universal. Social scientists used a similar term "rules" to describe *what appear to be* regularities in human (or animate) conduct. "Rules" in social science are far less determinate and far less quantifiable, than are "laws" in physics. Laws in physics, because they are, under definable circumstances inviolable, can be used to predict the behavior of physical objects. The same cannot be said of social rules. Here the viewer is floundering in the dark, and quite often, what seem to be firm rules, turn out to be special local cases which appear to exhibit regularities, but do so only in a limited number of cases. Rules are rules because people interpret certain behaviors, those that are expected from them, and those they expect from others.

To add to the problem, not all social scientists, even those from the same discipline, agree what these identified generalizations are. The variables are extremely difficult to define properly, the subjects are self-conscious, and able to violate the rules, and the samples, compared to the physical sciences, minute. A chemist, to take one example, makes a generalization based on the behavior of several million individual examples—about the number of molecules in a single drop. An anthropologist may make generalizations based on several hundred highly individualistic and variable individual examples—a tribe, a neighborhood, few of which she is able to observe fully, without bias, and without temporal restrictions.

There is one final difference as well. Physical scientists are able to demonstrate the truth of their assertions—their laws—by using those laws in an applied setting.

That is, by manipulating their materials and constructing artifacts that behave in a predetermined manner under controlled conditions, AKA experiment. From this perspective, a nuclear power plant, an automobile, and a computer are merely demonstrations of the stability of physical laws, and the scientists' knowledge of them. Social scientists are unable to demonstrate the truth of their propositions—their rules—by manipulation. Few social scientists has ever been given the opportunity to manipulate people in this manner. Thus whatever rules are adduced, remain barely "proven" by the standards of the physical sciences.

In the following chapters I will attempt to deduce some basic social rules and see how they can be utilized to discuss the issues of ETI. Some of these rules may prove to have only local applicability (that is, to humans). Others may prove to be false, even for human behavior. Nonetheless, there may well be a class of basic rules—meta rules—whose relationship to the physical laws of the universe is hypothesized here, and which may be universal for all ILF.

1.5 Organization of Book

This book proceeds from firmer fact to greater speculation. Not incidentally, it also moves from emphasis on facts and theories of the hard sciences—physics, chemistry and biology—to those of the behavioral sciences—psychology, anthropology and sociology in particular—whose rules are far less certainly known.

The first three chapters deal with the bases for intelligent life. The cosmic environment, or that aspects of it which are relevant to life in the universe, is followed by a description of chemical and biological factors that could lead to life or might hinder it. Chapter Four discusses the major law we do know about species: the issue of evolution, and whether it should be applied outside Earth. In Chapter Five the key concept of intelligence is analyzed. Overall the attempt here, as elsewhere through the book, is to detach the phenomenon from its roots in the human species and human experience, and to attempt and see if it can be applied systematically in some way to species other than human. Chapter Six presents the critical variables of language and communication. How it is brought about, and how and whether human concepts can be applied *without* the human factor. The subsequent chapters, Seven to Nine, deal with culture and civilization, in effect the keystone of this book. Chapters Ten and Eleven look at various aspects of contact, communication, and exchange and the difficulties they represent going well beyond the issue of language. Chapter 12 briefly discusses the potential effects of communicating with non-human intelligence on human civilization. Chapter Thirteen deviates slightly from the theme of the book, and addresses the question of potential human reactions to the proven existence of ETI. Chapter 14 briefly summarizes the main points of the argument.

1.6 A Note on Theory, Empiricism and Imagination

A great many theories by many people in many different disciplines went towards the making of this book. Theory, however elegant and exciting, is worth little without empirical facts. Usually such facts should precede the flights of theory.

It is in the nature of this particular subject that theory precedes empirical verification. As will be clarified later, some facts about the universe seem conclusive: those dealing with its physical makeup. Other facts are not well known at all: those about behavior, about organisms, about thinking. This book therefore relies, to some extent, on a marvelous human device, the imagination. Reasoned, imaginative thinking is as much a tool of the scientific enterprise as are theory and fact. Without it, science goes plodding along without the flights and leaps of intuition that give the scientific enterprise its beauty. I make no apology for the use of reasoned imagination—imagination based to the degree possible on what is known, in order to generate propositions about what has not been thought of—leaving it rather to those who have done so in their own work to lead by example.

1.7 In Summary

It has been argued, most emphatically by Tipler (1981), that spending money in trying to find and contact ETIs is money badly spent. Arguments for or against searching for ETI are not my concern here. I am convinced, for what I think are good reasons, that there are other intelligent species in the universe, perhaps in our Galaxy, possibly in communicative range. For some of the purposes of this book, that is not really material. The principles that are discussed here can, and should be examined by two means. First, as most of the features of the Drake equation have been considered, by picking at the logical and theoretical inconsistencies and problems in inherent in it. Second, by making an effort to verify (or falsify) the propositions empirically. This book will contribute to the debate surrounding ETI and SETI. I hope that it will also affect the future in two other ways. First, that it will encourage other behavioral scientists to enter the lists and consider problems related to SETI. Whether they enter to challenge some of the assertions made here, or to support them does not concern me too much. The mere fact of the debate will continue and reinforce humanity's SETI effort against short-sighted attempts to squash it in favor of 'feeding the hungry' to use the current political catch-phrase.

A second hope is more ephemeral: that at some time we will be able to test the assertions in this book, and others, against empirical data.

References

Billingham, J. A. (Ed.). (1981). *Life in the universe*. Boston: MIT Press.

Billingham, J. Heyns R. et al. (Eds.). (1994) *Social implications of the detection of an extraterrestrial civilization: A report of the workshops on the cultural aspects of seti*. Mountain View CA: SETI Press.

Brin, G. D. (1983). "The 'great silence': The controversy concerning extraterrestrial life." *Quarterly Journal of the Royal Astronomical Society, 24*, 283–309.

Denning, K. (2013). Impossible predictions of the unprecedented: analogy, history, and the work of prognostication. In A. V. Douglas (Ed.), *Advances in astrobiology and biogeophysics* (pp. 301–312). Berlin: Springer.

Forward, R. L. (1988). *Future magic*. New York: Avon Books.

Freudenthal, Hans. (1960). *Lincos: Design of a language for cosmic intercourse*. Amsterdam: North Holland Publishing Co.

Hart, M. H., & Zuckerman, B. (Eds.). (1982). *Extraterrestrials: Where are they?*. New York: Pergamon.

Jones, E. M. (1982) Estimates of expansion time scales. In M. H. Hart & B. Zuckerman (Eds.), *Extraterrestrials Where are they?* (pp.66–76) Oxford: Pergamon.

Marshall, L. (1976) *The !Kung of Nyae Nyae*. Cambridge MA: Harvard University Press.

Richard B. Lee, and Irven DeVore (Eds) (1976). *Kalahari Hunter-Gatherers: Studies of the! Kung San and their neighbors*. Cembridge, MA: Harvard University Press.

Regis, E. (Ed.) (1985) Extraterrestrials: *Science and alien intelligence*. Cambridge: Cambridge University Press.

Sagan, C., & Drake, F. (1975). The search for extraterrestrial intelligence. *Scientific American, 232*, 80–89.

Sharp, L. (1973). Steel axes for stone-age Australians. In T. Weaver (Ed.), *To see ourselves* (pp. 459–464). Glenview, IL: Scott, Foresman and Co.

Steward, J. H. (1968). The concept and method of cultural ecology. In H. F. Morton (Ed.), *Readings in anthropology* (pp. 159–171). New York: Thomas Crowell.

Steward, J. H. (1977). *Evolution and ecology: Essays on social transformation*. In J. H. Steward, J. Steward & R. Murphy (Eds.), Urbana: University of Illinois Press.

Tipler, F. J. (1981). Extraterrestrial intelligent beings do not exist. *Physics Today, 34*, 9–81.

Umesao, T. (1986). Keynote address: The methodology of the comparative study of civilization. *Senri Ethnological Studies, 19*, 1–8.

White, L. A. (1977) Symbol, the basic element in culture. In W. Goldschmidt (Ed.), *Exploring the ways of mankind* (pp. 20–26). New York: Holt, Rinehart, and Winston.

Wilson, E. O. (1975). *Sociobiology: The new synthesis*. Cambridge Mass: Harvard University Press.

Chapter 2
Space in the Background

The basic parameters for the continuing discussion involve three major sets of concepts. The first has to do with the basic question (which I admit is sidestepped throughout the rest of the book, but not in this chapter) of *whether* other intelligent beings exist anywhere else in the universe. Inseparable from that question are two others: talking about life on other worlds implies a necessity for understanding some of the basis of what is meant by "other worlds." That is, how is the universe to be conceptualized? What are the bases upon which a discussion of ETI can take place? The third question is related to the first, but it has to do with another level of life (and of the universe). How is the universe made up? What elements in the universe need to be kept in mind in order to be able to make the assumptions that will be made here and later in this book? Because we are to deal with life only remotely related (if at all) to our own, we must start with the microcosmic universal basics, even with the laws that make the universe up, in order to understand why we should assume life would emerge, and, whether there are any fixed compass points that will help us understand *how* an intelligent species would behave.

To start, however, it is useful to recapitulate the efforts that have been made to confirm the existence of intelligent life elsewhere than on Earth, and to try and contact it. I am not concerned here with all such efforts, but only with general approaches and their principles. It is however necessary to emphasize that searches for extraterrestrial intelligence operate, as scientific endeavors, under a set of principles and limitations that do not necessarily limit the discussion here. The physical difficulties in reaching other stars do not concern me: I leave that to the physicists. It is the *consequences* and modes of contact that are my major concern.

M. Ashkenazi, *What We Know About Extraterrestrial Intelligence*,
Space and Society, DOI 10.1007/978-3-319-44456-7_2

2.1 The Search for Extraterrestrial Life

Humans have been discussing the possible existence of intelligent life forms on other stars or planets for centuries. The earliest such speculations recorded were those of Democritus, but they have persisted in many cultures since (Dick 1999; Crowe 1986). These speculations have been motivated by scientific interest (SETI efforts in the twentieth century), religious discourse (Renard 1986; Lamm 1978), political polemic (Bergerac 1923), intellectual discourse (Kant in Dick 2013). or for entertainment (Verne, Welles, and subsequent science fiction). It is only in the twentieth century that technology has provided the tools necessary to formulate some questions and at least partial answers, to many of the questions that the problem poses.

2.1.1 The Drake/Green Bank Equation

Insofar as we (the human species) know, there is only one intelligent life-form in the universe: us. This fact, known as Fact One (Hart and Zuckerman 1982) is almost unarguable. However, the statement as it stands includes a number of semantic traps we ought to be aware of, and sometimes are not.

In this particular instance, the 'we' embodies not a particular group of people, but a process, a particular way of 'knowing' that we call 'science'. Science implies a definite set of rules *about* knowing and about knowledge. When we say the human species 'knows' what we mean is that insofar as the process called science has been able to determine, using the methodology it uses, there is no evidence for the existence of any ETI. This does raise two questions: what about signs of ETI existence that are *not* amenable to our current scientific discourse? The manifold reports of UFO sightings would tend to support this methodological position (though I hasten to say this is not an argument for UFO existence, let alone that UFOs are ETI, as I shall show in Chap. 5). Another dimension is the rather short reach of our science. As Forgan and Elvis (2011) argue, a likely location of ETI *artifacts* may be in the asteroid belt. But even those objects, astronomically speaking practically a part of our planet, are currently way out of reach. So Fact One, important empirical evidence though it is, should not constrain nor end our search. It is also important, therefore, to understand the processes by which we go about 'knowing' whether or not there are ETI. Part of this process is a technical one: the construction of physical means for deciding the issue. A more important part of the process is the logical examination of the possibilities, the rationale for the scientific endeavor.

This rationalization is presented in abbreviated form as a set of relationships, generally known as the Drake equation (for its originator, Professor Frank Drake) or the Green Bank equation (where Drake first formally presented the equation). The

Drake equation proposes that the number N of intelligent communicating civilizations in the Galaxy can be expressed as

$$\mathbf{N} = \mathbf{R} * \mathbf{f_g} * \mathbf{n_c} * \mathbf{f_l} * \mathbf{f_i} * \mathbf{f_c} * \mathbf{L}$$

where

N number of intelligent life forms
R rate of star formation
$\mathbf{f_g}$ fraction of stars with planets
$\mathbf{n_c}$ fraction of such planets within the 'habitable zone'
$\mathbf{f_l}$ fraction of such planets where life actually arises
$\mathbf{f_i}$ fraction of such planets where life gives rise to intelligence
$\mathbf{f_c}$ fraction of intelligent species that give rise to a technological civilization (detectable by us)
L Average lifespan of a technological culture

The precise details of the equation, its modifications, and the parameters of its variables are still being debated today (Drake 1967; Drake and Sobel 1992; Forgan 2009; Gleiser 2010; Maccone 2010). Note that the variables start from the physical sciences on the left, and move onto attempts to quantify biological, and later social unknowns on the right. Two points, however, are to be borne in mind.

First, as we proceed from physical science to the biological and then social sciences, the variables become more difficult to assess. Some of the independent variables have approximate solutions. The approximate rate of main sequence star formation in the Galaxy has been variously estimated at 0.68–1.45 solar masses/year (Robitaille and Whitney 2010). Other estimates are as high as 7 solar masses. These estimates are being updated as more information about the universe becomes available as the result of work by astronomers.

Some of the variables which had no data at the time the equation was stated, are currently being filled in: star systems with planets are apparently common, and as of the writing of this book, many planets have been discovered in the habitable zones (that is, the zone in which water can be found as a constant liquid 0C < $\mathbf{n_c}$ < 100C) around nearby stars. Though there is currently insufficient evidence to refine $\mathbf{f_g}$ to the point of making it a known variable, that is merely a matter of time, though it is currently estimated as quite high (Beckwith 2008; Beichman 2002).

In contrast, the rightmost end of the equation which contains biological and social variables is unknown. For all the variables f_l, f_i, f_c, we have only one example, ourselves and our planet, to generalize from. For L, the lifetime of a technological civilization (defined by SETI as a civilization capable of building and operating radio telescopes. Oliver 1976) we have a complete unknown: we have no knowledge beyond the fact that it *could* extend for more than a century.

Second, the various searches for ETI (SETI projects) by necessity, must have a very conservative definition of extraterrestrial intelligence, and of an ETI species. ETI must, for example, be a technologically advanced species, otherwise the effort is futile: there is no possibility of signal exchange across space. Moreover, an ETI species must also, to a great extent, think like humans and be interested in similar things, otherwise there will also be no likelihood of communication. Finally, because SETI projects must deal with current technological and scientific know-how, they must accept all basic physical implications of the known universe. The most prominent of those are the vast distances involved, and the fundamental universal velocity limit **C**.

Those limitations are not as rigid for this book. The variables of the Drake equation will here serve as a framework for discussion, because they are the currency of intellectual exchange in the SETI community. Yet it will also be necessary to consider exotic possibilities, as well as ignoring completely some issues arising from the physical limitations. In a later chapter, where I consider the consequences of contact, for instance, I will at times ignore the (current knowledge-based) absolute limitation on faster-than-C messaging and travel.

2.1.2 Arguments Pro- and Con-: The N = 1 Versus N > 1 Debate

Speculations about the existence of ETI fall generally into two opposing views. The first is that N (the number of intelligent, communicating, life forms in the galaxy) is greater than 1. We know of at least one ILF species: ourselves. The hypothesis on which ETI research is based is that there are more, thus N > 1 (Billingham 1981; Sagan and Drake 1975; Sagan and Shklovskii 1966). The opposite hypothesis (Hart 1982; Tipler 1981) N = 1, is, the idea that it is improbable that there are any more than one intelligent, communicating species in the galaxy.

So long as none of the independent variables in the Drake equation equal zero, the potential for number N is always greater than 1. There are some 300 to 600 *billion* stars in our Galaxy alone. It is sufficient that the product of the variables is a measurable fraction, for N > 1. Estimates of N have varied between highly optimistic, and probably unrealistic estimates of one million, to lower estimates on the order of one thousand to one hundred communicating civilizations in the galaxy. Keep in mind, however, that 2 is a theoretically sufficient number for purposes of contact.

There is an additional factor that must be considered. Its importance derives from the physical nature of the universe. As currently known (that is, to date unfalsified, and seemingly unfalsifiable) the limiting velocity of any object in the universe, light and other electromagnetic phenomena included, otherwise known as C, is 299,792 km/sec. Communication—by physical or radiation means—with any but the nearest stars, is measured therefore in centuries and millennia. The nearest stars are mere years away *at the speed of light*: a speed that is practically and theoretically impossible to achieve (Einstein 1955). What this means, among other things, is that two communicating civilizations must overlap in time for a considerable period, in order to carry on any sort of communication, making the variable L a crucial limiting number: *both* societies' L *must* overlap for there to be any contact.

Theoretically some complex ways of overcoming the C limitation, without violating the velocity and energetic requirements posed by the universe have been proposed (Hawking 1993; Wolf 1989). In practical terms no one has been able to suggest a way in which these (which have some severe limitations of their own) can be achieved. For the Drake equation, however, this means that the variable L is absolutely crucial: the longer the average lifespan of a communicating technological culture (assuming $N > 1$), in terms of *millennia at least*, the greater the odds such civilizations will overlap and thus indeed be able to communicate with others. If, for example, L averages one thousand years, the evolution of a technological species lasts one hundred million years, and of two species, one started evolving 1 % of that time before the other, no communication will occur.

2.1.3 The Fermi Paradox AKA the Great Silence

There is one aspect of the $N = 1$ party that *does* necessitate discussion here, though its full implications will be explored later. In 1950, the great physicist Enrico Fermi, during an evening discussing SETI, asked a simple question now known as the Fermi Paradox: "If there are advanced ETI in the universe, where are they?" In other words, why have they not contacted us? Why have we not even seen any sign of their presence?

This simple question (all great questions appear simple) represents a major problem for $N > 1$. For the $N = 1$ party it represents empirical demonstration of the $N = 1$ hypothesis. For the $N > 1$ it represents an even greater set of problems, highly relevant to the subject of this book, which David Brin has labelled "The Great Silence" (Brin 1983).

Let us assume, following Brin and others, that (a) there are contemporary ETI civilizations out in our galaxy; and that (b) they are avoiding contact with us. What then are the implications? Several such possibilities have occurred to SETI researchers, including the Zoo hypothesis (they wish to preserve us in our innocence, to allow us to develop at our own pace, cf. Ball 1973) to the Great Silence (they are hiding from some greater threat, Brin 1983), and the Transcendence theory (they have evolved so far ahead of us technologically (Bracewell 1981) that, to quote the Old Testament "We are as grasshoppers to them." Numbers 13:33).[1]

From the point of view of this book, these are all relevant but from a different perspective: all three theories constitute speculations on the nature of ETI civilization and culture. We shall address this core topic later on. For now, it is sufficient to say that the argument about the Fermi paradox by both the $N = 1$ and the

[1]This does beg the question of the two other $N > 1$ possibilities, that ETI are at our technological level, or behind us. *Most* SETI speculation, for obvious reasons, assumes that ETI are more advanced. Since we are principally discussing CETI in this book, however, we shall be ignoring the assumption of ETI technological superiority for the moment.

N > 1 parties are a side issue: we accept N > 1 as a given. And whether they are avoiding us deliberately or not, the principles to be discuss here remain the same.

2.1.4 SETI Programs

The possibility that other intelligent life may exist in the universe has led to a number of attempts to find ways of empirically demonstrating their existence. Establishment of organized searches for Extraterrestrial Life (known in the US as SETI and in the former Soviet Union as CETI) began in earnest in the late fifties. To sum up the results in terms of the final goal, they are zero so far. That is, there has been not so much as a hint of the existence of other intelligent life forms in the universe. This having been said, we must indicate that the 'zero' must be qualified. The searches *have* yielded a great deal of important information, not the least of which are the refinement of techniques for SETI. Given the universal realities discussed in the previous section, most searches have concentrated on using radio. Generation and detection of radio waves are both cheaper and more efficient than the transmission of physical artifacts.

The SETI Project (formerly the official NASA project searching for Extraterrestrial Intelligent life forms, currently a voluntary astronomical project run by the SETI Institute in Mountain View, California) defines an intelligent life form operatively as a life form able to construct large radio telescopes (Oliver 1986). This means a life form that we will be able to communicate with, given our own technological level. As an operative definition *for a SETI project* this is an admirable formulation, since it simplifies the discussion of what is intelligence, what is life, and so on. As a normative definition this does not do much good for the discussion here for two reasons. First, because the discussion is intended to examine as many possibilities as possible, and second, no less importantly, because the definition is ethnocentric: it assumes that all intelligence is roughly similar to human, which might not be the case.

It is useful to make a distinction between SETI efforts and SETI programs. A SETI program is a managed and organized project by some organization, to search for extraterrestrial life. Like all programs, any one of these might prove ephemeral for reasons of politics, financing, or technology. A SETI effort is the sum total of a species endeavor to search for intelligent life other than its own. This includes public opinion, intellectual debate, as well as SETI programs. A species with a SETI effort is different qualitatively from one that does not have one. Intelligent, even very intelligent lifeforms with advanced technology may well *not* engage in SETI efforts (Ashkenazi 1995). As we shall see subsequently, this will affect the SETI efforts of other species to some degree. No less importantly, reaching the position of "uninterested in SETI" is one that can be reached by several paths, many of them germane to the subject of this book. The absence of SETI programs or a specific program, on the other hand, merely means that a species has not gotten itself organized to engage in the enterprise. That might indicate shortsightedness, even stupidity by those responsible for organizing that particular species, but would still mean that the species is comparable, even comprehensible to us, at least in some fashion.

Organized SETI programs have been in operation world-wide for five decades (Tarter 1985; Drake and Sobel 1992). By and large these have been searches through the electromagnetic spectrum for indications either of intentional beacons or of leakage from broadcasts. There have also been searches for the possibility of physical arrival of ETI probes in the Solar System (Freitas 1980), known by the acronym of SETA (Search for Extraterrestrial Artifacts), and an ongoing search for signals in the visible spectrum: Optical SETI (Kingsley 2001). Other searches have been conducted for possible indications of ETI technology outside the Solar System (e.g. radiation leakage, stellar engineering. For some suggestions see Ćirković 2001).

The future of many of these programs depends on the will and enthusiasm of their various designers and proponents. These include private and philanthropic initiatives, ranging from the SETI Institute in California, to the 100 million dollar Breakthrough Listen project which started operation in 2016 (equivalent to the total budget of the government-sponsored NASA-SETI project). No less, it depends on the political and economic climate that exists in any country in which a SETI program takes place. These are not many, and with the impoverishment and demise of the Soviet Union, some of whose scientists and engineers were fervent supporters and maintainers of SETI program, some projects have disappeared (though the enthusiasm has not, as witness the Breakthrough Listen project, funded by Yuri Milner, a Russian businessman). We can only hope, if only for the sake of our remote descendants, that the SETI effort is sufficiently entrenched in the human species so that programs will be maintained and even strengthened by private or government funding.

2.2 The "ILF" Perspective

We start then by trying to picture and perhaps understand the relevant physical facts of the universe. Of those, the easier to grasp are those elements of the universe that have to do with size and dimension. We are a planet-bound species, though we have only begun to realize it in the past few centuries. Our planet is intimately tied to a star and the star is tied to still larger universal elements.

Understanding how these interrelate at some level, makes the following discussion easier. Understanding the implication in terms that make sense to an individual, even more so. We start therefore with a rough description of the major elements of the stage and the props. The obvious features within the universe that are of great interest to us are stars and their associations. We, the human species, exist as a consequence of factors inherent in physical activities at many levels. The most obvious, at least to the senses, is the star around which the Earth revolves. It is a reasonable supposition that *if* humanity came into being on a planet around a star, other species may do the same on other planets, around other stars. This not to prejudge the issue of what conditions need to occur for life or intelligent life to *occur*, but to suggest that similar conditions could bring about similar results. Given that the catalog of extra-solar planets in our near stellar neighborhood has swollen as of 1st July 2016 to 3443, partly due to the operation of the Kepler observatory (Exoplanet EU 2016) of which about 1/5 are in a habitable zone (Petigura et al. 2013), the odds of finding a habitable planet are growing daily. In other words, we can chalk up another victory for Drake's equation.

I must start by emphasizing that cosmology, as physicists understand it does not concern me here too much. Cosmological physics is a subject best comprehended by numbers and the relationship between them, not by humanizing the phenomena. But for our purposes, it is more important to humanize than to present an accurate mathematical description that a cosmologist would accept. The following description, wherever possible, humanizes the cosmos rather than abstracts it. Like all maps, it suffers therefore by being skewed. Accurate in some ways, it is misleading in others: if in making the universe comprehensible for human purposes it also makes it incomprehensible and unacceptable for the professional physicist and cosmologist. They have my apologies.[2]

2.2.1 Space and Time

The interrelationship of mass, time, and distance are complicated enough as physical phenomena. Our concern is however with ILF perspectives, which makes the issue more complicated still, because we need to add a *subjective* component to these issues: what, for instance, does time mean for an intelligent species? For example, the critical time-span for most people, over most of history, is one's own life span. Which is itself subjective: consider what 'life span' means to a ten year old and to a ninety year old. The same is obviously true for societies, and, we can extend that, for civilizations and species. Objectively (from a cosmological perspective) and subjectively (from an ILF perspective) time is a crucial dimension.

It is necessary to understand, first and foremost, that the universe is huge. Getting to our closest neighbor, Proxima Centauri, would take over 4.2 years traveling at the speed of light. To put it materially in ways that can (perhaps) be comprehended, if there were a highway stretching to Proxima Centauri., and you could drive at about 100 km/hr all the way, it would take you 34,560,000 *years* to make the trip (provided of course that you did not stop along the way for a snack or to relieve yourself). Most of that distance would be through 'empty' space.

---

[2]I could add, in consolation, that I find much of the discussion of social and cultural phenomena by my astronomer and cosmologist colleagues in SETI rather incomprehensible as well. Each to his knitting.

For a technologically developed civilization, the first stage of both space and time is its solar system: the cradle, and possibly the entire domain, of any non-exotic ILF. Since we presently adhere to the Principle of Mediocrity, we can assume, at least *pro tem* that ETI populated solar systems will be similar to our own in most ways.

2.2.2 The Solar System

The Solar System consists of a star, Sol, two torque-shaped planetary zones nestling one inside the other, a wide cometary belt of frozen ice and gas particles called the Oort cloud, and finally, a very diffuse solar atmosphere of charged particles which extends as far as the solar atmospheres of the nearer stars. Because of the rotation which initially formed them, most (well over 99 %) of the matter we are talking about is on the same plane, rather than existing as a sphere (with the exception of the charged particles which are emitted from the spherical surface of the sun).

Sol is a G2 V type star: mid-range, rather small, neither very hot nor very cool by stellar standards. This has implications for considerations of searching for extraterrestrial life. Life is an aspect of the cosmos, that is, a product of physical forces including energy and time. If Sol is an average solar system in physical-cosmic terms, then it is possible, for cosmic reasons and in the absence of another data point, to assume that it is average in all respects. This means in the existence of life on a planet surrounding it. If the Solar System is average, than on other stars that are close to the average we will find similar processes, and perhaps similar results. Other stars will also have life, other biospheres will also have intelligent life-forms. *If* the assumption of mediocrity holds true throughout.

Sol has a number of effects that are relevant to life *beyond* considerations of the evolution of life around it. First, and that is its most apparent characteristic to humans, it produces energy. All energy available to life on Earth is a product of the Sun. This has an implication as well: in a sense life on Earth is 'habituated' to certain types and ranges of energy from the sun. For example, we need liquid water, which implies a certain degree of insolation. Too much and water evaporates, too little, and it solidifies into ice. But is this a *universal* requirement for life?

Sol has mass, and effects on mass: the particles that stream away from it in the form of solar wind, and, more importantly, its gravity. It is the gravitational factor which accounts largely for the formation and continued existence of planets, and thus, directly and indirectly, to our continuation as a species: the speed of Earth's passage neatly counterbalances the effects of Sol's gravity: as neat and elegant a balancing act as one could ever expect to see, in fact, a marvel of engineering and architecture. Again, is this a feature in the emergence of life?

Finally, Sol, like any other star, has a temporal effect on its environment. This temporal effects starts from Sol's birth, some six billion years ago, extends through its lifetime as a nurturing star possessing a zone around it where life is possible, and continuing into the future to some final date when Sol most likely will go nova. So long as Sol exists in its current form, so do its dependents and immediate neighborhood. When it moves on to the next phase of its existence, as inevitably it must, this will affect its dependents. Lest one become complacent, it must be recalled that *we*, *Homo sapiens*, are one of Sol's dependents. We have, as presumably do all other stellar species, a "window" of existence. That is, from a certain time t_0 to a certain other time t_1 we can exist, in current form, as dependents of Sol. Before and after that window things may not be so simple. The lifetime of a star as it exists on the main sequence of stars is, effectively, the basic parameter for the existence of life on planets around that star.

Of greatest concern to us (because we live in it and it is most obviously accessible to our examination) is the planetary system. It consists of Mercury, Venus, Earth, Mars (the inner planets) Jupiter, Saturn, Uranus, Neptune, the recently demoted Pluto (the outer planets) and possibly another outer planet. Between them are the orbits of smaller objects generically (and often mistakenly) called the asteroid belt: the asteroid orbits are a "belt" to astronomers, but largely empty space (science fiction films to the contrary) for the rest of us (see Petit et al. 2001).

Insofar as we know, the only body capable of evolving and sustaining life in the solar system is Earth. There are some indications that other planets, e.g. Mars, may have had *some* of the conditions for life in the far past, and *may* still hold simple life today (Levin 2010; Yung et al. 2010), but there is no concrete evidence of the existence of life on Mars. One of the reasons for this consideration is the lack of flowing water. Water in liquid form may be a natural prerequisite for the evolution of life. Some recent studies (Battison 2011) indicate that other locations, notably some of the moons of Jupiter, may harbor free-flowing water. If that is the case, Ganymede, and possibly Europa which *may* have free standing water hidden under a crust (Keszthelyi 2011) are candidates for the existence of life.

There are some claims that life could originate in the methane clouds of a large gas giant such as Jupiter. The changing colors of Jupiter's bands and great spot have been suggested to mean that there is biological activity on Jupiter. If that is indeed the case, then necessary changes will have to be made in human calculations about the likelihood of life. So far, however, there is no real evidence that Jupiter possesses life.

The other planets—Saturn to Pluto—would seem to be too cold for the evolution of life. Outside their orbits there are belts of cometary material, largely ice, about which little is known from direct observation, perhaps other planets. No life would seem to be present there. Some suggestions have been made (e.g. by Shapiro and Feinberg 1982) that life may possibly emerge under conditions approaching zero absolute, but such life would be too strange for us to consider at this point.

Solar System Resources

Like a kid before a candy store, we can, at present, merely look at the material riches of the solar system, but not touch (the odd Lunar or asteroid rock notwithstanding). The solar system as a whole represents a bank of minerals, metals, and energy, as well as specific physical qualities that have major relevance to certain types of civilization, notably those that can overcome the gravity well. The human species is at the threshold of access to this cornucopia.[3] Other intelligent species may well have passed that threshold, a point worth keeping in mind.

2.2.3 *Why Habitability Zones*

Our life consists of a complicated chemical interaction between chemicals whose breakdown and reassembly provides us with our energy and the material of our bodies. There is good reason to suggest that a liquid is crucial for the emergence of life, since suspension in a liquid medium allows exchanges of chemicals, and later of cellular and genetic material which facilitates evolution. Not every liquid will do.

Empirically speaking, we have evidence that life evolved on Earth from, or within, a liquid or semi-liquid medium. This is true both for the emergence of very simple prokaryotic cells, and for the development of their more complex eukaryotic descendants into colonies of specialized multi-cellular beings, about a billion years later. Water has two relative advantages over other liquids as a cradle for life. First, both hydrogen and water are found in relative abundance in the universe, as they are extremely archaic and relatively light and they have a marked affinity for one another, creating H_2O molecules relatively commonly in cosmic terms. Second, water ice, peculiarly enough, floats *on top* of water. In our individual experience, most solids sink through liquids. Ice forms crystal-like structures whose volume is greater than the water the ice is formed from, and as a consequence, floats on top of the water. The same is not true for other potential liquids such as methane or sulphur, which have been suggested as candidates for a liquid medium for life. In those materials, the solid, ice form sinks to the bottom.

Why is this important? Ice forming as a crust on top of water serves as a shell allowing life to continue in the liquid that remains below. Lake Vostok, currently being explored in the Antarctic exemplifies that principle, which also raises hopes for the existence of life in the liquid under-ice oceans of Europa and Ganymede. Under normal weather conditions, atmospheric temperatures can drop below zero centigrade. This should not affect the existence, nor the exchanges of life processes necessary for life, so long as the bottom of the liquid body remains liquid. Any additional ice would accumulate on top, and eventually melt by solar insolation and heat, or crack, forming access corridors between the atmosphere and the liquid below. In contrast, methane ice is denser than methane liquid. The liquid, in effect, serves as a shield protecting the ice from heat and since the ice rests on the bottom of the liquid column, it is not subject to many stresses. Over time, the ice builds up, displacing liquid, and eliminating life processes.

---

[3] Whether the species turns out to be bright enough to access this wealth *sensibly* is another matter.

For many astrobiologists, the 'zone of habitability' is defined as the zone where water temperature never drops below 0 degrees Centigrade (at which stage it would be more or less a solid) or rises above 100 °C (in the form of steam or vapor). This *may* be 'carbon-centric' thinking (the Carbon-Oxygen exchange upon which our physiology is based works well with water), but may also be a universal truth. Very hot, very large (O, B, A types) stars suppress the chemical interactions by the sheer violence of their emission of heat and charged particles which destroy such interactions before they can start. The habitability zone—the distance from the star where water is liquid—is also for physical reasons very narrow, lessening the odds of a planet of suitable composition orbiting within the zone. Life *may* emerge in other chemical environments, and some thought has been given to that possibility as well, which changes somewhat the definition of 'habitability zone' (Shapiro and Feinberg 1982).

2.2.4 The Solar Neighborhood

There are some 14,000 stars within 100 light years of Earth, of which around 500 are spectral type G (similar to Sol in luminosity). There are in addition some 1300 F (yellowish) and K (reddish) stars within the same volume. Such stars—yellowish (hotter) to reddish (cooler) stars—seem to offer the best bet for the existence of life *as we know it*. The reasons for this have to do with the nature of stars.

As a general principle, the hotter the star and the larger, the newer it is, and the shorter its lifespan is likely to be. From the ILF perspective this provides serious obstacles to the emergence of life, let alone intelligence. On Earth, life took several billion years to emerge, and another billion to evolve into multicellular organisms that could devote enough energy to develop nerve tissue and brains. The evolution of culture—that is, for our purposes at the moment, the ability to develop artifacts to the point of being able to explore our speculations on the existence of other ILF (such as e.g. computers and radio-telescopes)—took several million years from the point at which our genetic ancestors first picked up a stone and shaped it to some purpose.

The other end of the spectrum—the M and N type stars—suffer from the reverse problem from a life perspective. The habitability zone is closer to the star, as it emits less energy. Planets are less likely to form or survive that close, as they may be ripped apart by solar tides. Nevertheless, life on planets around red dwarf M stars may be possible (Heath et al. 1999). M-class stars in the solar neighborhood, ignored until recently, are now (2016) being studied by the SETI Institute. For other reasons, other types of stars such as brown dwarfs and the fascinating array of exotic stars—quasars, neutron stars and the like—are even less likely to support planets, let alone those where 'mediocre' forms of life may have emerged.

2.2.5 The Galaxy and Beyond

The distances between stars within galaxies vary. Again, we can assume Sol is average in this regard, lying about two thirds of the way from the center to the rim along the Sagittarius arm of the galaxy we call the Milky Way. The Galaxy is composed of about 300 to 600 billion stars of varying sizes and temperatures. It is assumed to be about 1000 ly thick, and 100,000 ly in diameter. Gas and dust clouds veil most of its structure from our eyes, and much of the discussion of its structure relies on comparison with other, more distant galaxies that are visible. Some recent evidence suggests that the center of the Galaxy may be collapsing into a giant black hole (Genzel et al. 2010). In any case, the center of the Galaxy is the realm of closely packed stars. This too has implications for the existence of life. The closeness of stars in the galactic center may well increase radiation to the point that organic life may not be possible. Moreover, if as has been suggested, the center of the Galaxy is an enormous black hole, then the odds of life existing there go way down.

The Milky Way is not the only galaxy known. Two closer small galaxies, visible in the southern hemisphere of Earth are the Magellanic clouds. These two are much smaller than the Milky Way. Their structure is also more amorphous. The closest galaxy apparently similar to our own is the great Messier M31 Andromeda galaxy, composed of some one trillion stars. It too is a spiral galaxy like the Milky Way though twice as large. There are some thirty galaxies in our local cluster, and there are numerous clusters throughout the observed universe. In addition, space is rife with a multitude of observed physical material. Clouds of gas darken and shroud the light of stars and galaxies. There are large star clusters, too small to be considered galaxies on their own, which constitute nebulae. There are several types of hypothesized bodies, ranging from black holes (stars that have collapsed in on themselves to the point that the physical laws we know can no longer account for them), dark stars or protostars (large dark bodies that have not-yet ignited), brown stars, solitary wandering planets, neutron stars, and so on.

The point, again, is not to enumerate the wonders of the observable universe, but to indicate the ILF dimension: the universe, insofar as our understanding of physics goes, cannot be traversed, but it can be observed.

2.2.6 Universe

The universe is difficult to conceive of, let alone describe. For all ILF intents and purposes, the universe is infinitely large. That astronomers may *perhaps* be able to see to the limits of the universe means little to us or other ILFs, except insofar as it yields information regarding the universe itself. The ends of the universe are so far that they are out of reach, physically as well as conceptually.

What is important for us about the universe, is that it is the matrix for the operation of a set of physical laws. Whether those laws are inevitably supportive of life, as some (Margulis and Lovelock 1974; Gardner 2003) would argue, or whether other laws exist locally in the universe, is something to consider. But the truly important fact is that, insofar as we are able to tell, the physical laws under which the universe operates, are applicable throughout it. That is, the laws of gravity, light, atomic and subatomic interactions are as true on Earth as in Andromeda and far beyond.

In discussing universal rules an intriguing possibility must be mentioned. There is some theoretical evidence that other universes may also exist, in forms that appear inconceivable except to mathematical physicists (Wolf 1988). What this implies for ILFs is hard to conceptualize. The possibility is mentioned here merely to indicate that there may very well be *no* limit to the universe, if it is conceived of as a universe of universes: of possibilities that exist side-by-side without end.

2.3 Physical and Non-physical Contact

What the preceding discussion of space means, in terms of communication, is that there are a number of possibilities. Though interstellar physical contact is enormously—many astronomers and physicists would say "prohibitively,"—expensive, some cultural quirk, dire need, or driving urge might cause ETIs to contact us physically. This is rather far-fetched, but we must consider the possibility. In point of fact, STL contact, whatever conservative physicists say, is not prohibitive *provided* a civilization is prepared to pay the cost (Hodges 1985).

Still faster alternatives may exist. At constant 1 g acceleration, a vehicle would reach 99 % of light speed in 3,265,3061 s, or 9000 h: a bit over one year. The problem of providing power for such huge acceleration, and then deceleration is something we will not consider directly. The mass considerations *if a reaction motor is used* are horrendous. In practice, and notwithstanding time dilation for the passengers at that theoretical velocity, other stars are out of reach for immediate purposes.

The preceding very brief description of the physical universe must return us to my original point: the cosmic factors are important in this discussion only inasmuch as they fit ILF dimensions. The great distances between the stars, or between the galaxies, are only important (not 'interesting' which is a different kettle of fish) because of the value, or the difficulties, that they present to ILFs.

Now, interstellar distances may well be insurmountable—socially, economically, politically—for humans. But for an ILF with different motivations and a different economic persuasion, perhaps not so. Keep in mind too, that other human societies have indeed done things which are roughly comparable: Columbus is one example, Polynesian voyagers sailing to Easter Island another (Finney and Jones 1985). In each case a large amount of economic wealth were invested in an attempt to go someplace else. The economic return for the endeavor was always in doubt, but the motivation for the effort was not necessarily economic: while the Spaniards were anxious to be rewarded with the wealth of the Indies, they also wanted to spread religion. Polynesians had limited land resources and expanding populations, so long-distance travel was the only solution they could come up with. And though it can be demonstrated that energy considerations would be a major barrier for any given civilization from spreading into interstellar space, there is no reason to suppose that this is true for *all* species at *all* stages of technological development and knowledge.

For all ILFs anywhere in the universe, three factors are important: energy, matter, and distance. All three of these should be measured in ILF, not cosmic terms. Energy refers to those types of energy the ILF can use or enjoy: not to hot *for it*, not to cold *for it*; usable. Matter refers to solid points: star systems, planets, moons, dust clouds, providing both reference points and the very material of life, as well as fuel for exploration. Obviously this is related very closely to the issue of energy. The reasoning is the same for the discussion of energy: we are interested in the ILF, not the physics. Finally, distance is a factor that too is measured in ILF terms. Put very simply, the distance between Greenland and Africa is very different for an ant walking it, a man canoeing it, or a bird flying it on its yearly migration. Life span, other activities, energy needs, all conspire to define that elusive factor called "distance" for different life forms. Certain life forms (very simple ones, such as bacteria and viruses, and to a lesser extent, some more complex ones such as lungfish and cicadas) have the ability to aestivate (suspend their natural processes of anabolism and catabolism) for a period ranging from a season to several years. The technology of slower-than-light star travel is not too complex, theoretically, and given the desire to do so, we could probably send humans to the nearest stars. Yet we define that as 'impossible'. Why is that? Because in human socio-cultural terms it is indeed impossible. Under some configurations, the trip to one of the nearest stars (Barnard's star: 6 ly, the Centauris: 4.3 ly) would take only several years subjective. *Objectively* however, the time would be so great that few individuals would be willing to contemplate the trip. This is because in *our* terms, both social, psychological, and physiological, there are too many barriers. But what if we could, and did aestivate? Much of the cultural generational notions (i.e., I am younger than my parents' generation and older than my children's) would not exist. Moreover, we would not have the physiological limitations on life span (we actually would, but they would undoubtedly appear different to us) and therefore the two hundred years spent traveling would not be that much of a burden. Thus 'distance' for aestivators and for non-aestivators represent two subjectively different things even if physically they amount to the same distance in kilometers, light years, or inches.

The various arguments pro- and con- physical exploration and colonization of the universe are not provided here in order to enter into the debate of whether 'we' or 'they' can visit another star system. They do suggest that one must carefully examine the assumptions that are used to discuss CETI. We *cannot* assume, from the start, that ETI are likely to affect us in any particular way (see for example the

Zoo hypothesis, of which more subsequently). What we can do is suggest what the parameters and limitations of certain sorts of behaviors *might* be.

The major point of this chapter has been to describe very briefly the background against which the issue of xenology must be discussed. It is also important to reemphasize that the dimensions and background *must* be seen in ILF terms, not those of a physicist. That having been said, it is necessary to identify what 'ILF-terms' are. For that, it is necessary to attempt the construction of some parameters of ILFs, starting from the simplest building blocks and what we know about them, and progressing gradually, with ever-growing uncertainty, to some assumptions about the parameters of intelligent life under various circumstances.

2.4 The Microcosm

The stars and their dependents are the consequence of interactions between features at atomic and sub-atomic levels. These interactions have an effect on life as we know it, and certain features are obviously of interest if we are speaking of life elsewhere in the universe, insofar as those features are truly universal.

Stars are composed of elements, largely hydrogen and helium, and also are the furnaces in which heavier elements, including metals and the very heavy transuranics, are created by fusing lighter elements. The physical laws by which this activity takes place were determined, for reasons that are still debated, at the initial moment of the universe (cf. Hawking 1988).

These universal features affect us in a number of ways. First, they determine the form of the universe, and thus the forms we take and *can take*. Certain possibilities we could conceive of are simply not possible, given the laws of the universe as they are. This is important for the continuing discussion here, because these laws provide us with a boundary for considering the possibilities of life, and thus of ETI.

Second, the basic laws may well determine factors ranging from gross morphological features of a species, to the ways in which a species may or may not organize itself into intelligence. We need not consider the possible intelligence of a rock if it lacks certain organizing characteristics that are the result of truly universal rules.

Finally, the basic laws of the universe determine the possibilities, and the means by which we can communicate with ETI. For example, one derivative of the basic microcosmic laws of the universe is the limitation on velocity C. Grossly stated, a physical object cannot travel faster than 299,792 km per second. This is an absolute limit *insofar as we know*. It places limitations on communication, physical visits, travel in the universe and so on.

This final range of features will become less important as the discussion here progresses. "Insofar as we know" allows for an escape clause, one that will allow me to ignore the issue at some points. Nonetheless, the effect of these basic universal rules is profound and overall, unavoidable.

2.4.1 Evolution in the Universe

The universe—whether considering the entire cosmos or merely our part of it—is in constant movement and ferment. In essence, the universe is undergoing a persistent process of change at all levels. In fact, this process is often indistinguishable from a process of evolution. Evolution, in its grossest sense, means the development of more, complex forms, from fewer, simpler ones. This happens in the universe at the level of atoms, where complex hydrocarbons are formed through a random process like the one that affects biological evolution. It also happens at the levels of stars. Older population stars, formed "soon" after the Big Bang which initiated our universe, are giving way to younger stars, some of them complex and variable. Evolution is thus not only a biological rule. To the contrary. It would appear that biological evolution, the evolution of species that we normally think of when we consider the term, is but a special case. Under the proper circumstances, such as the hearts of stars, heavy elements are evolved from lighter (and simpler ones). Interactions between some of the molecules of these elements cause the creation of complex molecules in space. On Earth (and possibly elsewhere) elements stimulated by energy can create long molecule chains: amino acids that are the precursors of life as we know it.

This universality is crucial for ETI. If evolution is indeed the universal law it seems, then we have the basis for some comparisons that go beyond the Earth we know. We can speculate intelligently on possibilities elsewhere.

2.4.2 The Chemical Basis

The interaction of atoms, brought together in larger groups of molecules, constitutes what human science arbitrarily defines as a specific discipline: chemistry. The interaction of atoms requires energy, as does any other interaction, and certain inviolable laws of chemistry derive from the basic laws which compose the universe: certain elements interact better, or to a greater degree with some, and worse, or to a lesser degree with others. These differences determine the conceivable building blocks of organisms that can be termed living. Insofar as we know, these laws apply throughout the universe. That is, an oxygen atom and a carbon atom are the same throughout the universe, as are their features. Consequently, under the same circumstances, their interactions will be similar, whether on Earth or on a planet across the Galaxy. This is important for it allows us to predict what combinations are impossible, which are possible, even which are likely.

Life on Earth is based on the reaction and combination of several elements, the principle of those being oxygen and carbon. Other elements—sulphur, potassium and metals such as iron—are essential for our form of life, and may be essential for the energy exchanges that power all life. These interactions, in themselves, are sufficient for the creation of very varied life forms, as they have done on Earth. At the molecular and unicellular level, life could have risen through several distinct pathways which are not mutually incompatible (cf. Russell and Hall 1997; Lahav et al. 2001). This has significant implications. If life has arisen several times *and succeeded* on Earth, then given similar conditions, it may well arise elsewhere. Whether it becomes complex enough to form multicellular organisms, even intelligence, is a separate issue (Bainbridge 1983).

2.4.3 That Thing Called Life

Well, what then is life? Or, again, to use our particular lens of the ILF perspective, what can we identify as life? The question has, of course, many ramifications, ranging from the purely taxonomic and scientific, to the religious and theological, even artistic. The abstract question of 'what is life' is less critical for the issue of ETI, than the concrete issue of the *product* of life.

However. Roughly speaking, for biologists, life is identified by the presence of three characteristics:

- Self-reproduction (that is, some copying of the parent organism)—without which genetic information would be lost after each generation.
- Mutation—without which genetic information is 'unchangeable' and hence cannot even arise.
- Metabolism—without which the system would regress to equilibrium, from which no further change is possible (Murphy and O'Neill 1995).

We therefore come closer to accepting that evolution is a natural law for the universe: the development of complexity from simplicity.

2.4.4 Biological Basis and Biological Preconditions

Once life has arisen, it starts a process of interaction, with other life forms and with its environment. On Earth that interaction *apparently* has been one of competition for resources, and consequently of the evolution of more complex forms. This process, the Darwinian process, may or may not be universal. What is necessarily the case is that life will, wherever it emerge, start interacting with its environment in more-than-mechanical ways. On Earth, the competition for resources led to the evolution of eukaryotic cells. Unlike their predecessors, the prokaryotes, basic eukaryotes have a cellular membrane which protects the cell from its environment, and a cellular nucleus which contains genetic information. This means that eukaryotes are better able to survive environmental changes than prokaryotes, and that their survival strategies (e.g. the instruction 'develop a membrane') can be 'taught' to other eukaryotes *if* a method of exchanging genetic information can be developed as well. Which it has been. Once these principles—'develop specialized cellular elements', and 'exchange information for survival'—have been implemented, one can also learn how to specialize, aggregating individual cells into simpler, then more complex colonies. The rest, as they say, is history.

Emergent Properties
A significant issue in the emergence of life is what is now being called "emergent properties" a concept formalized through the study of complexity. Roughly put, the properties of a complex body of any sort are not predictable from the properties of the components that make it up. Thus, while living things on Earth are made up of proteins carbohydrates and fats (all types of molecules) mediated by DNA and RNA (other types of molecule) the molecular interaction is insufficient to predict, let alone explain, the American Declaration of Independence, the Forbidden City in Beijing, or the works of Beethoven. The complex specialization-cooperation-competition of colonies of eukaryotes (you and I included) is insufficient to explain cultural products, or even social practices. From our perspective, the critical implication of emergent properties that needs to be kept in mind is that *it will be impossible to predict the complex features of ILFs—their social arrangements and culture—from elementary principles*. In other words, familiarity with an ETI's molecular or cellular composition may tell us something about what that being is *not*, but very little about what it *is*, let alone what it thinks, produces, or communicates. On the plus side, however, knowing that ETI are most likely formed as we are—some arrangement of cooperating, specialized cells and elements—means that the evolution of complexity is a universal law. This principle is important, because *maybe* we can assume the same principle is true for more complex behaviors as well.

2.5 Conclusions to Chapter 2

One major important fact emerges from this chapter. *If* there exist intelligent life forms, they *can*, and perhaps will come into contact. This is the premise of this book: that indeed contact and communication will occur, and that initial steps must be taken to ensure that such contact is fruitful and successful.

All of the physical and cosmological issues discussed above are the stage for ETIs. They also point out some of the difficulties we are likely to face in the search for ETI. They do not, however, indicate with any convincing data that there are, or there are not ILFs beside us. The question is, and will remain for some time, a moot issue. In the final analysis it can only be decided empirically, and, in the absence of proof to the contrary, we are free to assume that the phenomenon exists. First, negative evidence do not prove the existence or absence of a phenomenon. The fact we are *unable* to find ETI is not the same, scientifically, as the fact of their nonexistence. Second, no less importantly, the SETI effort is in a sense an insurance against the positive case: that there are ETI somewhere in the universe. The potential benefits of searching for them far outweigh the possible risks. Finally, there is also the issue of serendipity. Whether SETI does or does not achieve its ultimate goal, along the way, as is the nature of scientific research, many other sets of valuable data, of techniques, of theories, are likely to emerge. These may well create undreamed of-as-yet technological spinoffs.

References

Ashkenazi, M. (1995). Will ETI be space explorers? Some cultural considerations. In G. S. Shostak (Ed.), *Progress in the search for extraterrestrial life*. San Francisco: Astronomical Society of the Pacific Conference series#74, pp. 507–20.

Bainbridge, W. S. (1983). Attitudes towards interstellar communication: An empirical study. *Journal of the British Interplanetary Society, 36*, 298–304.

Ball, J. A. (1973). The zoo hypothesis. *Icarus, 19*, 347–49.

Battison, L. (2011). Niche habitats for extra-terrestrial life: The potential for astrobiology on the moons of Saturn and Jupiter. *Journal of Cosmology, 13*, 3890–3903.

Beckwith, S. V. W. (2008). Detecting life-bearing extrasolar planets with space telescopes. *The Astrophysical Journal, 684*(2), 1404–1415.

Beichman, C. (2002). Captain cook, the terrestrial planet finder and the search for extraterrestrial intelligence. F - arXiv:astro-ph/0212336.

Finney, B. R., & Jones, E. M. (Eds.) (1985). *Interstellar migration and the human experience*. Berkeley CA: University of California Press.

Bergerac, C. de. 1923. (1662). *Voyages to the moon and the sun* (R. Aldington, Trans.). London/New York: Routledge & Sons Ltd/E.P. Dutton & Co.

Billingham, J. A. (Ed.). (1981). *Life in the universe*. Boston: MIT Press.

Bracewell, R. L. (1981). Manifestations of advanced civilizations. In J. Billingham (Ed.), *Life in the universe* (pp. 343–350). Boston: MIT Press.

Brin, G. D. (1983). The 'great silence': The controversy concerning extraterrestrial life. *Quarterly Journal of the Royal Astronomical Society, 24*, 283–309.

Ćirković, M. M. (2001). Is many likelier than few? A critical assessment of the self-indicating assumption. F - arXiv:gr-qc/0110002.

Crowe, M. J. (1986). *The extraterrestrial life debate 1750–1900*. Cambridge: Cambridge University Press.

Dick, S. J. (1999). *The biological universe: The twentieth century extraterrestrial life debate and the limits of science*. Cambridge: Cambridge University Press.

Dick, S. J. (2013). The societal impact of extraterrestrial life: The relevance of history and the social sciences. In D. A. Vakoch (Eds.), *Advances in Astrobiology and Biogeophysics* (pp. 227–57). Berlin, Heidelberg: Springer Berlin Heidelberg.

Drake, F. D. (1967). *Intelligent life in space*. New York: Macmillan.

Drake, F. D., & Sobel, D. (1992). *Is anyone out there? The scientific search for extraterrestrial intelligence*. New York: Delacorte Press.

Einstein, A. (1955). *The meaning of relativity*. Princeton: Princeton University Press.

Exoplanet.eu. (2016). *The extrasolar planet encyclopedia*. http://Exoplanet.eu. Accessed July 4, 2016.

Forgan, D. H. (2009). A numerical testbed for hypotheses of extraterrestrial life and intelligence. *International Journal of Astrobiology, 8*(02), 121–131.

Forgan, D. H., & Elvis, M. (2011). Extrasolar asteroid mining as forensic evidence for extraterrestrial intelligence. *International Journal of Astrobiology, 10*(4), 307–313.

Freitas, R. A. (1980). A search for natural or artificial objects located at the earth-moon libration points. *Icarus, 42*, 442–447.

Gardner, J. (2003). *Biocosm: The new scientific theory of evolution: intelligent life is the architect of the universe*. Makawao Maui HI: Inner Ocean.
Genzel, R., Eisenhauer, F., & Gillessen, S. (2010). The galactic center massive black hole and nuclear star cluster. *Reviews of Modern Physics, 82*, 3121–3195.
Gleiser, M. (2010). Drake equation for the multiverse: From the string landscape to complex life. arXiv:1002.1651.
Hart, M. H. (1982). An explanation for the absence of extraterrestrials on Earth. In M. H. Hart & B. Zuckerman (Eds.), *Extraterrestrials: Where are they?* (pp. 1–8). New York: Pergamon.
Hart, M. H., & Zuckerman, B. (Eds.). (1982). *Extraterrestrials: Where are they?*. New York: Pergamon.
Hawking, S. (1988). *A brief history of time: From the big bang to black holes*. Toronto [u.a.]: Bantam Books.
Hawking, S. (1993). *Black holes and baby universes and other essays*. London: Bantam Press.
Heath, M. J., Doyle, L. R., et al. (1999). Habitability of planets around red dwarf stars. *Origins of Life and Evolution of the Biosphere, 29*(4), 405–424.
Hodges, W. (1985). The division of labor and interstellar migration: A response to 'demographic contours. In B. R. Finney & E. M. Jones (Eds.), *Interstellar migration and the human experience*. Berkeley: University of California.
Keszthelyi, L. P. (2011). Planetary science: Europa awakening. *Nature, 479*(7374), 485–485.
Kingsley, S. A. (2001). *Optical SETI observatories in the new millennium: A review*. http://www.coseti.org/pdf/4273-06.pdf
Lamm, N. (1978). "The religious implication of extraterrestrial life. In A. Carmell & C. Domb (Eds.), *Challenge: Torah views on science and its problems* (2nd ed., pp. 354–398). New York: Feldheim Publishers.
Levin, G. V. (2010). Extant life on Mars: Resolving the issues. *Journal of Cosmology, 5*, 920–929.
Maccone, C. (2010). The statistical Drake equation. *Acta Astronautica, 67*(11-12), 1366–1383.
Margulis, L., & Lovelock, J. E. (1974). Biological modulation of the Earth's atmosphere. *Icarus, 21*, 471–489.
Murphy, M. P., & O'Neill, L. eds. (1995). *What is life?: The next fifty years: Speculations on the future of biology*. Cambridge: Cambridge University Press.
Noam, L., Nir, S., & Avshalom Elitzur, C. (2001). The emergence of life on Earth. *Progress in Biophysics and Molecular Biology, 75*(1–2), 75–120.
Oliver, B. M. (1976). Proximity of galactic civilizations. *Icarus, 25*, 360–67.
Oliver, B. M. (1986). Technological civilizations. *Journal of the British Interplanetary Society, 39* (7), 327.
Petigura, E. A., Howard, A. W., & Marcy, G. W. (2013). Prevalence of earth-size planets orbiting sun-like stars. *Proceedings of the National Academy of Sciences, 110*(48), 19273–19278.
Petit, J. -M., Alessandro M., & John C. (2001) The primordial excitation and clearing of the asteroid belt. *Icarus, 153*(2), 338–47.
Renard, J. B. (1986). La Croyance Aux Extraterrestres: Approche Lexicologique [Belief in Extraterrestrials: A Lexicological Approach]. *Revue francaise de Sociologie, 27*(2), 221–29.
Robitaille, T. P., & Barbara, W. A. (2010). The present-day star formation rate of the milky-way determined from Spitzer detected young stellar objects." F - arXiv:1001.3672.
Russell, M. J., & Hall, A. (1997). The emergence of life from iron Monosulphide bubbles at a submarine hydrothermal redox and Ph Front. *Journal of the Geological Society, 154*(3), 377–402.
Sagan, C., & Drake, F. (1975). The search for extraterrestrial intelligence. *Scientific American, 232*, 80–89.
Sagan, C., & Shklovskii, I. S. (1966). *Intelligent life in the universe*. New York: Dell Books.
Shapiro, R., & Gerald, F. (1982). Possible forms of life in environments very different from the Earth. In M. Hart & B. Zuckerman (Ed.), *Extraterrestrials: Where are they?* (pp. 113–121). New York: Pergamon Press.

Tarter, J. (1985). Searching for extraterrestrials. In E. Regis (Ed.), *Extraterrestrials: Science and alien intelligence* (pp. 167–189). Cambridge: Cambridge University Press.

Tipler, F. J. (1981). Extraterrestrial intelligent beings do not exist. *Physics Today, 34*, 9–81.

Wolf, F. A. (1988). *Parallel universes: The search for other worlds*. New York: Simon and Schuster.

Wolf, F. A. (1989). *Taking the quantum leap: The new physics for nonscientists*. New York: Perennial Library.

Yung, Y. L., Russell, M. J., & Parkinson, C. D. (2010). The search for life on Mars. *Journal of Cosmology, 5*, 1121–1130.

Chapter 3
The Physical Parameters of ILFS

We do have some examples of life evolving, and, moreover, have some idea under what conditions life on Earth *has* evolved. The objective of this chapter is to "… examine Earthlife and consider which of its large scale features are essential and which are merely local solutions to a general problem" (Feinberg and Shapiro 1980: 402).

The question of what life *is* has some bearing on the entire question dealt with here. Numerous definitions have been proposed. "An organism will be considered living if it is able to select between several possibilities. It will select that possibility that is least harmful" (Fasan 1986: 4), seems appropriate until one examines it semantically. Surely life is somehow self-organizing, possibly reproductive, and changes its environment. Beyond that we run into theoretical realms that have the usual problem: theorizing from a single point: life on Earth (Cleland and Chyba 2002).

3.1 The Physical Parameters of Ilfs

Whatever else they may be or may do, ILFs must obey the basic physical laws of the universe. This is based on the assumption (which may *not* be correct, but which must be accepted *pro-tem*) that the gross physical laws of the universe that we experience on Earth exist throughout the universe, and that we know, at least in rough terms what they are. At least initially, I am forced to ignore "exotic" possibilities which deal with the existence of life on, for example, neutron stars (Feinberg and Shapiro 1980; Shapiro and Feinberg 1982). Life elsewhere must obey the same basic physical and chemical rules as do our own. This does not mean, of course that ILFs are or must be physically similar to ourselves. The contrary is virtually certain (Gould 1983). Rather, we must assume that the make up of their biochemistry and physical construction: atomic, molecular, and mechanical, must obey the same laws.

At a basic level, therefore, when we search for a common denominator, we must consider the issue of energy. Life, even artificial constructs of life such as computer

M. Ashkenazi, *What We Know About Extraterrestrial Intelligence*,
Space and Society, DOI 10.1007/978-3-319-44456-7_3

programs in operation, consist of organized regularized exchanges of energy. It is to these exchanges that we must direct our attention. Simply put, what forms of energetic exchange are *possible*, and perhaps which are, given our knowledge of the universe *probable*. Secondly, we should consider the mechanical aspect. This is not because we are attempting to create a possible 'zoo' of life forms, but so we can ask whether the mechanics are likely to have second or third order effects that can be extrapolated about the effects of form on intelligence and culture.

3.1.1 Chemical Energy Creation and Exchange

At a basic level, life consists of an organized, replicable exchange of matter and energy where complex molecules serve as intermediaries between simpler ones. That is a complex way of saying that simple molecules such as oxygen, carbon, iron, and others perform a complicated quadrille with intricate figures, in which certain molecules serve as channels for conveying energy created by these exchanges.

With the exception of exotic life, it seems probable that life, if it evolves at all in the Galaxy, will have evolved on much the same basis as life on Earth. That is, since there is a high prevalence of organic molecules in space (Ehrenfreund and Charnley 2000), and since it has been proven experimentally that archaic planetary atmospheres, water, and energy input from lightning and ultra-violet radiation can create organic molecules (Miller et al. 1976; Cleaves et al. 2008; Trefil et al. 2009), it is reasonable to assume that the basic principles of life can occur elsewhere. Therefore it is highly probable that life has emerged elsewhere, even though we do not have to accept Christian's inference that life is necessarily common in the universe (Christian 1976: 20).[1]

There are also good reasons to suggest that the commonest, and perhaps only form of life is that based on a carbon-oxygen cycle (Chela-Flores 2007). There are a number of alternative candidates to carbon-oxygen that have been proposed over the years (Seckbach et al. 2008) as a basis for life. Unfortunately, all of them suffer from major drawbacks. The carbon molecule is demonstrably the most versatile and tenacious in its ability to form long molecular chains: a necessity for the creation of life, and no less importantly, for the retention of long strings of information. Without these long carbon-based chains, it seems that reproduction would not be possible, because long chains imply the ability to store information, thus to reproduce. Oxygen reactions (in combination with sulfur and potassium) yield more energy than any other comparable chemical reaction we are able to conceive of.

[1]It must, however, be noted that this is a contentious argument. Astrobiologists and others have come to the opposite conclusion which is also valid (see Gould 1983). Given the perspective of this book, we choose to ignore the reefs of caution. Sail on.

Thus the carbon-oxygen basis for life still seems to be the leading, perhaps only candidate.

3.1.2 *Significant Physiological Parameters*

Any organism has a vast number of physiological parameters that can be measured. These can yield important clues about the animal's behavior, some of the parameters do so better than others. For the purposes here, many of these parameters are irrelevant, or their importance and effect are not well documented, or they lie in the realm of speculation. But some physiological "realities" are very important, and perhaps can be derived from an interaction of the understanding of physical processes, and our understanding of environmental, in this case, the grossest planetary, effects.

Roughly, the physical parameters we are concerned with are those that could have an effect on the generation of intelligence and culture. Here we are forced to rely more on terrestrial analogues than I would like, and the limitations of the model will soon become apparent. The parameters of interest fall into two groups: gross physiological characteristics that have to do with intelligence and its generation, and physiological characteristics that have to do with the sensorium: the means by which an organism gathers information on the outside world. Both of these groups of characteristics are likely to be affected systematically by environmental, that is, planetary factors.[2]

3.2 Planetary Effects

Life forms we are familiar with have evolved in a planetary environment. Whether this is an absolute prerequisite for life depends on whether one entertains the possibility of 'exotic' life discussed in a later section. It seems likely, on the face of it, that planetary existence is a conservative option and will be the common one, and therefore it is necessary to examine the effects of the planetary environment.

There are, of course, a large number of variables to consider. Some may well be hidden, and we can only speculate on many. A point to start from is Dole's seminal work on planetology. He suggests several important affects a planet might have on *humans* trying to live there. To enlarge his argument, I will use the same parameters to suggest factors that might affect life in general. The important effects seem to include *gravity*, *temperature*, *light*, *atmosphere*, *liquid*, and *radioactivity* (Dole 1970). Some more recent studies have added other parameters, including ultra-violet radiation (Buccino et al. 2006).

---

[2]Again, leaving exotic life forms apart for the time being.

3.2.1 Gravity

Gravity, that mysterious force, is crucial in two senses. Gravity as a constant from the ILF perspective. It allows resources—chemicals and energy—to coalesce so that life can occur and once occurring, can go on. In the absence of gravity, molecules would move about, dissipating energy, attenuating the fabric of resources, and probably disallowing life. Different gravities (larger or smaller, less or more dense planets) are likely to affect the gross structures of organisms. There have been a number of experiments examining the effects of micro gravity on growth and development (Dole 1970; Matía et al. 2010). The data from these are invalid for our purposes however. All of the biological material used in these experiments had its origin on Earth. It must be assumed, therefore, that it has been conditioned by generation after generation of development in a particular gravitic environment. The same is true with data from higher animals.

Certain physical effects of gravity must therefore be considered in the abstract. The higher the gravity, the faster things fall and the greater their weight (*not*, I trust I need not add, their *mass*). This means that high gravity will probably cause the emergence of organisms that have lower centers of gravity, greater support, possibly evolve body architectures that will reduce body weight, and faster reflexes (or more forethought?) to deal with the greater speeds of falling objects. Complex organisms that evolve on heavy gravity planets are likely to be smaller in absolute mass, have a complex support structure for their internal architecture (e.g. hollow bones or struts), and will have faster reflexes. It is also possible that they may never leave the sea, or if life evolves on land, will return to a marine habitat, since the effect of suspension in liquid allows mass more freedom. This may imply that technological intelligence may never evolve on higher gravity planets (how high is at present unknown and unguessable) *if* assumptions about the land origin of intelligence are correct. Another factor that should be considered is reproduction. Human babies are born embryonic. That is, less well-developed in terms of motor activity than other primates. This is because of the large head size that encloses an enlarged brain. As a direct consequence humans are more social than other animals (they learn by acculturation, rather than instinctively or purely by imitation). The large size of the human infant's head affects the size of the human pelvis and the human stance. High gravity life is likely to carry babies to shorter term (or perhaps be strictly oviparous). In such a case, if intelligence *does* develop, then it will likely be even more reliant on socialization, and could be a more social animal than humans. Whether this implies a less aggressive intelligent animal or not, is worth considering.

Lower gravity complex life may well be larger, weaker, and slower. That is, reverse effects to those in high gravity may operate on life evolving under lighter gravity. Another effect pointed out by Dole (1970: 129) has to do with radioactivity. In lower gravity planets, there would be less fractionalization of rocky materials during formation, which would mean a higher percentage of radioactive materials in the crust, thus higher local radiation and consequent greater rate of mutation. This supports the possibility that body forms might be varied, and that evolution might well be speeded up. It is possible that the evolution of an intelligent life form, if at all possible, would take less than the 3.5 billion it took on Earth.

3.2.2 Temperature

The effects of temperature must be considered in terms of absolute temperature, and temperature fluctuations (Dole 1970). Planetary temperature is a complex function whose basic variables include star radiation, distance from star, and planetary inclination. Greater inclination means that the planets temperature and seasonal variations will be much greater. The absolute limit for ILFs, assuming carbon-oxygen life form, would seem to be the range 0 < to < 100 centigrade. At or below an average temperature of 0, liquid water cannot exist, and life will probably not emerge. The same is true of temperatures above the boiling temperature of water. This is not because of the temperature itself (some terrestrial organisms live in higher temperatures), but because most of the water will turn to vapor, which means many of the characteristics of water that encourage life will be lost.

As a general rule, higher temperatures will require more mechanisms to disperse heat (mechanical, cellular or chemical), colder temperatures the reverse. High temperature organisms will likely have large surface areas relative to mass, and perhaps lesser bulk. Lower temperature organisms will likely have smaller surface areas relative to mass and some thermal control over their epidermis. They might also have lower blood circulation and thicker circulatory fluids.

Highly variable temperatures are of course going to upset this neat scheme. Great variability in a planetary climate, because of large orbit inclination or planetary tilt will not only make the habitable zone smaller. Other effects will be possible too. Very great, long term perturbations, such as those caused by a long highly elliptical orbit, might bring about two separate kingdoms: hot climate and cool climate, that will alternate in some form of estivation or seed form until the conditions are ripe, much like desert plants on Earth. In such a case, the emergence of intelligence is highly doubtful, since the possibilities of cultural transmission will be low. Another alternative might be some form of cold weather/hot weather dimorphic pattern that is seen in superficial form on Earth. Ermine, snow hare, and snow owl all display seasonal dimorphism in their coloration. It is at least conceivable that a similar mechanism may cause more radical changes in a species, or, indeed, an entire biosphere.

3.2.3 Light

The insolation from a star is going to have an effect on the food supply (assuming, which is likely, that photosynthesis—the conversion of basic molecules into storable and usable sugars—of some form evolves) as well as on the development of organs such as eyes. Adapting to the range of electromagnetic spectrum we term visible light has some advantages. Visible light is a compromise. On the one hand, the wavelength is short enough to yield a great deal of information, that we interpret as color, shape, and so on. On the other hand, the wavelength is not *too* short, so that biologically generated receptors have fewer problems with it. A visible spectrum may well extend into the infra-red, as it does for terrestrial vipers, or into the ultra-violet, as it does for bees. Extending it well beyond might create insuperable problems for a biological organism.

Light is not a prerequisite for the emergence of living organisms, yet it will affect the morphology. For example, organisms evolving in a light-poor, cooler environment, will require senses that will operate in low-light conditions. These may be (based on Terrestrial analogues), magnetism, heat, sound, pressure. Each of these possibilities would require specific *physio-mechanical* evolution to support the sense. Reliance on sound would require large receptors, on magnetism some organ like a shark's ampullae of Lorenzini, on heat the control of heat exudation for protection, rather like camouflage among Terrestrial animals.

3.2.4 Atmosphere

The composition of the atmosphere will affect biology as well. Provision of more or less oxygen will have an effect, as well as trace gasses, CO_2, atmospheric dust, and so on. The availability of the combustive gas used—probably oxygen, though we can keep methane in mind—will affect energy levels. Up to a point, the presence of more of the combustive gas will mean greater possible energy levels. Less of the gas will mean lower energy levels for the organisms in it.

Atmospheric particles are also important. Their composition, size, and saturation of the atmosphere will affect methods of breathing and thus physiology. Beyond the senses, atmospheric particles might well also determine which parts of a planet may be habitable or uninhabitable in particular times. Moreover, atmospheric particles block certain wavelengths of light, which in turn have implications for the wavelength relied on for vision.

3.2.5 Liquid

As noted above, the presence of free-flowing water is considered critical for the evolution of organisms and their ability to continue surviving. Other liquids have been proposed as well, but water still seems to be the best candidate. In addition to flowing water, the amount of water vapor in the atmosphere is probably important. What role it would play in structuring life can only be guessed by examining possible poles. Roughly speaking, the dryer the atmosphere, the more organisms will have to expend on maintaining their internal moisture. This may mean the evolution of moisture traps in ET physiology, possible insulation from the atmosphere, radiant cooling (if necessary) rather than respiration cooling which is common among Earth mammals, and so on.

Very moist atmospheres with a saturation of close to 100 % will not have those problems, but may well require defenses against airborne spores and microorganisms. ET will also not be able to cool themselves through respiration and will likely have to evolve other means of maintaining steady body temperature, or alternatively, very complex means to allow delicate nervous tissue (or its analog) to deal with poikilothermy.

3.2.6 Radioactivity

Generally speaking, the higher the incidence of radiation (from the star or from the planetary crust), the greater the rate of mutation. Mutations, in general, are not beneficial. For life to succeed on a high radioactivity planet it will have to evolve some mechanism to deal with the radiation. This may be protection of some form. An alternative may be high rates of reproduction and offspring mortality to ensure the continuation of the genetic line. It may also means that the entire concept of 'species' so useful for terrestrial biology, will no longer be applicable. Terrestrial speciation is a matter of identifying and ordering relatively stable characteristics: this will not be possible. Life on a highly mutagenic world might simply emerge as variant shapes, without any apparent taxonomical ordering, and, if this is the case, with genetic reproduction spread among all organisms. That is, the biota will consist of organisms of different shapes and physiologies which share only in their ability to reproduce. Evolution, as understood on Earth, may not occur: greater ordering of traits will be as likely as lesser ordering in any generation, even after the non-viable mutations are culled.

Lower rates of radiation will lead to the opposite effect: lower rates of change and mutation and possibly slower biological changes over similar spans of time. We do now know that radiation is not the sole mechanism for Darwinian genetic variability. Nevertheless, it constitutes a significant mechanism and cannot be ignored.

3.3 Physiological Limitations

It should be clear by now that each of these specific physical environments imposes certain burdens, just as it gives certain advantages. To give one example from one planet. Given our gravity (9.8 m sec/sec) it can be demonstrated *empirically* that there are certain mechanical limitations on size: African elephants and the larger dinosauria are about as large as a land organism can get. Even so, such animals are "designed" by nature very carefully in terms of the mechanics of their skeletal and muscular structures: dinosaurs had to have hollow bones and latticework skeletons. They also have, or had, major limitations in terms of their ability to move those structures. In other words, there is a size constraint which presumably goes up as gravity goes down. This has to do with physical/mechanical factors. Assume an elephant whose bones were constructed of some light, strong alloy such as titanium, rather than calcium. Could the elephant be larger? Possibly yes. However, another limit would soon impose itself: the muscular needs would also increase with size and the doubled weight of the titanium skeleton. That would require a more efficient blood supply, use of enriched oxygen, use of more efficient food system, possibly still more weight. As each limit were overcome, another limit would present itself. So, there is a series of interconnected limitations which define the size of land animals *on Earth*. If Earth's physical parameters were changed, presumably so could the size of it's largest organisms.

ILF physiology is a product of many planetary, or at least, initial development site factors. We cannot assume that an ETI will have any particular shape, but we *can* make some assumptions about the parameters under which it would have to operate. Given a certain gravity, certain prevalent elements in a planetary makeup, we can predict what the basic physiological parameters of an extraterrestrial biology will be like.

Lighter gravities would permit the emergence of larger elephants, that is, of larger life forms. Higher gravities would impose size restraints much more quickly. However, high oxygen content might offset *to some degree* problems of gravity. On two planets with a gravity of 1.5 g, that with the higher oxygen content in its atmosphere would permit larger animals.

The same is true for intelligence. A minimum cortical size (which means some minimum flow of active gas for energy exchange), a minimum mobility, the ability to evolve some senses, all depend on the specific planetary conditions. These may allow the evolution of simple organisms, but not allow the evolution of other traits which, finally, will allow the evolution of intelligence.

The multiple effects of these planetary factors can be roughly estimated, even if, at present, we have only one data point: our own environment. A second data point —some knowledge of the crude features of an ETIs planet—would provide us with better extrapolation, and, parenthetically, better understanding of the ETIs nature.

The various planetary effects are not intended as an exhaustive catalog of the kinds of planets where ILFs *could* evolve. In many ways this simplistic approach is misleading. Much more goes into the development, evolution, and makeup of any multi-celled (or even single celled) animal than mere physiological effects. As we shall see later when we discuss perception, direct planetary parameters — > physiological features relationships are likely simplistic and thus misleading. What this comparison *does* do, is alert us to the fact that the planets on which ILFs evolve impose certain limitations on origins and evolution. It is possible, even likely, that we can transcend those limitations, *even in our biological selves*. But the fundamental impositions will likely last well into our development of non-biological, cultural evolution.

3.4 Sensing

Sensing, as noted earlier, is one of the two critical variables when attempting to communicate with ETI. There are some obvious issues, such as for example, whether they would be able to see our raster images if we are communicating by some form of television, and some less obvious ones. The issue of the sensorium also allows us to start tackling another, wider problem: basically, we want to ask, are there physical peculiarities that are *derivatives* of factors in a particular chemical-physical environment and will *affect* communication. The problem can be illustrated by reference to three of our major senses: vision, hearing, and smell. All three of these are important on Earth for the survival of complex animals, be they canaries, sharks, or humans. Subsequently, all three senses (to different degrees) are important for the existence and transmission of intelligence, and, finally, for the creation and maintenance of culture *on Earth*. Though the interaction of sensory stimuli and thinking is not perfectly understood, still, there *must* be some connection, if only because we are products of our environment. We *may* assume that that general rule will apply elsewhere: senses and thinking are intimately related in some way. Change the sensorium, and the thinking is likely to be affected too.

It can be assumed that under similar (not identical) conditions to those on Earth, similar organs will have evolved, if only because of what seems to be the universal principle (insofar as we know) of convergent evolution (Arbib 1979: 11; DeVito and Oehrle 1990). Thus, animals on an oxygen-carbon cycle, are likely to evolve some analog to each of the senses we know of, and are *unlikely* to evolve other, more esoteric senses *unless* there is some special condition that obtains on that planet. Ear-analogs, eye-analogs, smell-analogs, touch-analogs are highly likely. They will not look *precisely* like those we are familiar with, but since they fulfill a particular function, they must fulfill certain physical requirements. Under similar physical conditions, they are likely to be easily identifiable as to their function. Under slightly different conditions, they will, presumably, vary in a way predictable (at least generally) according to some basic rule deriving from the physical properties of the environment. To give an example, terrestrial animals have protuberances intended to enhance their hearing. Convergent evolution has seen to it that rabbits, foxes, owls, and humans, have identifiable external ears, even though these vary considerably in shape and composition. Fishes, living in the sea where sound propagates differently, have evolved a different method of hearing, and we see that the cetacean ears are externally similar to fish ears. We can therefore assume that under specific conditions, ears will evolve, and that differences in their shape will be the result of differences in their functioning, importance for the organism, and the degree of maladaptiveness they create. Moles, snakes, lizards, and other burrowing animals have lost their external ears for the same reason as the Cetacea: they are at the same time non-advantageous (they catch on things underground, which is maladaptive) and the medium (ground, water) does not require nor allow external ear cones. In this section we shall explore, superficially, some of the implications of various senses, keeping largely to the assumption of mediocrity once again.

All intelligences must have some sensory access to the universe. It seems, at least until some further evidence is in, improbable that intelligence would not have any input from the surrounding universe, and still be able to function intelligently. DeVito and Oehrle (1990) have argued that eyesight may be a universal. Whether this is the case or not, it seems that some form of efficient sensory input is a must, even though the only basis for saying so is logic and the example of Earth life.

3.4.1 Vision

Our vision relies on two factors. First, the existence of light that ranges from about *390 to 750 nm*. Some animals can see deeper into the infra-red using special organs (snakes, for instance) or the ultra-violet (bees, for example). The other factor is the relative transparency of oxygen, in its various forms, to those wavelengths. Obviously the two are related. The visible light window which permits certain wavelengths of solar light onto the surface, also provides more scatter for blue (thus coloring the sky blue) than for red.

What happens under another, hypothetical gas mixture? Higher rates of dust, moisture, or trace gasses in the atmosphere will affect the form of the visual receptors, just as different rates and spectra of starlight will. Greater shift towards the red (because of atmospheric particles which block shorter wavelengths AKA blue; or solar type) will mean that visual receptors will have to be larger. That having been said, we should consider two things: the effects of vision, and the morphology of the receptors.

Vision allows an organism to identify food (whether prey, welcoming flowers, or ripe fruit) and threats, including movement, coloration and shape, which most Earth animals are attuned to. The silhouette of a small, fast moving black body alerts us to the presence of a fly which we might want to swat, but for a frog means dinner. The silhouette of a flying raptor—a cruciform shape with the wings close to the direction of flight—causes chicks to freeze, whereas the same shape flying 'in reverse'—the wings close to the back—means 'goose' and is ignored (Bressan and Pizzighello 2008). Given the speed, accuracy, and multiple values that light provides (shape, movement, color) it would be unsurprising to find some form of vision (that is, light reception) evolving on other planets, though, as we note above, not in all planetary environments would vision necessarily be as useful as on Earth.

The morphology of the eye derives from the nature of light and of biological requirements. Eyes have evolved in various organisms and in different terrestrial phyla. Interestingly enough, there is good evidence that certain forms are particularly effective. Humans (phylum *chordata*) and octopi (phylum *Mollusca*) who may not share recent animal ancestors have nonetheless evolved eyes that share many characteristics (Halder et al. 1995). This seems to indicate that certain abilities such as an eye lens and retina have a positive evolutionary value: the ability to control the amount of light impinging on the retina is obviously an advantage for surviving in changing light intensities. This leads to the development of the iris.

Long-range/short-range sight requires some analog of the flexible lens, or multiple lenses/eyes. Eyes also serve for quick target acquisition and homing (in predators) and for obstacle and predator avoidance in prey, and therefore must be concentrated as close as possible to the nerve cluster to be of benefit. Two eyes allow for stereoptic vision. Three frontal eyes would not increase that ability, but would cost more in terms of increased tissue, if nothing else. A rear eye might be a good idea for prey species, but it would require greater investment in nervous tissue to handle the added information, which might be a drawback, thus three eyes, or over specialized eyes (one set for distance viewing, another for closeup, for example) are unlikely because they are, biologically speaking, too expensive. That having been said, such forms *can* evolve, as witness Earth spiders.

3.4.2 Hearing

For animals, sound, or sensitivity to airborne vibrations, serves as a general warning system. Overall, though with many exceptions and counterexamples, vision specifies, hearing generalizes. What is no less important, and for very good reasons, sound is the major medium of general communication in most higher species. There is no question that sound is critical for the transmission of intelligent communication among humans. Hearing loss among small children, for instance, is far more of a developmental 'brake' for an individual than is vision loss since they are unable to access language cues.

Would another environment affect hearing? We do know that the denser the medium, the faster and better sound propagates. This is one of the reasons whales rely so heavily on sound (the other, of course, is the absence of light at great depth). Some ETI may be more heavily reliant on sound than we are, notably if they live wholly or partly in a liquid environment. Sound pulses from some species of whales are exceedingly complex. Thus we may be relatively benighted in the realm of sound, to offset our relative acuity in vision.

The ability to detect airborne vibrations is important for both predators and prey. It is likely to evolve in some form in any environment where vibrations can be carried. The degree of reliance on sound, and the ability to make sense of it, will differ to a great degree. Animals using sound as a sight analogue will be likely to evolve organs to help them identify small sounds as well as large, and to focus their attention on particular sounds. Because it is in many ways as good as sight, and in some ways even superior (easier to generate by oneself, for instance) it is possible that some species, even entire phyla or planetary biota, will evolve to take greater advantage of sound then of light. This will inevitably change their way of behavior, of thinking, and definitely their morphology.

If a generalized warning system such as sound were *not* available, could some other sense fill the place of sound, that is, would there be another general warning system as effective, one that could evolve to a specific, detailed communication device? Electrical fields are sensed by many fish, some birds, and apparently some mammals as well (platypuses, moles). Note, however, that in the mammalian case, this is very short-range. Even in water, that supreme predator the shark, identifies primarily by smell and vibration, and only later homes in using sensors called the ampullae of Lorenzini. This implies, at least, that electrical-field sense would offer a smaller range, and consequently other, social factors may also be different.

3.4.3 Smell

Smell is an important sense for most animals. Like sound, it is a general identifier, but in addition, it serves two other purposes. First, it has storage capabilities. Any owner of a dog knows that dogs are constantly leaving delayed messages to other dogs. In addition, smell provides unambiguous physiological information for intra-species communication.

Smell is generated by air-borne micro-particles which fit receptacles in the olfactory organ. The presence of such factors depends on the relationship between their size/mass ratio and that of the atmosphere. Would the same effect of atmosphere-carried complex molecules exist under different atmospheric conditions? Could it exist under different gravities? The effect of gravity upon a body depends ultimately upon the size of the body. That is why ants are able to lift grains many times their own weight, and fleas are able to jump several hundred times their body lengths: less affected by gravity, they are able to devote more of their muscular strength to physical tasks, rather than just to support their own mass. Presumably, therefore, gas-borne particles, inasmuch as they propagate rather rapidly, are good candidates for sources of information under whatever conditions. However, being airborne, they are far more dependent on external conditions, far less controllable. Because their frequency is less, they can only convey messages at a frequency that is far less than sound or light. It is *unlikely* that any species will rely *solely* on smell as its major sense. Animals that do so on Earth are usually restricted to specific, and limiting ecological niches where those with better generalized sensoria are at more of a loss.

One could suggest that such (for Terrestrial species) unique niches may be the norm on some planets. Even so, the inherent limitations of smell (inability to focus it, high degree of disturbance by external environmental factors), make it rather unlikely a candidate as a focused sense.

3.4.4 Other/Exotic Senses

One consequence of life is that it will have a major effect on the environment in which it lives. This has occurred on Earth as prokaryotic organisms converted Earth's atmosphere from a reducing to an oxidizing atmospheres some time after their emergence. The effects may well extend to crustal elements as well. A. Smith has suggested that the presence of life forms on Earth has been responsible for the existence of metal concentrations that are of importance to technological civilizations (Smith 1982). In a broader context, this view is known as the Gaia hypothesis (Margulis and Lovelock 1974). Briefly stated, the Gaia hypothesis claims that biological factors constantly modulate the Earth in order to produce the conditions necessary for life. If this hypothesis is confirmed, or only partly confirmed, it suggests that other planetary environments where life has arisen will also be modified to support that particular kind of life. How this principle (if it holds true) will affect the evolution of organisms is not clear at all.

Nevertheless, we must accept the possibility that senses that exploit other physical features of a world are inherently possible. The senses that have been discussed above illustrate a basic requirement from all life forms: a means of receiving information from the environment. The need for reception of such information is so important (much as we take it for granted) that life has been defined in reference to that. In fact, put together sensory input and organization of that information, and we may well be on the way to defining the next step, intelligence.

Since I am concerned here with possible alternatives to humanity, it is necessary to ask a further question: could there be, or are there, other senses that could be used by an organism? If so, under what conditions are they likely to occur? And, would such different senses have implications—for intelligence, for culture—that we can conceive of and justify?

3.4.5 Subjective Experience of Time

A similar and even greater problem is that of time and time sense. All Earth living creatures are subject to circadian cycles. These are, for some reason, crucial to the existence of the organism (Ayensu and Whitfield 1982; Pittendrigh 1981). I would hazard a guess and suggest that, just as computer needs a clock to regulate input and output, so do living things and much for the same reasons: to regulate information flow, and thus their entire 'living' existence. Our (living beings on Earth) time sense is the product of a number of factors. Day-night alteration, lunar cycles, seasonal change, and possibly even longer cycles. On another planet, it is only to be expected that, given the different astronomical factors, there will be different time cycles. Now, what does that mean in ETI terms? It is highly unlikely that 'different time cycles' implies animals that are physically slower, or faster, than those on Earth. Chemical reactions take place without regard to the astronomical cycles that affect complex organisms. However, a computer with a quicker inner clock reacts faster (for that, but also for other reasons having to do with the structure of its processor) than one with a slower clock. Perhaps having a quicker cycle base might imply that the *organism* is quicker? Even on Earth, some animals have a notably quicker set of reactions than others.

These questions are progressively more difficult to answer because we know so little of the factors ourselves. Consider the electromagnetic spectrum. We are sensitive to a relatively small portion of it, even if we consider some of our more esoteric (and thus less useful senses) such as the sensitivity some people (and many animal species) have to magnetism (homing pigeons are one such example in the animal world) or to electrical fields (such as various kinds of fish). Now, there are very good reasons for that. The particular range of the total spectrum we use is fairly coherent, requires little energy to use. If we used, say, extremely long waves, our receptors (eyes, and other less specialized cells and cell colonies) would have to be much larger, which would have an effect on our size, thus our mobility, and so on. The same is true of shorter wavelengths. There is no direct correlation between body size and sensitivity to particular wavelengths, but the wavelength/body size equation does produce certain limitations.

3.4.6 Interaction of Mind and Senses

To summarize, we need to go back to some basics. "Senses" is a shorthand description of the various modes we have of receiving information from our environment. Obviously, the ability to receive such impressions is absolutely crucial for any animal. Without these, at even the simplest level, any organism is little more than a rock. Several other generalizations appear to hold true as well.

The more complex the organism, the more likely it appears to rely on a sense *set* rather than on only one sense. Organisms that exist in relatively unchanging, simple conditions—moles, for example—can afford to specialize evolutionarily on one, or a few senses. But for those living in challenging and varied environments, a full suite of senses is a must. This would also seem to have bearing on another very important subject: the evolution of intelligence. Evolutionary pressures imply the ability to deal with changing and unexpected situations, and a full, even a redundant sensorium is a necessary prerequisite for that. We are, I believe safe in assuming that any ILF is likely to have 'redundant backup' senses, which sense various environmental stimuli.

There is a further point to be made. Sensing the environment *also* implies being able to manipulate it for other purposes, such as communication. Many, indeed most, terrestrial multicellular animals make use of other animals senses. They send olfactory, visual, and (certainly at our size and body mass and above) auditory signals, which do much to establish social and collective, as well as individual behavior.

We *know* at least some of the parameters of chemical interaction in different solutions, even if there is no complete agreement on the possibility of creating life under those parameters. We do *not* know what the limits would be on organs that might evolve under those conditions. Keep in mind that all of the many adaptations of organs we know of have evolved under the same physical conditions. These range from a particular part of the spectrum (affected both by star color and by oxygen atmosphere), through questions of gravity, to the effect of atmospheric density on vision and hearing. For some of those problems we can find some sort of analogy on earth, for others we are more restricted.

3.5 Exotic Chemistries and Physics

So far, the discussion has upheld a conservative approach. The biochemical issues discussed above have the smell of ordinariness about them. They are, in effect, possible variations of what life on Earth is, or could be. We in effect followed the assumption of mediocrity, which essentially argues that Earth is well-placed in the median of possible intelligent life development. The assumption of mediocrity is, however, merely an unproven assumption. If only for the sake of completeness, it is useful to take the discussion several steps forward. Rather than considering alternatives to what we see on Earth, with all the limitations it involves, it should be asked whether truly different possibilities may occur in the development of life.

In order to do so, it may well be useful to modify our definition of life itself, or, at least, to broaden it in such a way that it encompasses more than the oxygen-carbon forms familiar to us. Arguments have been offered, as we saw above, to suggest that given the chemical structure of the universe, where there is life, oxy-carbon life-forms is the most likely to emerge. This is true provided we entertain only the possibility of life based on a familiar set of electro-chemical properties. To counter that, we ought to keep in mind that both matter and energy act differently under widely divergent sets of circumstances. These circumstances may include different temperatures (very hot, very cold), gravities (e.g. extremely high or low), and energies. Broadly speaking, I will term these environments and life/intelligence possibilities 'exotics'. Under exotic circumstances, is it possible that 'life' may have different meaning, and that its functions may change?

In order to answer that question, it is useful, as noted, to change the definition of life. In essence, life consists of complex organization of matter and energy, that is capable of reproducing this order by some means. For our purposes, such a simple definition is insufficient. We also need intelligent life, or at least its potential. Therefore life could be defined to be a complex organization of matter and energy, capable of reproducing its order, and capable of imposing other forms of order, not innate to it, on its surroundings.

Feinberg and Shapiro (1980) have proposed a number of intriguing possibilities for the bases for life. These speculations are based on the extrapolation of what is currently known about the physics of the universe. Plasma life may exist in interiors of ordinary stars, which depends on the interactions, at magnetic field levels, between ionized gasses. The mutual effects of charged particles and magnetic fields (each affects the other) is vaguely similar to the reciprocal influences of proteins and nucleic acids. It could, theoretically, lead to greater order and to rapid growth of such order. Energy is supplied by the flow of radiation within the sun. Since the temperature differences at higher levels of stars are very great (even one hundred meters apart) there is effectively an energy sink and free energy can be used. (Feinberg and Shapiro 1980: 383–386). In complete contrast, at an atmosphere only ten or so above absolute zero, there are no chemical reactions, oxygen and nitrogen are frosty solids. Helium and hydrogen would form the atmosphere, freezing occasionally to hard masses which would float on a hydrogen sea. Hydrogen has chirality and appears in two forms: o and p, in which the different forms rotate in different directions (clock or counter clockwise). Incident radiation—at very low intensity to be sure, but nonetheless available—would power interactions between the o and p types of hydrogen and would provide the life organizing process (Feinberg and Shapiro 1980: 383–393).

At a even greater extreme, in a high radiant environment in an interstellar dust cloud, a collection of atoms or molecules (in the case of a gas cloud, in which case it could also use infra-red energy) might become ordered under a specific pattern of excitation. These, under radiation from outside, emit radiation in an orderly pattern. This radiation would interact with the same group of atoms and excite them to greater order (e.g. it reinforces the original order). Every once in a while the group would reach a stage of 'flashing' excess energy and exciting a nearby unexcited group into excitation as well, thus 'growing'. Again, this is like information exchange between nucleic acids and proteins. Because atoms tend to emit on the same wavelengths and in the same order as their neighbors, order is going to grow. Evolution would be fueled by the removal or accretion of molecules that would be hindering or increasing radiation propagation and atomic excitation (Feinberg and Shapiro 1980: 393–400).

Though these examples are highly speculative, they do, nonetheless, propose acceptable hypotheses, given what we know of physics. For our purposes here, this poses an important problem: would we share any congruency which would allow us to interact with such exotic beings? Keep in mind that given the different nature involved, we might not even be aware of one another, and the only communication that might emerge could be the destruction of one by the action of another, as by the presence of what would amount to a superheated body in a hydrogen sea if a non-exotic spaceship were to attempt to land on the exotics home planet.

3.6 Possible ETI Forms

On Earth, an animal's form is a derivative of its life style, basically the combination of its place of residence and activity, and its way of providing itself with food and reproduction. The factors interrelating in a complex, generations-long dance.

Whatever assumptions we make about an ETI species, must bring into account not only those aspects of its physiology that derive from physical and chemical factors, but also the ecology that has produced it, and the ETI's role, *as an animal species* within it.

Human beings, one end product of the hominoid line, which includes the extant and extinct apes, the live (*Homo sapiens*) and extinct (*Homo erectus*, *Homo habilis*, *Homo neanderthalis* or *Homo sapiens neanderthalis*) hominids, and the various extinct pithecanthropenes, emerged as a specific form of animal. Hominidae ancestors were tree dwellers, who adapted over generations to a terrestrial mode encompassing seeds, roots, and carrion. The biological features of the Hominidae derive from differences in lifestyle. Most apes, presumably originating as, or returning to an arboreal lifestyle, have long grasping limbs. All have binocular vision, for reasons that differ considerably than the binocular vision of predators such as cats and hawks. Predators need to sight prey from a distance, primates need to estimate the distance to tree branches. A primate that later turns to hunting can use the binocular vision developed in the forest to sight prey as well.

We would therefore expect that an ETI (with some exceptions such as artificial intelligences) would have atavistic characteristics that would be the result of a species' history. An ETI, like a human, could not be understood as "a predator" or as "a burrower" but it could not be understood fully without keeping in mind its basic species history either. The forms its intelligence takes would be constrained, to *some degree* though not absolutely by the species history.

3.7 Summary of Chapter 3

From reading this chapter it might seem as if I am advocating homomorphism among ILFs. After all, if our sense cluster can be rationally generated from simpler principles, then the same must be true for other ILFs, who, logically, at least, should evolve the same way. This is not the case. Any of the general principles above is based on *limiters*. These simply claim that some forms are inherently less efficient then others, and given Darwinian evolution (which is addressed later) are unlikely to survive in a species. Within these limits, however, many other possibilities may emerge. For example, the problem of a single eye that must focus at different distances may be solved by the equivalent of bifocal glasses: a lens that is divided into zones for distant and close viewing. This would tend to make certain demands on the organism that differ from those made on flexibly adjustable principles.

Exotic possibilities for solving the problem of seeing at variable distances, and for estimating distances are also at least theoretically possible. David Brin has suggested the possibility that naturally generated coherent light could be used by an organism instead of mechanically changing focus. He has even suggested a mechanism whereby an organism could create such a laser effect. That life form would not have to be binocular: the laser would provide all the distance measurement it could need.

References

Arbib, M. A. (1979). Minds and millenia: The psychology of interstellar communication. *Cosmic Search, 1*(3), 21–25.

Ayensu, E., & Whitfield, P. (Eds.). (1982). *The rhythms of life*. NY: Crown Publishers Inc.

Bressan, P., & Pizzighello, S. (2008). The attentional cost of inattentional blindness. *Cognition, 106*(1), 370–383.

Buccino, A. P., Guillermo, L., & Pablo, M. (2006). Ultraviolet radiation constraints around the circumstellar habitable zones. *Icarus, 183*, 491–503.

Chela-Flores, J. (2007). Testing the universality of biology: A review. *International Journal of Astrobiology, 6*(03), 241–248.

Christian, J. L. (1976). The story of life: Earth's four-billion-year beginning. In J. Christian (Ed.), *Extra terrestrial intelligence: The first encounter* (pp. 15–31). Buffalo, New York: Prometheus Books.

Cleaves, H. J., Chalmers, J. H., et al. (2008). A reassessment of prebiotic organic synthesis in neutral planetary atmospheres. *Origins of Life and Evolution of Biospheres, 38*(2), 105–115.

Cleland, C. E., & Chyba, C. F. (2002). Defining 'life'. *Origins of Life and Evolution of the Biosphere, 32*(4), 387–393.

DeVito, C. L., & Oehrle, R. T. (1990). A language based on the fundamental facts of science. *Journal of the British Interplanetary Society, 43*(12), 561–568.

Dole, S. H. (1970). *Habitable planets for man* (2nd ed.). New York: American Elsevier.

Ehrenfreund, P., & Charnley, S. B. (2000). Organic molecules in the interstellar medium, comets, and meteorites: A voyage from dark clouds to the early Earth. *Annual Review of Astronomy and Astrophysics, 38*(1), 427–483.

Fasan, E. (1986). *Discovery of ETI: Terrestrial and extraterrestrial legal implications. International Astronautical Federation. IAA-86-476*. Oxford: Pergamon Press.

Feinberg, G., & Shapiro, R. (1980). *Life beyond Earth: The intelligent earthling's guide to life in the universe*. New York: William Morrow and Co.

Gould, S. J. (1983). The wisdom of Casey Stengel. *Discover*, 62–65.

Halder, G., Callaerts, P., & Gehring, W. J. (1995). New perspectives on eye evolution. *Current Opinion in Genetics and Development, 5*(5), 602–609.

Margulis, L., & Lovelock, J. E. (1974). Biological modulation of the Earth's atmosphere. *Icarus, 21*, 471–489.

Matía, I., González-Camacho F. et al. (2010) Plant cell proliferation and growth are altered by microgravity conditions in spaceflight. *Journal of plant physiology, 167*(3), 184–93.

Miller Stanley, L., Urey, H. C., & Oro, J. (1976). Origin of organic compounds on the primitive earth and in meteorites. *Journal of Molecular Evolution, 9*(1), 59–72.

Pittendrigh, C. S. (1981). Circadian systems: General perspective. Handbook of behavioral neurobiology. In J. Aschoff (Ed.), *Biological rhythms* (Vol. 4, pp. 57–77). New York: Plenum Press.

Seckbach, J., Aharon, O., & Julian, C.-F. (eds). (2008). *The extreme environments and their microbes as models for extraterrestrial life*. Munster, Germany: European Planetary Science Congress.

Shapiro, R., & Feinberg, G. (1982). Possible forms of life in environments very different from the Earth. In M. Hart & B. Zuckerman (Ed.), *Extraterrestrials: Where are they?* (pp. 113–121). New York: Pergamon Press.

Smith, A. G. (1982). Settlers and metals: Industrial supplies in a barren planetary system. *Journal of the British Interplanetary Society, 35*, 209–217.

Trefil, J., Harold J. M., & Eric, S. (2009). The origin of life: A case is made for the descent of electrons. *American Scientist*, 206–213.

Chapter 4
Evolutionary Parameters of ETI

It is impossible to discuss the question of life, on the one hand, and of intelligence on the other, without considering questions of change and evolution. By its very nature, life must be a dynamic system. Current theory has it that intelligence derives from the dynamics of the life system, though exactly (or even principally) how this comes about is not clear.

This chapter is intended to discuss some of the aspects of life as a dynamic system leading up to intelligence. The question of how life evolved on Earth, which is more convincingly discussed by biologists and paleontologists, is part of a larger issue: is this the *only* way life could evolve? Evolution for the purposes here, means change in a species leading to greater complexity.[1]

Evolution does not only occur in the biological world. There is good evidence to suggest its occurrence at other levels of the universe. Stellar and universal evolution appears to be a fact. This leads to the suspicion that evolution is a general rule in the universe. Evolution occurs when a "System possesses some degree of freedom and some rules. Freedom leads to the appearance of new forms which are then tested by the rules of the system. If they fail the test, they vanish. If they succeed they are accepted into the system. As cosmic evolution advances, the complexity increases, and the constraints newcomers have to face multiply" (Papagiannis 1983: 1). We can conclude, if only as a strong hypothesis, that evolution is a consequence of very basic rules in our universe. Therefore we should expect to find it in other biological systems than ours as well as on Earth.

The common scientific paradigm for evolution on Earth is the Darwin-Wallace theory, also known for its dramatic aspect as "survival of the fittest." To simplify (oversimplify?) Darwin's theory, the theory proposes two mechanisms which, in tandem, work to change species. First, a process of mutation, whereby new traits in

[1]Evolution need not, of course, lead to greater complexity. The Darwinian paradigm speaks of 'fitness', not greater complexity. However, we are not truly that interested in attempting to communicate with bacteria (or even more complex animals) unless we are going to get some form of intelligent response, which in turn, arguably, requires some form of complexity.

M. Ashkenazi, *What We Know About Extraterrestrial Intelligence*,
Space and Society, DOI 10.1007/978-3-319-44456-7_4

any given species come into being essentially on a random basis.[2] Second, a process of selection, whereby only those traits that do not have negative value for the survival of individuals in the species to the age at which they breed, are perpetuated in the species. Gradually this process leads to a greater fitness of a particular species to a particular environment.

The specifics of the Theory of Evolution have been under scrutiny since the theory has been proposed. A number of modifications and refinements have been proposed and accepted. Currently, it is the dominant theory of how complex life emerges on Earth. Possibly, similar mechanisms operate elsewhere under the same rules.

It is this last statement that will be challenged here. Not because I wish to challenge the dominant paradigm for changes in life forms and the development of complex forms, but because the presentation of alternative possibilities for change and evolution may yield us more clues about possible forms life and intelligence may take elsewhere.

4.1 The Dominant Evolutionary Theory and Its Alternatives

A useful step forward in defining ET life forms lies in Darwin's basic formulations. Though it has been often said (e.g. by creationists) that Darwin's ideas are no more than an unproven theory, they are nonetheless accepted by the large majority of biological scientists as the only theory going. This is because it is the only theory proposed to date which meets the demands of science: falsifiability, parsimony, and empiricity. All other theories proposed involve more elements, and need more complicated variables to account for the same set of phenomena. While we could posit a number of alternative biological developmental schemes, none of them are likely, given our physical and chemical knowledge and the rules of scientific enquiry, to have the elegance and simplicity of the Darwinian paradigm.

We will therefore start by assuming (a somewhat dangerous assumption, I'll admit) that the basic Darwinian laws apply in ET situations as well as on Earth. A question we also need to address, is whether physical/chemical differences are likely to cause alternatives to the Darwinian paradigm common on Earth? While this is of course possible, it does not, in the final analysis, matter much. We are interested in the end product: an ETI and its doings.

Is Darwinian evolution likely under other suns? That depends to a large degree on the similarity of basic of life systems. Given the realities of the observable universe *outside* the Earth, the question is emphatically "yes." From what physicists

[2]It must be noted that Darwin and Wallace did not know how the process of mutation came about as they had no knowledge of genes and their effects, nor of radiation. They did, however, an excellent job of observing the *effects* of genes and therefore hypothesizing their presence.

and cosmologists have been able to observe, evolution derives from some very basic causes. There are no reports, yet, from biologists. Yet, if we can posit a system that is as 'elegant' in the mathematical sense, and will answer the requirements of an evolutionary system, then we will only be able to judge by empirical evidence (of which, I emphasize, we currently have none) or by the effects of Occam's razor.

4.1.1 Alternatives to Darwinian Evolution?

We shall start with a negative, as a thought experiment. The objective here is to propose an alternative mechanism, the end result of which will provide (a) species diversity, (b) adapting species to various biological niches/opportunities (c) some advance or evolution when alternative opportunities become available. Could such mechanisms be conceived of? If they are logically coherent, and provide responses to these three basic issues, then given the variability of the universe, we ought to consider them seriously.

A Darwinian Alternative One way of examining the possibility of Darwinian alternatives, is to question the entire basis of biological and physical foundations that underlie the Theory of Evolution. I cannot pretend to have the expertise to do so. However, it has been done, most recently, by a biologist, Rupert Sheldrake. Sheldrake's theory of 'morphic resonance' (Sheldrake 1987, 1996, 2011) suggests that natural laws are the consequence of a form of "memory" inherent in the fabric of the universe and of its constituents. Now, like UFOs, Sheldrake's claims of 'morphic resonance' may be nothing more than pseudo science. However, like UFOs (see below, Chap. 5), Sheldrake's position provides an important point: we cannot take our scientific assumptions for granted. Which, in the evolutionary context means that we must be able to conceive of other forms of evolution, nature currently unknown, that could take place on other worlds.

Although the Darwinian paradigm is the only one for which we can adduce any reasonable proof, the example suggested by morphic resonance is instructive. Unproven or not, it suggests that Hamlet's warning to Horatio ought to be taken seriously, at the very least as a frame of mind. If species adaptation is not the simple matter of genes being arranged and rearranged by natural selection, than the universality of the survival of the fittest paradigm is in question. If that is the case, than it is only reasonable to entertain the possibility that *given different initial conditions*, the Darwinian subset of evolutionary possibilities is not the only one.

In fact, there has been some indication of less spectacular alternatives than formative causation in the discussion among biologists of "altruistic genes" (Taylor 1992; Smith 1974). However, altruistic genes—where an individual in a genetic group gives up some fitness advantage to increase fitness for the group—are merely another expression of Darwinian evolution. The colonies of single-cells we call complex organisms are of this sort. Most colony animals (including you and I) are composed of specializing, interacting cells which, quite obviously, are not competing fiercely with one another. On physiological-taxonomic grounds alone, there are as many differences between blood-cells and nerve cells as there are between various species of amoebas, and yet *they do not compete.* This implies, therefore, that even in a Darwinian world, competition is itself selective. We must modify the Darwinian alternative and suggest, instead, that Darwinian evolution is one in which competition at all levels is *in most cases* a useful strategy for exploitation of the environment. This axiom has a corollary: 'In some cases, cooperation between units is a successful competitive strategy.'

Cooperation-Cooptation Paradigm: Cooperation as an Evolutionary Alternative
Let us posit (and for the moment, the genesis of such a system need not be considered) a system in which cooperation is the *norm*, and competition is the *alternative*. In such a case, which is not a mirror image of our own[3] you are likely to find large scale cooperatives existing, perhaps in competition with one another under *some* circumstances but not others. These cooperatives will be composed of diverse units, at any level, most probably organized in hierarchies. It is also possible that dominance will be established by competition and selection. There are examples such as that on Earth, such as humans and their commensal/parasite, the Norwegian brown rat (*Rattus norvegicus*). That is, the evolutionary success of the rat has been almost completely dependent on the evolutionary success of humankind. It is also dependent upon the evolutionary success of one particular culture—European maritime culture—but that is a different story. Note, that the rat does not 'cooperate' with humans. Nor, for that matter, is it a true competitor. Instead, there is a factual hierarchy existing between human and rat. Individuals on both sides are killed, though not in evolutionary competition. It is likely, that given the absence of humans, the rat would be far less successful, having to compete with other predators far better evolved (and willing) to deal with it. Moreover, it is also likely that other competitors for the same type of foods, would be better adapted in many environments to deal with the brown rat's competition.

The biosphere and life in general are products of the initial rules by which these systems are organized. Under the Earth system, a 'free-market biosphere' in which each kind competes against each other, leads to developmental change and evolution of, effectively, the entire economic/ecological system, even at the risk of entire species going down the drain. Alternatively, a 'socialist' type is possible in which there is some sort of unified shared interest, and everyone is closely interlocked (Feinberg and Shapiro 1980: 402, 403). Neither of these possibilities are 'pure' examples.[4] In the case of Earth's biosphere, competition *appears* to be

[3]The mirror image of our biological system would be competition between cells in an organism, and cooperation among complex organisms. What is proposed here is a mixture of competition and cooperation with the latter predominating.

[4]Which is, of course, true of the social forms from which this simile is drawn: even the most capitalist of human societies has a large touch of socialism, and vice versa.

dominant. This appearance may only be a cultural bias on our part. Each case of the evolution of competition *between* complex organisms, is also a case of cooperation *within* cellular organisms, and examples of cooperation and sharing abound.

A 'socialist' paradigm might evolve as the *major paradigm* on other planets. Organisms fulfilling different, though overlapping ecological niches might emerge as cooperating wholes. They will have reached some form of co-existence internally, and compete, as a unified whole, with other such colony organizations. What differs from Earth would be the inclusiveness of 'internally'. On Earth, this is circumscribed by specialist boundary cells: skin or other armor. On another planet, the interface 'skin' may allow different multicellular organisms to cooperate with a larger frame of some sort.

At the level of intelligent organisms, it should be noted that as organisms, we prey on other life forms. Without that predation, we die individually and as a species. This is as true of those of us who eat meat, as those who prey on vegetables. A purely cooperative ecological system is thus unlikely, since each organism would be required to find its energy resources at the lowest common denominator—basically solar energy—a long and time-consuming process which does not leave much time for other activities, as most bacteria could tell you.

4.2 Reproduction: Sex in Many Ways

Sex—gamete exchange between members of a species—is a function of the need to exchange genes. That is, to ensure variability: a part of the evolutionary paradigm. A weak hypothesis, one we can adhere to here, at least, is that any species will require a means to ensure the production of diversity, hence, sex. It is likely, if only on the basis of simplicity, that ETI will be sexual beings, though of course they might be partially asexual, have only one intelligent sex, etc. (Ruse 1985: 66). The likelihood of a non-sexual species—one where genetic exchange *does not take place*—is remote: at the species level, lack of diversity kills. A species unable to adapt to changing times and places is a dead species. Sex would not necessarily be the same form as human or mammalian.

The exchange of genes (which is important) must not be confused with the creation of offspring (also important). Single cell animals, many plants, and even some fish such as sawfish, reproduce asexually, in whole or in part. That is, the exchange of genetic material is separable from the birth of offspring: not always, but enough for us to consider this as an option. Moreover, even sexual mating can take place without any sexual distinction or sexual act in the sense we think of it. Flowering plants broadcast sperm (pollen) and have receptive ova without a 'sexual' act taking place. That these alternative *on Earth* take place only in 'lower' taxa does not mean that sexual acts are solely and necessarily associated with 'higher' taxa such as vertebrates. ETI may well have some form of reproduction in which genetic exchange (necessary for species survival), birth, and nurturing (necessary for intelligence and culture) would be separate activities, having different biological imperatives and social or a-social loci. Before considering differences in sexual strategies and gene exchanges, however, we need first to understand the creation (or rather, survival) of offspring, after which we can consider sexual strategies as well via a small thought experiment.

4.2.1 K and r Reproductive Strategies

Reproduction [besides being fun, for humans at least (see Diamond 1997)] can be viewed from a number of perspectives. It is the way a species perpetuates itself. It is the way individuals perpetuate their genes. It is a physical and emotional drive for most individuals. It is a way genes have of ensuring their own continuity. Whatever the perspective taken, one element is true to all: reproduction requires some strategy to ensure that the succeeding generation attains sufficient maturity to breed the subsequent generation. How is this to be achieved?

Biologists have long noticed that there are, effectively, two general strategies a species (or an individual, or a set of genes) can adopt to maximize the likelihood that their genes will be reproduced for a third generation. An organism can produce as many offspring as possible, trusting to statistics that *some* of them at least will survive to perpetuate the third generation. This is generally known as an **r** strategy, defined formally as the production of numerous low-cost offspring. This is a very successful strategy, as animals ranging from insects through brown rats can attest. In contrast, some species adopt a **K** strategy, formally: increased parental investment and cost in offspring. Larger animals including cats, Cetacea, and all primates adopt this strategy. A rat female will drop a litter of eight or ten, trusting that half or less will attain breeding adulthood, and investing a few weeks at most before the offspring are on their own. Her material investment (food energy turned to milk, risk of discovery) is minimized. In contrast, an elephant female may drop a single calf once in five years, and then will feed that calf for a year, and protect it until maturity.

Humans, like most primates, invest heavily in their offspring. In fact, humans invest more in their offspring than any other animal we know. The parental investment is very high by all measures including calories from feeding, time, and even money: around US$225,000 to maturity per child in the US. K strategies are also often not always individual strategies. An entire group may be involved. Elephant and chimpanzee mothers can rely to some extent on 'auntie' behavior: other females in the herd or troop will keep an eye on younger children to ensure their safety, and even occasionally feed and groom them. This has three important outcomes: bonds are created between members of the group, the odds of an offspring reaching maturity are enhanced, and the child is socialized into the group's way of doing things, including learning the group's particular language, social hierarchy, proper behavior, and survival strategies.

As a general rule *on Earth*, the higher the intelligence, the more a species is likely to use a K strategy. In humans, this has a physiological reason: human babies are all born prematurely and physically underdeveloped. This is because relative to body size, human infants are born with enormous heads: any larger, and they could not be born at all, since they would not pass through the uterine space in their mother's pelvis. This means the human infant is helpless for a period of some two years from birth, and not very functional for a number of years afterwards. A baby deer, in contrast, is able to stand and follow its mother within minutes of being born.

Does this mean we should expect ETI to necessarily be organisms that have adopted a K reproductive strategy? Let us keep in mind that the objective of reproduction from the biological point of view is *not* intelligence (whatever that is). It is continued reproduction. So, we need to ask ourselves whether K reproduction strategy is the *only* way which might also, incidentally, develop and sustain intelligence in a species.

To answer that question, we need to look at how the difference strategies contribute, in their different ways, to the *individual*, not to the *species* (which has been the focus of most r/K selection theory thinking).

Elephants, because they are non-primates can serve as a useful analogy. The individual elephant child is born with a great deal of intelligence. By the standards of the African or Indian environment where it is born, however, it is, on its own, little more than a slowly moving larder: without the protection of its mother and herd, it would fall prey to a predator. Its social environment, however, provides alternatives:

- It is fed, until it can browse.
- It is protected against predators.
- It is likely to be assisted in case of accident by its mother or aunties.
- It will be trained to utilize resources *and* avoid dangers.
- It will be socialized as an elephant and as a member of a particular herd.[5]

As a general rule, we can say that a K strategy provides at the least food, security, safety, learning, and socialization to the infant (defined as an immature member of a species unable to care for itself). It is highly likely that these universal needs, probably in different mixes, are necessary for *any* K-strategy reproduction, *including* ETI. And, not coincidentally, K strategy reproduction favors intelligence.

What about r strategy reproduction? Most turtle species can serve as an analogy here. Once the turtle mother has selected a nesting site and laid her eggs, she takes no further interest in reproduction. A turtle clutch can contain between thirty and sixty eggs. The hatchlings must claw their way across the dry beach sands, through a hungry mob of predators, into the surf (which they recognize instinctively, but have not been taught about), through shallows crowded with sharks and other marine predators desirous of some turtle snacks, into the deeps where they can

[5]Different elephant herds are known to 'speak' different dialects of alarm, food, and emergency calls (Sukumar 2003). This is true of other mammals such as orca (killer whales) and primates as well.

presumably hide and grow. Fewer than 1 percent reach adulthood. Around 0.1 % manage to reproduce. Obviously, learned behaviors in the turtle case take a back seat to instincts and genetic programming.

What we need to ask is whether some mechanism could ensure the development and nurturing of intelligence in an r-strategy species? If this is *not* the case, then we know for *almost certain* one fact about ETI: they *will* utilize a K strategy for reproduction. If this is the case, and given whatever sexual reproductive structure they have *ETI will have a family analog: some social structure designed to nurture infants to maturity*. This has immense implications. Much of our social structure, our thinking, our behaviors derive, at least in part, from our reproductive strategy combined with our species sexual roles (inseminator/birther): consider kinship and extended families, "God the Father", and the terms 'maternal' and 'paternal' and how they are used in realms such as politics. There is no reason to suggest that something similar will *not* be the case of ETI, assuming a K strategy reproduction.

4.2.2 From Genetic to Social Reproduction

One consequence of the sexual system we have is a complex organizational form—the family—which centers around nurturance. The family among humans focuses around a mated pair and their kin and offspring. The family is the social locus where the functions of gene mixing, bringing offspring to term, and nurturing them to adults are brought about. It is unique to the human species, it is common in *all* human societies, and it has a wide range of variation, from monogamous two parents and children nuclear families to polygynous/polyandrous multi-generation forms (Fox 1967). But it is not the only way of going about it, once we separate the different issues of gene mixing, bringing offspring to term, and nurturance,

The important variable, when considering *any* form of social species, is not the question of the biological mechanics, but of the social pattern of nurturance that occurs subsequent to the biological process and in relation to it. This is not to say that the biology is an unimportant factor; to the contrary, it may be the determining one in the evolution of a species. But given our current state of knowledge, where N = 1, we cannot know the various permutations of gene-mixing among intelligent species, even if we only consider a maximum of two sexes. We can, however, make some very educated guesses about the consequences of nurturance on the type of society we might encounter. Moreover, since it is behaviors we would most immediately encounter, with nurturance directly feeding into it, and biology hovering in the background, it is substantially easier and more profitable to try and understand nurturance and its consequences than genetics, for in nurturance we do have a limited number of alternative models within the human species available for examination.

4.2.3 *Tri-Sexual Species Thought Experiment*

Now, to return to sexuality. The question of sexuality may well be important for the issue of culture, something to be taken up later. But would biology necessitate the bisexual system that is dominant on Earth for complex organisms? And how, assuming that is not the case, could nurturance come about? There are good reasons, as Bylinsky (1981) and others have stated, for a binary sexual system. First and foremost, it is the most efficient, allowing for a mixture of genetic material while permitting as many encounters as possible and ensuring a minimal configuration for nurture. Three or more sexes having to act together would reduce the number of matings (since the odds on finding an appropriate mate would go down from 1/2 to 1/3).[6]

Nevertheless, as a thought experiment, let us hypothesize an intelligent species which has, for good and sufficient evolutionary reasons, three sexes. Let us also assume they are impelled by biology into one of three roles: a *child-bearer* produces an embryo and brings it to term. Upon birth, a member of the *nurturer* sex takes on the task of child rearing to maturity. A *gene mixer* accepts genetic material from the nurturer, mixes that with its own genetic contribution, and implants the genetic material into the child-bearer. For simplicity sake, we shall ignore the significant (in modern human societies) differentiation between sex (biological designation) and gender (social role designation) in reproduction.

Assuming this scenario, how would association between individuals come about? There might be a single form of 'family,' or, as in human societies, different social forms would emerge in different places, different times, and for different individuals. In our tri-sexual reproductive model, we could hypothesize that three individuals, each of a different sex would form a family-like association, since the investment in nurturance would be higher, relatively than for humans (on average, at least one member of each sex needs to be born from each sexual association, for the species to survive). However, in contrast to human societies, what would be the glue that keeps this 'family' together? In human families it is a mix of sexual reward, economic interest (sometimes), and social norms. However, anthropologists have long been aware that the *critical* factor is a basic biological one: men cannot give birth. This means male humans (and chimpanzees and lions and other species) interested in perpetuating their genes (and most animals are, apparently, genetically programmed to do so) must 'borrow' fertility, or, to be more crude, the wombs of females. And if a male wants to ensure his offspring survive, he needs to hang around and make sure it happens. This simple, elegant, and very basic fact forms the foundation for many social issues, ills, and religions. So, in our tripartite scheme, we could expect, nevertheless, that immediate social roles would evolve *based* on the biological facts, but far more complex and far-reaching. *Both* genetic reproduction and the need to provide for offspring dominate multi-cellular

[6]This assumes of course that every individual a person met was sexually receptive, mature, etc., which in reality is of course not the case, reducing the odds significantly.

organisms' lives on Earth. The same is likely to be true elsewhere. And to maintain long-term viability of intelligent offspring, who learn culturally rather than genetically, is an expensive process. Obviously, this expense (or something like it) would need to be based on some form of social support mechanism, else the species would very rapidly disappear.[7]

To return to r strategy once again. It is possible that, for example, our tri-sexual species *could* have an r strategy, in whole or in part. For example, only one of the sexes might be intelligent, or the nurturer would select only some offspring to be nurtured, while the rest would have to try and survive in some form in the wild. Culling could produce some offspring that are nurtured to adulthood based on some biological principal while others are not. Or intelligence in a clutch of offspring would be nurtured among a small number when the nurturer changes environmental factors, in the same way a crocodile mother selects for sex by varying the temperature in her nest. Nevertheless, and this is an absolute we know about ETI: *some* mechanism *must* exist to ensure the inculcation, in some or all of the young, of social and cultural behaviors. Obviously, we cannot predict what the social form will be in detail. What we *can* predict is that this basic form will be of extreme importance, consciously and unconsciously. Consciously, because intelligent beings are likely to spend a great deal of effort in nurturance, and they will therefore see its locus as an important factor. Some culturally driven mechanism will exist to provide nurturance (including socialization for new members of the species). Unconsciously, because analogy is one of the best ways to learn, and therefore many of their social relationships *outside* the breeding locus, may be modeled on it to some degree, just as they are among humans (see for example Ashkenazi 1991). However, nurturance can be either collective (a large group collectively nurtures all infants of members of the group, disregarding parenthood) or *specific* (a specific individual or set of individuals cares for specific infants). One of the forms—specific nurturance—is most common among humans. The closest humans come to an enduring normative collective form is the artificial environment of the Israeli kibbutz where children have been raised mostly by the group (Aviezer et al. 1994). Thus we cannot expect the ETI nurturance mode to be similar to the human one.

The Reproduction of Intelligence The question of nurturance ties directly into the question of intelligence. Whether an offspring leaps fully armed from parental brow or not, biologically intelligence is merely a potential: it needs be filled with knowledge, and practiced to make it function. Even a species whose offspring are fully intelligent upon birth, is going to need to have some form of nurturance to

[7]To complicate things further, it is not inconceivable that instead of sexes, we could have a three *species* reproductive system, in which each of the reproductive roles were played by members of different *species* who need the other two to bear their young and to raise them to maturity. The principles would be the same, as would the consequences. However, in the interest of simplicity, let us stick to a single tri*sexual* species.

ensure that knowledge, and no less importantly, social norms are acquired by the infant (conceivably this could be acquired in the womb, but that does not change the point). True, some very complex behaviors may well be transferred genetically. How complex we do not know, since in the case of Earth species, control of complex social behavior by instinct is largely limited to animals such as bees and other colony insects. However, it seems doubtful whether we could call a species whose behavior is purely genetic and instinctive—whether or not it may be able to build radio telescopes or even spaceships—intelligent. So in any intelligent species we are going to find some form of nurturing. This may take any of a number of conceivable, and probably some not-as-yet conceived of forms. There may be placental-equivalent nurturing in which offspring are nurtured and taught *before* being born, or time or situation triggered metamorphosis in which not all individuals develop intelligence, but only those under certain situations. These situation-responses require greater neural volume, they do not, in themselves trigger intelligence, except that, as on Earth, intelligence has become part of unused brain capacity. There must also be a mechanism inhibiting continued creation of the ETI 'intelligent individuals' variation, else these will evolve to the exclusion of the non-intelligent-until-triggered variety. Nurturing may take place without connection to gene transmission or bearing: Dawkins (2006) has noted that genes, like people, tend to be selfish, but a proper fostering system, in which parents broadcast offspring widely, only to have some of them fostered by others in return for acting as servants, soldiers, political supporters, economic clients or mates, is just as possible. Now these examples, which are definitely not exhaustive, have been provided to illustrate the simple point that we should *not* expect, that because biological requirements of gene transmission, birthing and nurturing are probably universal, the modalities by which they are expressed should be universal as well. However, we *should* anticipate that ETI have some mechanism which serves to reproduce social norms and behaviors, and bring infant intelligence into its full potential. We *can* call such a system a 'family' if we are willing to extend our term to somewhat strange extents.

The Domination of Shape: Homomorphic and Polymorphic Species In the previous chapter I discussed the possibility that animal forms on planets with high radiation and perhaps low gravity (the two may correlate), or rapid environmental changes (temperature, atmospheric composition) may bring about life forms that are polymorphic. On Earth, all highly complex animals are homomorphic. That is, humans have roughly the same shape from the day of birth to death. This may not be the case with ETI. If so, does this have implications for intelligence and social behavior? The likelihood of polymorphic species cannot be predicted, but the consequences are examined here. There are two possible polymorphic alternatives: punctuated polymorphs and general polymorphs. The former is common on Earth. It is exhibited in postpartum form by the entire insect order. All insects are initially hatched as asexual larvae. In some species (e.g. ants) these grubs are helpless and undeveloped. In other species, (e.g. dragonflies) the larval form is a formidable and intelligent predator. Later, the larva encysts, morphological changes occur, and the individual emerges in completely different form from its chrysalis. In many cases the change is so radical that neither form is predictable from the other. To add to the confusion, the different forms may exist in very different environments. Dragonfly larvae are aquatic predators. Their adult forms, though living close to water, are aerial predators.

While polymorphism *will* undoubtedly have social and cultural effects, it is unlikely to mean that polymorphic species will be radically different from humans in their possession of a culture or the ability to think. There is, however, an advantage to being a member of a species that change form throughout life in some way: unless one or some particular forms are notably aggressive or threatening, polymorphic species are likely to exhibit far less inherent xenophobia than the human species does. They may well be puzzled (notably if their polymorphism is punctuated) by the absence of different morphologies among humans, but by and large it is unlikely that the strange shapes of other ILFs would cause any more than wariness of the unknown, rather than the almost visceral fear of those different that haunts many human interactions.

Individual Polymorphism and Species Polymorphism and Their Implications Human beings change rather slowly as a species. The physical shape of humans (superficial differences such as skin, hair and eye shading aside) has been roughly fixed for all of recorded history, and morphologically modern humans may have emerged as early as 40,000 years ago. On other planets, however, rate of physical change may well be much higher: recordable in the memory of cultures, perhaps of individuals. If this is the case, there will be implications for the structure of the society concerned, something that will factor in our ability to communicate with them. Polymorphic ETI, particularly if the polymorphism is relatively random, are likely to be far less xenophobic than human beings as has been said. One the other hand, it is also possible that knowledge of cultural forms would be far more important to species that has a wide range of forms, or whose history includes change from one physical form to another. The primary anchor for membership would be *behavioral*, not appearance, somewhat like in Chinese society (Anderson 1988), where knowing or not knowing the proper cultural forms labels one very strongly as member of the group or an outsider. Chinese exclusion of outsiders is also a form of exclusion, of xenophobia, but unlike the form of xenophobia we call racism, it allows, for an outsider to *learn* how to become an insider. This could be of great consequence for our ability to communicate with such a species. The inverse would be a great difficulty members of such a race might have in understanding many of our social peculiarities, xenophobias, and prejudices.

A *caveat* is of course in order. Natural polymorphism could not be too extreme. It seems unlikely that any given species, no matter how plastic it is, will be able to adapt itself with sufficient rapidity to all environments, or environment changes it is likely to find itself in during a planet's lifetime. Moreover, the effects of polymorphism, on the one hand, must be balanced against the issue of intelligence on the other. The brain, where intelligence appears to reside, functions as such partly because of a particular three dimensional geometry—a set of ganglion connections

—which would seem to be crucial to its functioning (Langs et al. 2010). Thus any polymorphic intelligent species will be limited to some definite, if even currently unknown degree, in its bodily changes. The same is presumably true of its sensorium: senses, as we noted in Chap. 3, rely on physics, and thus have very strong engineering constraints. For a particular environment, therefore, even if a species is polymorphic, we must assume there will be limits of variation, which, once we have sufficient data points (more experience in contact) we will be able to predict. A note of caution: all bets are off, as we shall see subsequently, when a species engages in cultural evolution in preference to natural evolution.

4.3 Selection

Are there behavioral implications for different environments? We know that, roughly speaking, the *principles* of selection work uniformly, though their *expression* differs. Cold-weather animals on earth are selected for insulation (fur or blubber), lower radiation surface, and in many cases, white camouflage. Those in the tropics are selected for the reverse: low insulation, large radiation surface, and appropriate coloration. To a very small degree this is true of humans as well. While in reality things are more complex than described here, in principle, the different environments produce effects through the same principle of selection.

Given these principles, and short of the discovery of a non-Darwinian set of principles, it seems likely that ET organisms, including ETI will be products of generations of selection for their environment. Unfortunately, this bald statement does virtually nothing to help us understand ETI any better. We need to look once again at selection and environment to find a clue.

4.3.1 Natural Evolution and Cultural Evolution

In the past 50,000 years evolution has been charging ahead with great vigor.[8] We know of the rise and extinction of the woolly mammoth and woolly rhinoceros, giant monitor lizards and sabertooth kangaroos, elephant birds and Haast's eagle. Some have been replaced in their ecological domains by other species. Many species have increased their ranges. Paradoxically, from the perspective of a zoologist, one species has not really changed much physically (local puerile physical differences aside) over the past 40,000 years or so: *Homo sapiens*. Yet again, paradoxically, from the pure zoological perspective, that species has successfully spread into virtually all ecological zones, from the arctic to the equator,

[8]Though not in the way many of us like, and which the older generation, as is usual, generally disapproves of.

and seems to do quite well there. How has this happened? Have the rules of evolution been suddenly suspended?

As we know, of course, evolution has not taken a holiday. What has happened is that human evolution now dances to a slightly different tune: cultural, rather than natural selection is now the driver of human evolution. What this means precisely is difficult to say. But there is very good evidence, anecdotal and superficial in most cases, but thoroughly documented in others, that this is the case. For the latter, we need to turn to JR Flynn.

Flynn (2007) noticed a peculiar historical effect in the 2000s on ordered, regular intelligence testing of the male young adult Dutch population: IQ score, on well-respected tests, have been steadily climbing since the 1950s, which *seemed* to indicate that Dutch males today are almost a third again as smart as the same cohort in the 1950s (Flynn 2007). Data from other countries supported this observation. Obviously, since such an increase in intelligence would not go unremarked, the rational suggestion is that the IQ tests used were inaccurate, inappropriate, or just junk. Effects such as greater familiarity with tests and testing were also discounted statistically. Through a rather elegant analysis, Flynn shows that this is indeed the case *for some, rather outdated tests*. However, the samples he kept are those of people who were tested by the most modern, unbiased, and neutrally constructed IQ tests known. Flynn (and others, e.g. Wicherts 2008) argue quite convincingly, that there has been a substantial shift in intelligence, but that is a shift I would call 'sideways'. Overall, Flynn argues, the difference between the 1950s and today is down to a cultural shift. In comparison to thought patterns taught today, thought patterns taught fifty and a hundred years ago were far more concrete, far less abstract, far less formally logical. In 1916, the trick to getting a cow into a barn was knowing cows, what would move them, how to tie a knot in a lead rope. In 2016, the trick to getting a cow into a barn is knowing how many bytes your computer has, how to move your mouse on the mouse pad, and which path you lead your electronic cow through will yield the most bonus points: quite a different proposition.

For our purposes here, I want to do some violence to Flynn's formulation. I argue that what he is touching upon is a radical shift in human evolutionary patterns. In 1916, let alone 1816 and every century before that, those who could not handle a cow in a barn (or a hoe, a spear, a demi-lune knife, as appropriate) did not eat. They rarely reproduced. Those who knew how to use *the tools and social obligations their society offered*, lived, and sometimes prospered by the only measure Darwin allows: their ability to produce viable offspring, and nurture them to maturity to produce offspring in their turn. Since the 1970s, this has changed radically. Fitness is now measured in one's ability to deal with the exigencies of modernity, computer dating, Googled information sources, and all.

Consider K reproductive strategies and nurturing once again. Inherent in this, in the modern world, is searching for toys and information about the best food for your infant; searching out good schools on the Internet; mobile phones to keep the infant safe; ordering consumer items and comparing prices from half-a-dozen vendors. Leading cows around? Interesting, doable (I'd probably check the Internet on the best way to do that, wouldn't you?), but not essential for survival. In effect, while evolution is not in abeyance, some specific selection pressures no longer have weight. Welcome to cultural evolution (Currie 2013; Ekstig 2007; Matsui 1996; Stull 1977).

Cultural evolution is no different *in principle* from natural evolution (Ekstig 2007). Those who are fit, will grow to adulthood and produce offspring of their own. There is likely to be a selection: some will do it better, some worse. The parameters for selection are, however different: they derive from the environment produced by an intelligent organism's culture, rather than conditions imposed by nature. These effects may be devastating—nuclear war would certainly have some sort of effect—and they operate at individual and collective levels. The Japanese state evolved from a militaristic to a peaceful one as result of a form of selection engendered purely by culture imposed from the outside.

All natural evolution *whatever the physical parameters of the world concerned* is likely to be similar: some form, competitive or collaborative, of selection of the fittest individuals of a species, and of a species as a whole, to survive the *natural* environment and reproduce. Cultural evolution, on the other hand, seems to have some different rules going for it. Less a matter of physical fitness, and more a matter of a mix of *social*, *technical*, *and* environmental fitness. Thus a different, more complex or nuanced, form of selection comes into play, overshadowing natural selection. Cultural evolution *seems* to imply that those who can adapt themselves to the complex interplay of social actions (such as e.g. new forms of communication), technological flexibility (e.g. using a variety of complex tools) and natural selection, are those who will breed, and successfully bring their offspring to adulthood and breeding status.

We will discuss the concept of culture and its implications below in Chap. 8, here where we are discussing evolution, the question we need to ask is whether this makes a difference to communicating with ETI?

Let us consider the fundamentals of the two types of evolution. Natural evolution is an effect of the aggregate of physical, chemical, and biological effects of a planetary environment. It is, by definition almost, unguessable. When exobiology and improved astronomy supply us with more detailed data, we could perhaps reach a stage of making some generalizations about how ET planetary ecologies function. For the moment, this is unknown, and we must assume, subject to the realties we have uncovered such as the need for nurturance of the young of the species, that the nature of evolution and its products on an ET planet are unknowable at present.

Cultural evolution is a somewhat different kettle of fish. Let us assume, as a hypothetical argument, that in some identifiable way, the development of all cultures exhibit some commonalities.[9] From what we know about changes in human history, 'modernity' is characterized by: very rapid technological change (which is not necessarily the same as improvement, since *some* of this technological change is

---

[9]This is a dangerous assumption: much of the European disdain for other cultures derived from assuming that there was some inevitable path to development: the European way.

driven by fashion and mode. Appadurai 1996; Featherston 1990); rapid social change as social regularities erode due to technological innovation, economic fluctuations, and population shifts (Giddens 1990); a change in intelligence so that logical and abstract thinking, connectivity, and innovation have advantages (Flynn 2007); and an intensification and higher frequency of modes and fashions (Bikhchandani et al. 1992). From the perspective of the individual, bringing an offspring to maturity is also a feature. This means that being able to pass on the abilities to thrive in such an environment is a form of selection: those who do, live to produce offspring who can do so as well.

Since we are currently experiencing the first century of modernity, it is very hard to see ourselves objectively. Nevertheless, *if* this type of change, and its drivers—IT, digital technology, globalization—are universal (and it appears they are *very* powerful so that might be the case), then other species may develop in roughly the same direction. In such a case, a large number, many, or all ILFs may the product of the same sort of forces. This has the curious effect that we *may* be closer to ETI, as individuals than we think. While the *biological* processes would differ very greatly, the other two evolutionary forces—society and culture—may be in many ways similar, producing much more of a closeness to ETI than, say, to one's stone-age ancestor. This in turn implies that *some* problems of communication will be lessened.

We finish this section by suggesting that, just as we cannot see our own process of evolution/cultural evolution, so perhaps we would be unable to see a further stage: there may be some form of 'post-cultural' evolution, unseeable, even unguessable from where we stand here. Thinking about such 'transcendent' societies is well-documented in SETI scholarship. For the moment, we need merely entertain the possibility, that cultural evolution could well imply a convergence, of some sort, perhaps nebulous, between us and ETI.

4.4 Intelligence, Evolution and Ecology

Finally, we must come to the question of the evolution of intelligence. In order to discuss intelligence, it is useful to consider a theoretical perspective from Terrestrial animals. Crucial to an understanding of intelligence in land animals, is the question of reproduction. Humans (or at least, zoologists) can conceive of intelligence as merely one in a series of strategies designed to ensure species survival (Gould 1983). Intelligence is not, one may also argue, one of the most successful strategies (Lovejoy 1981: 323). Of the highly intelligent species of extant primates, only one has actually "made it". The rest are either moribund and on the verge of extinction, or else have actually died off. Vaunted intelligence is not necessarily as good a strategy as that of the rat, whose species continuity is ensured more by sheer numbers than strictly by intelligence.

The **r** and **K** nurturance strategies were discussed earlier. Here we look at it from the perspective of the end product of one of the **K** strategy. In this case, the parents invest heavily in their offspring, doing their best to ensure their viability. As a direct consequence, there are fewer offspring, but the offspring must needs be able to last longer *as offspring* and absorb much more learned information than offspring of **r**-strategy species. In brief, therefore, there is a tradeoff: greater intelligence, longevity, and fewer offspring, or less intelligence, briefer life-span, and more offspring.

We must therefore ask ourselves what strategies are likely to occur among ETI. Now, it is important to stress that **K** and **r** reproductive strategies are consequences of certain basic realities, but that only some of those realities are immutable. One, the need to allocate resources by an animal: energies used for foraging, evading predation, learning, and reproducing must be somehow balanced. We should probably assume this to be universally true. The second truism is that all land animals on Earth are bisexual, with one species nurturing the embryo, the other merely passing on genetic material. This places a heavy burden on the female. In most **r** selection species, the female bears almost the entire burden of reproductive cost, and is most vulnerable to its deprivations. In **K** selection species, it is possible to evolve cooperation (which is probably the basis for social behavior) and sharing of the parental post-parturition burden.

But suppose that ILFs were not bisexual, as in our thought experiment? What would the consequences be in terms of intelligence, since intelligence is, on Earth, one aspect of **K** selection? We can make certain basic assumptive principles.

First, we must assume that genetic variation is an evolutionary advantage (keep the need for adaptability as a survival trait in mind). As a consequence, any method of species reproduction must be built around, or include, a means for splitting and splicing the genetic code, by whatever means possible.

Second, such a system must be relatively simple, to allow, at the pre-sentient level, for a species to reproduce itself to the point it can improve its intelligence.[10] Thus, as a matter of principle, the fewer sexes, the better. A member of a single-sex species has a far better chance of mating than members of a bisexual species, not to mention tri-sexual, and so on, providing it can solve the problem of transferring genetic information. Two, or more, looking for a third, or more, would make reproduction practically intolerable unless genetic material were broadcast, something rather unlikely in evolutionary terms, since genetic material is *expensive* in terms of energy investment.

This is however only true *if* all sexes are exclusive (that is, males are only, and always males and so on). It is not true if each individual is two or more sexes at once, or, if there are three and no more than three sexes, if any two will produce a third. That is, $A + B \rightarrow C$, $A + C \rightarrow B$, $B + C \rightarrow A$. Such a system seems unwieldy (and thus unnecessary and unlikely) but at least possible.

Third, the reproductive method must include a variant of **K** selection: some way by which offspring can be nurtured (or, more likely, must be nurtured), and thus,

[10]Without, for the moment prejudicing the question of what intelligence *is* which is left for later.

will be able to develop intelligence *and have information transferred to it from a parent or other species source.*

This latter brings us to the question of how intelligence evolves. Briefly put, in the only example of which we have any experience—the genus *Homo*—seems to have evolved in what anthropologists call a process of punctuated evolution. The size of the brain (probably a primary prerequisite for intelligence to develop), evolves not as part of the process of evolving intelligence, which, in any case, is not a successful evolutionary strategy for a species, but as required for other activities (e.g. longevity, which has developmental consequences, or manipulation and binocular/color discriminate vision). The presence of a large brain allowed the development of intelligence as well as the brain's original, purely behavioral functions. The same may well be true of dolphins. That is, large size in the dolphin brain evolved as response to a need to process and organize marine sonar signals. However, because of the lack of evolutionary pressures (or for so0Ome other reason), this 'excess storage and processing space' never was used among dolphins for 'intelligence' whatever that particular label may imply.

Evolution does not necessarily lead to intelligence. In fact, as Lovejoy (1981: 323) points out, intelligence is not particularly survival positive for a species. Of the various descendants of the Dryopithecinae (the ancestral ape line including all hominids and hominoids), all except humans are either extinct or relict. Thus intelligence per se does not appear to offer any great survival advantages. In fact, it successfully developed only once on Earth, out of several billion existing and defunct species. Increased nervous capacity and its ability to encode relational experience is only useful if it enhances an organisms other capabilities. It also leads to increased reproductive liabilities, such as risks for the mother, both from predators while she is pregnant and giving birth, and from malnutrition caused by feeding the infant. Most successful species exhibit far lower brain/body ratios than humans. Cognition is likely to be extremely rare, because even on earth it was selected in only a few species (Lovejoy 1981: 327)

Intelligence may represent a qualitative change in a species but it seems more likely, that 'intelligence' is extant in all animals with a nervous system to some degree or another. Certainly all the mammalia, avia, and reptilia exhibit behaviors which appear, to different degrees, to be problem solving. One consequence of that is the view that intelligence derives naturally as a consequence of greater complexity in an organism. Biological evolution towards intelligence occurs as different organisms compete for survival (Smith 1976). This does not seem to be the case, given the limited sample on Earth. Rather, what seems to be more likely, is that intelligence is one of the possible strategies for survival given a species becoming involved in **K** selection. That is, given a preference of long-nurturance/fewer offspring, one possible strategy is to take the larger nervous system allowed by long-nurturance strategies, and evolve it into the use of greater neural functioning, and thus intelligence.

The development of intelligence is only one strategy, and it has costs in terms of the need for **K** reproductive selection rather than **r** reproductive selection. It is likely that it is not the end product of a process, but a particular historical happening, since it is not much more useful for predator avoidance than other means. It is a good, though not exclusive device for increasing locomotive and food acquisition ability. (Lovejoy 1981).

Some attempts have attempted to asses the possibilities of the evolution of intelligence in other faunal families or orders. One question, for instance, is why intelligence has not emerged in orders such as the dinosauria—an extremely successful order—since they survived for about 160 million years. Recall that cytodonts, the mega dynasty that preceded the dinosauria and was very similar to dinosaurs in some respects (complete domination of the land mass, homeothermy, infant nurturance in at least some cases), did not develop intelligence. Smith's (1976) idea that given time, any suitable environment should produce an intelligent creature, appears to be fatuous, since unprovable: until the universe ends there is still "time" to evolve. Moreover, if we consider each of the major land mega-dynasties on Earth to be a separate environment suitable for development, then his idea is empirically falsified. Neither the age of cytodonts, nor the age of dinosaurs, both longer than the age of mammals, produced intelligence. Russel's attempts to construct a model of a possible intelligent dinosaur, Stenonychosoid (Russell and Seguin 1982), while interesting and well reasoned, still fails to convince that intelligence will have eventually emerged among dinosaurs, particularly since the dinosaurs were so successful *without* any increased intelligence (Raup 1985). Lovejoy's (1981) perception of the ecological influences on intelligence being essentially accidental may be extreme, but we have at present no way to unambiguously falsify it.

4.5 Summary

Bylinsky (1981) has argued that given the probable limitation of the creation of life to carbon-oxygen cycle worlds, extreme differences between ILFs are unlikely. That is, since the functions required of intelligence will be similar, ETI form will be required to approach the human. Therefore, he argues, ETI will likely be bipedal quadrupeds (since quadrupedality is more efficient than either tripedality or multipedality), bilaterally symmetrical, two eyed (for minimum provision of binocular vision). This argument of convergent evolution makes sense and seems to be reasonable. However, this forces us to ignore the effects of alien chemistries and physics, not to mention ecologies. As I have shown above, the range of variation is far too great, even if we do not consider exotic contexts, to assume that efficiency equals bipedality or that depth vision is best achieved with two complex eyes.

Even the lessons from the most basic realities of our own ecology seem to be in doubt when we attempt to apply them outside the Earth. That is because we often confuse surface expression of a particular universal law with the law itself. Genetic mixing is a useful means of ensuring variability, which is itself necessary for continued and long term existence. That, however, must not be confused with the particular bisexual form of gene mixing that is practiced widely on Earth. Similarly, the evolutionary path of humans to intelligence was the result of specific accidents, but that must not be confused with the possibility of a different path working just as well for another ILF under different constraints. True, in both cases, nurturance is probably absolutely necessary. But that does not mean that mother love is universal: some other form of nurturance might do just as well.

From the discussion above also flow some realities about ETI behavior, and possibly society, subjects that will be addressed once we have a grasp of a concept intrinsic to the entire discussion, something we have taken for granted until now, the question of intelligence.

References

Anderson, E. (1988). *The Food of China*. New Haven: Yale University Press.

Appadurai, A. (1996). *Modernity at large: Cultural dimensions of globalization*. Minneapolis: University of Minnesota Press.

Ashkenazi, M. (1991). Traditional small group organization and cultural modelling in modern Japan. *Human Organization, 50*(4), 385–392.

Aviezer, O., Van Uzendoorn, Marinus H., et al. (1994). "Children of the dream revisited: 70 years of collective early child care in Israeli Kibbutzim" *Psychological Bulletin, 116*(1), 99–116.

Bikhchandani, S., Hirshleifer, D., & Welch, I. (1992). A theory of fads, fashion, custom, and cultural change as informational cascades. *The Journal of Political Economy, 100*(5), 992–1026.

Bylinsky, G. (1981). *Life in Darwin's universe: Evolution and the cosmos*. Garden City, New York: Doubleday.

Currie, T. E. (2013). Cultural evolution branches out the phylogenetic approach in cross-cultural research. *Cross-Cultural Research, 47*(2), 102–130.

Dawkins, R. (2006) *The Selfish Gene*. Oxford: Oxford University Press.

Diamond, J. (1997). *Why is sex fun? The evolution of human sexuality*. New York: Basic Books.

Ekstig, B. (2007). A unifying principle of biological and cultural evolution and its implications for the future. *World Futures, 63*(2), 98–106.

Featherston, M. (1990). Introduction. In M Featherstone (Ed.), *Global culture: Nationalism, globalization and modernity*. London: Sage.

Flynn, J. R. (2007). *What is intelligence?*. Cambridge: Cambridge University Press.

Fox, Robin. (1967). *Kinship and marriage*. Baltimore, Maryland: Penguin Books.

Giddens, A. (1990). *The consequences of modernity*. Cambridge: Polity Press.

Gould, S. J. (1983). The wisdom of Casey Stengel. In *Discover* (pp. 62–65). March 1983.

Langs, G., Golland, P. et al. (2010). Functional geometry alignment and localization of brain areas. Advances in neural science. http://people.csail.mit.edu/polina/papers/Langs-NIPS-2010.pdf. Accessed February 15, 2016.

Lovejoy, C. O. (1981) Evolution of man and its implications for general principles of the evolution of intelligent life. In J. Billingham (Ed.), *Life in the universe* (pp. 317–329). Boston: MIT Press.

Matsui, A. (1996). On cultural evolution: Social norms, rational behavior, and evolutionary game theory. *Journal of the Japanese and International Economies, 10*(3), 262–294.

Papagiannis, M. D. (1983). *Natural selection of stellar civilizations by the limits of growth*. IAA-83-272.

Raup, D. M. (1985). ETI without intelligence. In E. Regis (Ed.), *Extraterrestrials: Science and alien intelligence* (pp. 32–42). Cambridge: Cambridge University Press.

Ruse, M. (1985). Is rape wrong on Andromeda? An introduction to extraterrestrial evolution, science, and morality. In E. Regis (Ed.), *Extraterrestrials: Science and alien intelligence* (pp. 43–78). Cambridge: Cambridge University Press.

Russell, D. A., & Seguin, R. (1982). Reconstruction of the small cretaceous theropod *Stenonychosaurus inequalis* and a hypothetical Dinosauroid. *Syllogeus, 37.*

Sheldrake, R. (1987). *A new science of life: The hypothesis of formative causation.* JP Tarcher/Houghton Mifflin.

Sheldrake, R. (1996). Seven experiments that could change the world. *Complementary Health Practice Review, 2*(2), 93–99.

Sheldrake, R. (2011). *The presence of the past: Morphic resonance and the habits of nature.* London: Icon Books.

Smith, J. M. (1974). The theory of games and the evolution of animal conflicts. *Journal of Theoretical Biology, 47*, 209–221.

Smith, R. (1976). The abdication of human intelligence. In J. Christian (Ed.) *Extra terrestrial intelligence: The first encounter.* (pp. 147–159), Buffalo, NY: Prometheus Books.

Stull, M. (1977). Cultural evolution. In P. Morrison et al. (Eds.) *The search for extraterrestrial intelligence.* (pp. 47–52). Moffit Field: NASA.

Sukumar, R. (2003). *The living elephants: Evolutionary ecology, behavior, and conservation.* Oxford: Oxford University Press.

Taylor, P. (1992). Altruism in viscous populations: An inclusive fitness model. *Evolutionary Ecology, 6*(4), 352–356.

Wicherts, J. (2008) What is Intelligence? Beyond the Flynn Effect. *Netherlands Journal of Psychology, 64*(1), 41–43.

Chapter 5
Defining Modes of Intelligence

We come now to the major problem in discussing any sort of ILF: The question of intelligence. Effectively, this question breaks down into three separate but related issues. One of these is the question of intelligence itself. What is it? How do we measure it?

A second question that must hover at the background, when discussing possible intelligence universals, is how likely is it that intelligence will evolve?

No less important, is a tricky question: Would we recognize intelligence if we met it? This is not a puerile question, as we shall see subsequently using three concrete examples.

This chapter will concern itself first with trying to define the parameters of intelligence as humans have worked them out, from various points of view, and second, with how these insights can be applied to non-human intelligence. I emphasize humans, because those are the only unambiguously (pardon my ethnocentrism) intelligent beings the psychologists know (or think they know) of.

We do need to stipulate that all animals, and perhaps even plants, have some form of 'intelligence' in the sense that they are able to react in complex ways to their environment. Planaria, simple worm-like creatures, are able to navigate their way through a maze, and even remember what they have done. Certainly this does constitute 'intelligence' for some, very limited, value of intelligence, and presumably, there is some likelihood that planaria-like beings (in the intellectual sense) might exist off Earth. However, what we are looking for, is either some cutoff point on a gradient of intelligence, from, for sake of argument, planaria, to what we mean by the colloquial term 'intelligent' being. An alternative possibility is that some quality or collection of qualities in a discontinuity, would be defined as 'intelligence'. In essence, the difference is between increasing quantities in a bundle of characteristics that make up intelligence, or the addition of new characteristics.

However we address this issue, we need to keep in mind that we need an understanding of 'intelligence' that helps us to communicate with ETI, *as well as* some quality that distinguishes the species *Homo sapiens* + ETI from other

M. Ashkenazi, *What We Know About Extraterrestrial Intelligence*,
Space and Society, DOI 10.1007/978-3-319-44456-7_5

terrestrial and extraterrestrial species. For the moment, let us leave it at that, until we can get some sense of the issues.

5.1 What Is Intelligence? Definition and Problems

The question of what intelligence *is*, is difficult to disassociate from the question of what intelligence *does*. Yet the two questions are quite different. The one allows us to proceed via a process of induction, searching for traces of the phenomenon on the basis of a constructed model. The second allows us by a process of deduction, to identify the phenomenon by its resultants. Both of these processes have much to recommend them, but both may have different consequences for the issue of ET intelligence. The first course makes possible the creation of an abstract model which, hopefully, will fit both humans and other ILFs. The second, derived as it does from observations of our own species, may identify only those intelligences that are similar to us, thus defeating the main purpose of this exercise.

In concrete terms, approaches to intelligence have been either empirical (usually from the viewpoints of behavioral science or neurology) or else philosophical (from the viewpoint of theoretical psychology). Both of these approaches may be fruitful here, provided we keep the limitations of each in mind.

To start, we need to keep in mind that notwithstanding over a century of debate and research, there is no agreed on definition of intelligence (Legg and Hutter 2007). The ins and outs of the argument are not discussed here. However, some different parties in the debate can be identified in the form of contrastive questions.

1. Do we define intelligence based on what we see it does or based on what we think it is (and then find examples 'in the wild')?
2. Is human intelligence unique and qualitatively different from animal intelligence, or is intelligence (whatever it is) a continuum, from, for sake of argument, planaria to human?
3. What *is* intelligence? Problem solving? Associative learning? Fit to the environment? Social or emotional abilities which ensure survival?

Each of these different perspectives has its partisans and opponents. For us here, the crucial point is whether any definition (or characterization) or cluster of them helps us in the potential to identify, communicate with, and understand ETI. Due to the fact that the span of the debate (over a century) and the literature (thousands of articles and books) there will be no attempt to cover all the issues, or to settle them satisfactorily (which I am not competent to do in any case). Instead, we shall look at those elements of the issue that have, or appear to have, some bearing on the central question here. So, before trying to define intelligence in any way, it is worth examining the implications of answers to the three questions posed above.

5.1.1 Intelligence as a Philosophical Concept

Intelligence comprises a spectrum of phenomena. It can be defined as a "property of advanced information systems which allow such systems to analyze their environment then engage in processes which enhance the survivability and/or the reproducibility of the system." (Stonier, 1988) but it is also a process independent of the peculiarities of the structure originating the intelligence (Angeles 1976).

Intelligence may well be a componential system, one that can only be understood through what makes it up. Thus Coffey (1980) would term an action 'intelligent' only if it involves 'understanding': a far more complex issue. Moreover, it is possible to take this view to some extreme. For Coffey, 'understanding' is intimately tied to creativity, and thus thinking is not strictly a logical process, but is characterized by spontaneity and irrationality, selecting from imagined possible scenarios that that seem to model the outside world and its outcomes most closely: there is no fundamental distinction between perceiving, thinking and imagining (Coffey 1980: 273).

Intelligence also seems related to the process of information storage and recall, or knowing. There are three basic ways of knowing; empiricism (based on sensory experience: we know to the extent we perceive correctly); rationalism (based on logical consistency); and metaphorism (based on the degree to which symbolic cognitions lead to universal awareness [Royce 1976: 179]). Each may lead to error, but each may also lead to truth. None of these operates independently of the other.

One crucial aspect of intelligence is the ability to be adaptable, flexible, and to be able to predictably model the environment (Arbib, ms: 10). When this becomes a learned, rather than genetic system, it would seem to be an aspect of intelligence. For learned systems, unlike genetically programmed ones, offer two advantages. First, learned systems are infinitively expandable. Second, no less importantly, they enable the receiver to critique the message. That is, given the opportunity, the recipient of a model can say "This does not match my experience" and correct it *without* having to be destroyed, and its genes eliminated. This flexibility may be a consequence of the fact that "intelligence" is a compound phenomenon, in which the intelligent organism engages is the creation of perceptually triggered hypotheses (Coffey 1981). Interestingly enough, as Coffey notes, there exists no absolutely reliable connection between experience and hypotheses. The latter derive from delusion, fantasy, illusion, or hallucination. The sheer *irrationality* of the thinking process are what "create" intelligent thought.

5.1.2 Intelligence as a Consequence of Behavior

A very different view of intelligence proceeds from a different set of methodological assumptions. Simply put, if we are able to categorize the results of intelligent action, then we are able to identify intelligence. This of course presupposes that we know at the outset, what the results of intelligence are supposed to be.

For Minsky, intelligence is simply the ability to solve difficult problems (Minsky 1985). This seems like a reasonable approach, worth pursuing. At a species level, "… intelligent behavior produces solutions to problems which enable the individual or the species to survive or perform better. Problem solving ability is a key part of most definitions of human intelligence. An intelligent species is able to react successfully to experience and to respond to new situations. We need not be concerned with how this is done and can thus avoid words like reason, understanding, thought, and conceptualization." (Raup 1985: 38–39). In effect, Raup attempts to circumvent the problems posed by defining intelligence in terms of what it is. Coker (1984: 130) goes even further in claiming that intelligence may have no independent existence. Conceivably, it is simply apparent to the beholder when the behavior she studies reaches a degree of complexity within an arbitrary range of complexities: presumably when it is the same order of magnitude as her own.

5.1.3 The Varied Forms of Intelligence

From the xenology perspective, we also need to keep Raup (1985) and Gould (1991) in mind: high intelligence is not necessary for survival, so it is unlikely to come about as a matter of course. We deal with this conundrum later in the chapter when we examine three cases of non-human intelligence. For now, however, let us first catch our tiger.

Legg and Hutter (2007) attempted to distill a definition of intelligence from the wide range of such definitions that have emerged in several disciplines—psychology, machine intelligence—over the decades. They summarize their findings in three statements that all definitions agree on, or at least do not exclude:

Intelligence is:

1. The ability to interact with environment or environments.
2. The ability to succeed or profit from setting goal(s).
3. The ability to adapt to different goals and environments.

If we add to that (4.): the ability to communicate these issues to others, we may have a useful beginning in identifying intelligence.

5.1.4 A Working Synthesis for Intelligence

A real problem emerges when we attempt to define an intelligence different from ours. In essence, almost all discussions of intelligence cited above, have extrapolated from human intelligence and Earth experience, and consequently, are based on human evaluative models. Simply stated, intelligence is seen as the ability to be adaptable, flexible, and to be able to predictably model the environment, usually by the construction of a symbol set, which we shall call Intelligence definition 1.0. This may indeed be the dominant, or the most salient aspect of intelligence in all ETI as well. On the face of it, at least, it seems a universal proposition, one with which it would be difficult to argue. But the danger is that this is a quintessentially *human* definition of intelligence.

There lingers a suspicion that by doing so we might exclude some aspect of intelligence that may well be vital, if only because some percent (ranging from 1/n to n-1) of all intelligences in the galaxy (or the universe) are constructed differently. To proceed in an orderly fashion with the discussion, I shall call all intelligences, whatever their biological or physical origin who fit definition (1.0) above, "normal" (from our point of view: returning again to the assumption of mediocrity) intelligences. Other types of intelligence, those that do not fit the definition above, are 'exotics', a label we have already applied for biologies and physical surroundings.

Two questions must be asked. First, is there any reason to assume that 'exotic' intelligence could exist (that is, beyond the question of'why not')? If such intelligences could exist, is there any way of judging what form they might take, or of seeing what parameters might be identified that they fit into? The answer to the first question rests on the answers to two other questions asked earlier. If we are able to posit the existence of a non-Darwinian evolutionary scheme, then possibly if we can identify some appropriate mechanism, we could have an intelligence that does not derive from the need to compete with other organisms. Is this possible?

The second question has to do with physical surroundings. Can life arise at all (and we are very close here to a tautological negative) in relatively benign, unchanging conditions? This is a challenge to the 'Darwinian' view of intelligence as the product of an evolutionary struggle.

It is possible to enlarge upon this basic list of prerequisites for intelligence. Indeed, Minsky (1985) insists that human beings and ETI will think in similar ways simply because all intelligent problem solvers are subject to the same ultimate constraints of solving problems as economically and compactly as possible. Angeles (1976) goes even further, in that he characterizes all higher intelligences by the quest for information (and thereby a search for interrelatedness) about (a) What is known about the world, (b) How it is known, (c) How this 'known' can be represented and/or replicated, and (d) What can be known, that is, in what forms or within what parameters. Other orders of intelligence should be pursuing common relational procedures for organizing systems procedures, and for interrelating "content-less" relational patterns (i.e. searching for generalizable rules about their environment).

Tool using and communication are the foundations of how *we* humans generally identify intelligent actions. These are the most obvious expressions of whatever it is that we call "intelligence." The first, tool-using, is however a physical manifestation of problem solving in general. The second, communication, is a manifestation of the retention and transmission of knowledge, which in turn is useful for solving problems, particularly complex ones over the long term. Collecting and disseminating knowledge is particularly crucial for solving *large* problems whose solution needs breaking the problem down into parts, and solving by what computer specialists call "parallel processing," in which several processes each deal with one aspect of the problem. Allied to these two, in practice, is the nurturance factor. That is, fairly long-term habituation and practice of problem solving and utilization of knowledge. In fact, for biological organisms, nurturance, in practice, is probably essential. The same is true of artificial organisms (of which more presently) if we perceive of the period of design and testing as a period of nurturance.

One indication of intelligence is the search for interrelatedness in different context and content patterns and factors. If you are looking for a relationship between apples and oranges and bananas, you may be an ape, but if you are looking (as a civilization) for a relationship between one bottle of gin, seven organisms, a bottle of bitters, a lemon, and forty-nine ice cubes, then I would have to assume you are aware of the possibilities of more complex issues, thus, perhaps, intelligent.

The implications of this are not so clear, partly because we have only one certain sample to generalize from. Because we are so constructed ourselves, we expect several things from intelligence, as a basis for the previous qualities: (1) an efficient system of information reception; (2) an ample information store; (3) an efficient system that can modify its storage, storage system, or even methods of processing as result of information it has received; (4) some means of communicating its decisions to the outside world; and (5) it will at intervals find it expedient to communicate these decisions in some form (Coker 1984). Perhaps we can even afford to ignore those intelligences that do not, or cannot live up to our standards. This will be examined in the next section, as well as in the chapter on communication.

In the ETI context, therefore, we should search, in an unfamiliar environment for manifestations of problems solving that are *varied*, that extend beyond the immediate experience of any one organism, and possibly which require the ability to transmit a description of the problem and possibly of its solution or parts of the solution, to others. We might also search for those species that engage in long periods of nurturance for their young, in combination with the intelligence factors cited, whether or not we can identify the transfer of knowledge *to* these young: I have already noted the different balance between genetic information and learned information that may occur among ETI.

5.1.5 Technology and Handedness

Tool creation was regarded for a long time as the primary indicator of intelligence. Indeed, one species of the genus homo was so defined: *Homo habilis*, apparently a human precursor, who existed 2.2–1.1 million years ago. *Homo habilis*'s defining characteristic was handiness, that is, the ability to manufacture and use tools, and that served as the definer for the species (Leakey 1971). Along came Goodall (1970) and demonstrated that other species, notably chimpanzees, are both tool makers and tool users. Later, other species (for example, crows and otters) were added to the list. The definition of a species as intelligent, cannot, therefore, be dependent on the identification of its tool-making capabilities, unless we define 'tools' in some way that restricts the term, which in turn could make it virtually useless for our purposes here.

Moreover, even the combination of large brain and handiness is not a sufficient criterion for the emergence of intelligent. In addition to man, at least one other species has both a large and complex brain and manipulative dexterity: the elephants. Yet no one can claim that elephants are intelligent, at least not by any of the criteria we commonly use.

The use of tools seems an obvious criterion: if an organism has the ability to construct complicated structural devices which use, transfer, and impart energy, then presumably that organism is also capable of conceptualizing these things. Therefore, such an organism is comparable to us, i.e. intelligent. That is not necessarily the case, as I have underlined above: consider diverse life forms such as termites, bower birds, and apes. In an ET environment, as in ours, tool using could emerge without "intelligence". And while tool-using among non-humans on earth is relatively simple, this may not be the case elsewhere: the problem of generalizing from a limited sample remains.

SETI effectively (and in some cases overtly) defines intelligence as "the ability to construct large radio telescopes for the purpose of interstellar communication" Billingham (Introduction in Billingham 1981). While this is an excellent pragmatic approach, it does tend to miss the mark. It uses a measure of the complexity of tool use, as a measure of intelligence. The two should not, however, be confounded. If the general principle that tool using and making is not a sign of 'intelligence' holds good, than defining intelligence in terms of an elaboration of the tools might not necessarily be a good one either. Organisms may evolve large-scale radio telescopes on the basis of three different processes: for some biological requirement inherent in their natural surroundings, as the consequence of some non-intelligent tool-making activity, and as the consequence of some purposeful intelligent activity. In either case a far-off listener is going to be receiving radio signals, perhaps even modulated in some coded form, but only in the third is that listener going to be receiving an intelligent communication.

Intelligence does tend to solve questions faster, but not necessarily better than evolution. (Raup 1985: 39). Whales, fish, mollusks, birds have remarkable technical-biological solutions to a host of problems they have encountered, but their development was measured in generations rather than years. The familiar process of adaptation is one alternative route to sophisticated technology and it has no requirements of intelligence. Some of the most elegant solutions to engineering requirements for survival have evolved many times, including sonar and the use of electronic fields (Raup 1985: 41). Thus, as Raup notes, such seemingly artificial effects as radio beams do not necessarily indicate intelligence, they may merely indicate a different evolutionary paradigm, such as bees (Raup 1985: 37). An illustration of the problems inherent in the "artificial"—"natural" dichotomy will be offered when we discuss UFOs below.

5.1.6 Communication and Memory

A definite prerequisite for considering something as 'life', whatever else there may be to it, is the apprehension and transmission of messages. That is, for an object to be 'alive' it must be able to receive information about its environment, convert it to some interior medium, and transmit it internally and perhaps externally. Life is, in effect, a medium for arranging an environment. An AI could do that as well as any biologically-based life.

This also presupposes, as a necessary concomitant, that there is likely to emerge some means of storing information, again, in a form different than the way the raw data exists in the environment. Pain is a particular datum associated with the organism's interpretation of external stimuli.

Now as the information set becomes larger and larger, some form must emerge to organize it, otherwise the information cannot be retrieved and becomes valueless to the organism (which, by extension of the two paragraphs above, we must consider as 'live'). The prerequisite for a sophont organism is a more complex, possibly hierarchical, information/message storage system. Perhaps the manipulation of such messages must be considered the minimal prerequisite for intelligence. Intelligence, rather than being an all-or-nothing phenomenon, is therefore a continuum, 'greater' intelligence associated, at least on this level, with greater organization of messages. Thus 'sufficient intelligence' for whatever purposes is defined by the observer as some point on the continuum.

In passing, this decouples the concept of intelligence from that of technology. Valid as it may be, useful though it is in defining operative conditions, the SETI (and our cultural) bias towards equating intelligence with tool using/technological capability, hampers xenology. Some animals on earth, with minimal intelligence, are excellent tool-makers and users. This ranges from insects, through birds, to mammals. There are therefore two continua which may touch at any point: Non-tool using—high tool using, and low message organization—high message organization. Bees deal very successfully with their tools, but their ability to organize and transmit messages about their environment is strictly limited and formal, and thus too their intelligence, irrespective of the geometry of a hive and its social organization. Theoretically, we could encounter planetary, perhaps (and this may be stretching a point) even interplanetary "bees", with very little intelligence. Conversely, non-tool-using species with a capacity to transmit and organize information are also possible. How likely these two extremes are going to be is something that has to be examined.

Communication, like intelligence, is a difficult term to apply, largely because we have forms of communication which may well fit into a broad definition, but such a broad definition becomes, by its very broadness, almost useless for our purposes. Consider that almost all living species communicate in some way, and that even if we take the sophistication of the expressive medium as a measure of the sophistication of communication, we end up with some very 'unintelligent' species. Bower birds (the family *Ptilonorhynchidae*) create complex patterns of objects in order to communicate with potential mates. Many species have complex call signs intended to communicate within the species.

Any message may be a composite beyond the ability of any individual to create, but which an individual can comprehend. There are two requirements to deal with those restraints, which require a symbolic system for their manipulation. Every intelligence must develop symbol systems; for representing causes and goals and for formulating and retaining the procedures to reach those goals ["economics of thought"]. Every intelligent problem solver will encounter the same ideas about arithmetic and causal reasoning, because those are much simpler and more mathematically elegant than other solution ["sparseness principle of thought"] (Minsky 1985).

5.2 How Likely Is Intelligence to Evolve?

Here too we can find a wide range of opinions, with some considering it a highly unlikely phenomenon, whereas others, e.g. Gould (1983) consider it a highly likely phenomenon as a class if not as an example. In other words, we are likely to find other intelligence, but it is not going to be the same as ours in many respects. Essentially, three questions need be asked (Lovejoy 1981: 318):

1. Do biological systems naturally evolve towards more complex states?
2. Does increased intelligence favor survival and reproduction?
3. Does increased intelligence generally evolve towards cognition?

Related to the question of the evolution of intelligence is another one: the question of communication. Does communication indicate intelligence? Does it presuppose intelligence? Is it necessary and sufficient, or insufficient, for intelligence? Can we dismiss an ILF because it does not, for whatever reason, show signs of communication, with us or with others? The latter set of questions will be addressed only partly here, because in some ways the entire question of communication is itself problematic. SETI in a sense 'solves' the last of those questions unambiguously. Intelligence is defined as the ability and will to construct large radio telescopes (Billingham et al. 1991). This is fine, insofar as it goes, for SETI. It is less satisfactory when one is considering the question of CETI.

5.3 Would We Recognize Nonhuman Intelligence if We Met It?

As noted above, one of the greatest problems for SETI, is being able to recognize ETI. The problem has two prongs: false negatives, and false positives. That is, we may not recognize an ET intelligent species, or we may see a non-intelligent species as intelligent, when that "... just ain't so."

5.3.1 False Negatives

We tend to assume that an intelligent species will have a technological base, *and* that the technological base will resemble our own. Is this realistic? The majority expression of all human intelligence is the development and use of technology of a particular kind. Physical, chemical, and mechanical knowledge have been at the forefront of our technology. Biological sciences are now catching up. Social and behavioral sciences are still far behind, technologically-speaking. And though these different areas of thought and knowledge are clearly intertwined, we still lay greater emphasis on the concrete and physical, than, for example, the biological or social: we build individual family houses of concrete and steel, rather than growing them, or teaching ourselves to live communally. Not that these other solutions to existential problems have not been tried, or are not used, but those are exceptional, rather than the norm. Suppose, however, that an ETI (for example, in a metal-poor environment) developed biological solutions to the same problems? Eventually, of course, they would develop physics and mechanics as well, as sciences do converge. But their habits of thought, would be to reach for a biological, not mechanical solution.

And, as Flynn (2007) convincingly argues, intelligence is deeply affected by context and environment, notably the technological environment. That is to say that the type of ETI described in the previous paragraph, would have their intelligence 'skewed' in the direction of biological problem solving, not mechanical/physical problem solving, in which, to us, they may appear less-than-intelligent. Now, this would not be much of a problem (after all, there are would likely be other dimensions of intelligence they share with us) but for the fact that for the majority of SETI studies, the definition of an intelligent species is precisely that: specific-technology-based. Thus if our hypothetical biologically smart ETI did not build radio telescopes, *even if they were space-faring*, conventional SETI would not see them as intelligent. Of course, in reality, if there was any reason to consider them intelligent otherwise, the SETI definition would go by the board. Nonetheless, our techno-bias could well lead us badly astray.

5.3.2 False Positives

Suppose a Martian (= a hypothetical, completely innocent ETI)'s first encounter on Earth was an ant mound. "These creatures have a low technology. They do use tools to a minor extent, understand construction for their environment, keep animals for food, and communicate among themselves. Very much like ourselves, in fact, though of course at much lower degree of sophistication. Conclusion: these creatures are intelligent" thinks the Martian.

Whether the Martian's musings are right or wrong, what concerns us is the case where we identify some creature and make assumptions that it is intelligent in some way similar to humans. In fact, let the record show, we have three examples from the last century where this has indeed occurred. Ignoring, of course, the long, convoluted, and ongoing argument about the intelligence of our hominid ancestors, from Homo neanderthalis all the way back to the Australopithecenes,(cf. Brace 1976; Carter 1011; Corballis 1991; Howell 1978; Strait et al. 1997; Stringer 1987; Szalay and Delson 2013; Tattersall 1986, 1992; Wood 1992; Wood and Collard 1999). These examples are instructive, and we shall address them below.

5.3.3 Examples of Non-human Intelligence?

An empirical strategy to identify intelligence requires, if at all possible, to find some alternative examples of intelligence, so that the extrapolation about intelligence will rest on more than one observed point. We have no examples that are unambiguously intelligent. Nonetheless we can use 'pseudo-samples' by examining cases in which intelligence, though not proved, has at least been suspected. Moreover, these examples may provide clues as to the issues, and pitfalls, involved when we discuss non-human intelligence.

The question of identifying intelligence has been crucial for three major popular scientific conundrums of the twentieth century. All three are obscured by popular beliefs, misconceptions, and partisan expressions. They have all been studied extensively, yielding reams of data which can be interpreted in numerous ways. All have at different times attracted the interest of the man on the street, science, the military, the press, and theologians. The three examples are the whales and dolphins, or more broadly speaking, the *Cetacea*, Flying Saucers, or more properly, UFOs, and AI (artificial intelligence). These three phenomena have an interesting factor in common: their importance as platforms for discussing the meta-question of how do we identify another intelligence?

What do whales, AI and UFOs have in common? In one sense at least they are equally fabulous. For the past fifty odd years attempts have been made to test or to prove whale intelligence. Lilly's (1975) enthusiastic experiments attempting to define dolphin intelligence have proved inconclusive, and the host of experiments after him have done little to dispel the mystery (Herman 2006; White 2007). Hynek (1977) and Condon (1968, 1969), who did impressive work have pointed out a similar problem for UFOs. Demonstrating that UFOs exist, and analyzing what they are is a separate question from whether they are intelligent (or are directed by intelligent beings: for our purposes either position is useful). AI too are somewhat fabulous, partly because people working on AI have been stumped with the same question we are trying to answer here: how do we identify intelligence if we see it, notably if it differs substantively from our own?

Dolphins as Intelligent Beings

Dolphins, a genus consisting of the smaller species of toothed whales, have intrigued humanity for some time. Besides preying on, and sometimes killing them for reasons of expediency, human beings have come to appreciate that dolphins apparently often find humans interesting. There numerous reports of dolphin-human interaction, to the point that many people, including some scientists, feel that dolphins are intelligent (cf. Lilly 1975). Here too, I do not wish to enter into the dolphin intelligence debate. What is important is the existence of a species which (a) displays many (but not all) of the physiological characteristics assumed to correlate with intelligence (i.e. large and complex brain, sophisticated communicative ability, socialization and acculturation) but no hands or manipulative organs, and (b) does not respond to persistent attempts designed to allow the species to display its intelligence, if any. Lilly has suggested a number of factors that might account for this. First, that dolphins are communicating using a system so radically different from the monolinear one ('speech') used by humans, that the discussion, and indeed, even the identification of communication are essentially impossible. This seems unlikely, given the fact that opportunities for fairly basic communication have persistently been offered dolphins (Lilly 1975; Connor 2007; Herman 2006; Hovelmann 1989; Reiss et al. 1997). Another alternative is that dolphins have imposed upon themselves some form of taboo about speaking to humans, though this seems rather far-fetched, and, moreover, needs pass the barrier of Occam's razor. Particularly given the stress the dolphins, as individuals and as a species have suffered at human hands, its seems unlikely they would be so saintly as to persist in their self-imposed silence. Finally there is the possibility that dolphins have a *potential* for intelligence, but have not evolved it in practice.

Now, we may posit other types of intelligence for the sake of argument. For example, suppose tool-using was impossible (though for counter-examples see Mann and Patterson 2013), and symbolization unnecessary? The catodonts (toothed whales) cannot of course use tools, and, with the recent exception of man, have few natural enemies. However, if, as at least Lilly has argued (though others, e.g. Mann et al. 2000; Reiss et al. 1997 found no evidence for this claim), they communicate impressions and ideas by using multi-channel sonar (which would transmit a sort of hologramic picture, rather than a symbol of the object) then they may have no need for symbolization either. Instead of compressing information within one channel diachronically (which would require some speech 'object' to represent complex concepts, such as e.g. the verbal symbol 'food'), they would create effectively a multiple message synchronically, using different bands for communicating the same message, or aspects of the same message, or multiple messages. The compression of symbols, and their very use, would make no sense to them.

This of course relates directly to the issue of intelligence. Scientifically, in fact, intelligence does not "exist" until we have means to measure it. Much of our intelligence, and our ability to identify it, is based on the ability to symbolize, which all humans share. Humans do not all symbolize in the same way, something which led to tremendous biases in the early (and perhaps even current) uses of standard intelligence tests. The standard Stanford-Binet tests used on Navajo Indian and on Nigerian children, seemed to indicate they were less "intelligent" than their middle-class white American peers.[1] The same results were found when testing black ghetto children in the early years of intelligence testing throughout the US. What the tests were purporting to measure was intelligence. What they actually measured, of course, was knowledge of a particular set of symbols: middle-class white American symbols.

Whales do not have manipulative organs, and therefore do not create tools. They also do not communicate with us for whatever reasons. They have large and complex brains, and an elaborate, and so-far, not well-understood communication system (Reiss et al. 1997). The sole remaining possible criterion we would be able to identify for intelligence is whether they symbolize: something we are virtually certain less intelligent organisms do not do. Thus, if whales are unable to symbolize they are 'unintelligent'. But. They are communicating in a complex and responsive way. And in addition, they are also doing something that we humans are unable to do: communicating on a multiple-channel network.

Nonetheless, if it is the case that (some) Cetacea are transmitting images, symbolization may not be necessary, since rather than expressing information in a compact form, whales may simply 'send' an accurate picture, or rendering, of what they see, feel, plan to do, or whatever. If this is the case, then even the prop of symbolization as a primary characteristic of intelligence is snatched away. We cannot communicate with whales because they communicate in a way that is extremely alien, in fact physiologically impossible, for us to make use of. A further objection has been voiced by Minsky. If we (or, in this case, whales) could see everything at once, we'd never be able to focus on the parts, and thus be unable to solve complicated problems (Minsky 1985).

Let us assume, for the sake of argument (which we can retract later) that whales are indeed intelligent. The only ILF we know well—*Homo sapiens*—can be conceived of as a species engaged in purposeful activity: the production of its civilization. This purposeful activity is the cumulative result of the actions of several billion humans over the period of many thousand years. Partly this activity was directed at specific instrumental behaviors such as making a living. Partly, at least, human behaviors were directed specifically at the creation and maintenance of broader aims: civilization in some form or another. The whales, if intelligent, may well be engaged in some purposeful behavior of their own, one whose final product is invisible to us. They therefore would have little need, or motivation, to join us in our endeavor, and to 'prove' their intelligence to us, notwithstanding the horrible individual losses they have suffered at our hands.

This sort of response leaves us on a cleft stick. Yes, the Cetacea may be doing something but is what they are doing intelligent? It is obviously a complex, 'non-animal' set of behaviors, yet, it does not fit the current idea of intelligence.

[1]A detailed discussion on the weaknesses of IQ tests (and their strengths) can be found in Flynn 2006.

This seems to indicate that the definitions of intelligence in common use are inadequate for our purposes here. This has bearing on the primary definition of intelligence: problem solving. It is necessary to assume that the whales want to solve the question of communication as much as humans. Moreover, it must also needs to be demonstrated that whales see their problems in the same ways as do humans, even their major supporters.

If dolphins (the most studied species of whales) are intelligent, and given that human intelligence is the result of a particular historical accident, or concatenation of them (choice of K strategy, choice of upright posture, choice of intelligence), then it is likely *homo* and *delphinae* acquired intelligence for different reasons. The first for coordination and control in an arboreal environment where eyes and ears needed to be coordinated and where multiple offspring were a survival disadvantage, dolphins because of the need for echo location in the pursuit of prey, hunting coordination, and a sophisticated analysis of the marine environment. Later, when each had adjusted to a new environment (dry plains, marine) each also was able to use the increased brain capacity for other things: K selection and, more importantly, social activity and communication. That, in turn, selected for speech once the appropriate physical characteristic were available

In the human case, intelligence evolved to occupy those parts of the enlarged brain previously used, probably, for arboreal locomotion. The use of the brain for problem solving may well have been aided by handedness (Falk 1980; Corballis 2003), or possibly by the need for cooperation in bringing down prey. Dolphins may either never have needed intelligence, given the low ecological pressure on them in the sea, or else, their enlarged brains are being used fully for vision/communication (for the Cetacea there may not be any difference between the two) and thus unavailable for "intelligent" thought. Given evolutionary pressures—the gradual[2] introduction of a predator over many generations, gradual drying up of the ocean, less food or greater need for an intelligent 'crutch'—they may well be forced, biologically, to use their complex brain for other factors than vision-communication.

Identifying Intelligence: Is It a Bird? Is It a Plane? no, It's … a UFO

UFOs pose a peculiar problem for any kind of work on ETI.[3] Reports about UFOs have increased in the present century with the realization that artificial objects could fly. Presumably, any kind of report before the age of human flight would have been

[2]Whales *do* suffer from the presence of a merciless and destructive predator—humans—but that predation has been very severe for less than three hundred years, hardly time enough for a species to adapt, let alone develop intelligence as a counter to the new threat environment.

[3]One of my tasks as visiting scientist at NASA's SETI project in the nineties was to intercept and deal with calls from UFO enthusiasts wanting to report their findings to the experts. Quite a few professional UFO enthusiasts were very outraged when I explained, as gently as I could, that SETI did not deal with UFOs.

labeled in a different way: angels, demons, or celestial events. Moreover, at least since the fifties, an assumption has been made by those interested in the subject that UFOs represent the artificial constructs of non terrestrials. The considerable arguments pro (e.g. Vallee 1993, Hall 2001) and con (e.g. Sagan 1986; Klass 1968) the idea of UFOs as ETI visitations are not as significant for the present discussion as a question that is raised by the phenomenon itself (on the assumption, of course, that UFOs are not merely optical illusions or poor reporting).

As a hypothetical, I will accept the premise that UFOs exist as empirical phenomena. I do not wish to provide an alternative explanation (nor to commit myself to their existence), so much as to use the UFO issue as a paradigm for the problem inherent in identifying intelligent from non-intelligent behaviors. To do so, we can examine the characteristic of UFOs sighted, without going too much into the provenance of the phenomenon, and, what is no less important, without assuming *a priori*, that an unfamiliar phenomenon must have an artifactual nature. So, ignoring for a moment the possibility of mass or individual hallucinations or mental breakdowns, let us assume that *something* was indeed seen by at least *some* of the more reliable observers. To date there are easily over ten thousand reported sightings, but not all are confirmed. Hynek's (1977) work ended four decades ago. Even adding an additional four decades of sightings, there has been no confirmed evidence of intelligent beings associated with UFO phenomena. On that basis, let us try to examine UFO phenomena as a thought experiment.

First, certain consistent features can be abstracted from the reports:

a. The objects fly freely through the air.
b. Dimensions vary from about one meter to one hundred meters in length or diameter.
c. The objects move erratically, wobbling or oscillating.
d. They move determinedly in specific directions, and are able to change direction extremely rapidly.
e. They are attracted to certain human artifacts, notably high-voltage wires, moving cars, and airplanes.
f. They chase when fled from, and flee when chased.
g. They sometimes appear to 'spawn' smaller replicas of themselves.

(Blake 1979; Condon 1969; Hendry 1979; Hynek 1977; Lewis 1995; Lore 1969; Sheaffer 1981; Swords 1993).

These characteristics have led Klass (1968) to believe that UFOs are atmospheric plasma events ('ball lightning'), whereas other observers (see Swords 1993) feel this indicates the existence of secret flying machines. For others, UFO observations are examples of psychological disturbances (Zimmer 1985).

Normally, our cultural bias (twentieth century version, comic book heroes aside) assumes one of two general things about flying objects. Either they have wings, feathered or otherwise, which, by a rowing motion or gliding, allow the creature to proceed through the air. Such a phenomenon is 'natural' and indicative of a number of more-or-less familiar biological organisms, ranging from insects through reptiles to birds and bats. Or, rockets, planes, helicopters, and balloons are clearly identified as 'artifactual' because, in our culture, we know that "flapping wings = natural" and "humming sound + fixed wings/gas bag = artificial". A corollary of that is that any object that does not fit the characteristics of natural flying object is, *ipso facto*, an artifactual flying object. For a less technological society, a different syllogism would have been: "Birds have wings. Angels have wings. Flying object that is not bird = angel" (cf. Partridge 2004).

Now, let us make an assumption, and examine UFOs with an eye to their behavior and as little prejudice as possible. Very roughly speaking, UFOs have varied sizes and shapes. They seem to posses intentional mobility, and objectives of that mobility. They seem to posses both a sense of curiosity (they follow planes and human artifacts) and a sense of self preservation (they flee when chased). What should be most significant, but has generally been ignored by UFO analysts, is that they appear to have young. That is, a number of cases have provided evidence of 'smaller versions of UFOs splitting of a parent body' (Mason 2008: Table II p. 10). Let us therefore ignore our cultural and natural biases for a moment, and ask ourselves 'What phenomenon is this like?' The answer by now should be clear to the reader: a biological organism, in other words, an animal is as reasonable an explanation as a technological artifact. Now, this might mean one of two things: either the observations are false, or our perception of the data, which equates large flying biological organisms solely with the possession of wings is wrong. If we ignore for a moment Klass's claim (1975) that UFOs are atmospheric phenomena (a strong argument indeed), then the next likely explanation on the basis of the synthesized data in our list, is not that they are artifactual, but that they are natural biological organisms of some sort. Yet many people persist in automatically assigning them to the 'artifacts' category.

The purpose of this exercise was not to prove that UFOs are some sort of unknown animals. Nor even that UFOs exist as an empirical, material object, or as a delusion, or as mis-observation. The object is to demonstrate the difficulties in separating intelligent purposeful behavior from non-intelligent purposeful behavior. This is a crux of the issue. The little experiment also shows the need to overcome our biases, expressed here in two opposing syllogisms 'flying with wings = bird/insect = natural' versus 'flying without wings = artificial = intelligent/directed by intelligence'. To the degree that non-intelligent behavior approximates what we consider intelligent behavior, we should be inclined to call the originator 'intelligent'. As the UFO phenomena demonstrate, we need to be careful in making such assignments, since the data, when viewed in an unbiased way, can point in many directions.

This is particularly true when we encounter strange life forms away from Earth. They may be 'super bees' constructing elaborate structures in complex forms, up to and including in space, and not excluding large radio-telescopes, and do this by a set of genetic compulsions and reflexes that have nothing, or little to do with what we have labeled intelligence. We also need to remember that space is huge, and the possibilities of emergent phenomena are virtually infinite (at least until we learn of cosmic rules that serve as limiters). Just as the probability of N > 1 is as great or as small as one wishes to make it, so to is the likelihood of complexity-building non-intelligent life forms.

Artificial Intelligence: When Is It?

The discussion in the two sections above implies that we need to identify ETI in some way divorced from both artifactualism, and from human-centric prejudices. We do not *know* that dolphins and UFO are intelligent. (Once again: without prejudice as to whether the phenomenon is 'real' or otherwise.) Or are *not* intelligent, though so far there has been no evidence that they are either. We do know that alternative explanations exist both for their behavior and their inherent natures. We do not know, for instance, that both dolphins and UFOs are not simply refusing to deal with us for reasons of their own. Philip Klass correctly points out that the examination of the UFO phenomenon requires the ability to change one's view in light of evidence: the same may well be necessary in our assessment of the intelligence of both these two phenomena.

'Artificial Intelligence' (also 'machine intelligence') is a term used to minimally designate the emulation of intelligent responses by a machine (Alper 1990; Karelis 1986). A number of strategies have been proposed for solving the mechanics and problems associated with this. A fairly simple (theoretically) test was devised by Alfred Turing, and is known by his name. The Turing Test examines the proposition that if an intelligent human being were to communicate with a computer, using a free-discussion format, the measure of intelligence (that is, whether the computer program is indeed intelligent or not) would depend on the degree it (the program) would be able to fool the human at the console into believing the machine to be a human. If the keyboard operator were to ask a sufficient number of complex questions, only a sophisticated, that is, truly intelligent program, would have the flexibility to answer them, otherwise, the keyboard operator would realize there is something fishy.

There are a number of problems, from our point of view, with this approach. First, what if the keyboard operator only spoke and wrote Chinese, while the keyboard was in English? Alternatively, what if the computer spoke Chinese and the operator English? The problem was first propounded by a philosopher, J. Searle and has remained a cogent critique of the Turning test since (Preston and Bishop 2002).

A separate question derives from the issue of consciousness. Is an artificial intelligence conscious, sentient, that is, does it have an "I"? Marvin Minsky, father of the term "artificial intelligence', using a black box approach, argues that if it acts like intelligence, and its results are intelligent, then it is intelligence. His approach denies altogether the validity of the question of "consciousness" in AI research, and is generally labeled 'weak AI'. However, this poses a problem for the equation of artificial intelligence with any kind of 'higher' organism. Put simply, if it is only emulating the results of intelligent behavior, then it may well not be intelligent at all (Coffey 1980; McClelland 2009).

It also raises another question: perhaps the operator is simply not intelligent enough to figure out the computer program's errors? From the point of view of analyzing ET intelligence, the whole issue is very important for two reasons. First, we are still unable to determine if a particular set of behaviors is the result of intelligence or 'something like intelligence' if the end results from our point of view are the same (cf. Edmondson 2012). Second, there is good reason to argue that AI is one of the best ways to explore the galaxy, and that therefore, the first ETI we will meet might well be an artificial intelligence (Dick 2004).

Coffey represent an alternative to Minsky's view: since computers are unable to perceive, they lack consciousness, and therefore are not truly intelligent merely cleverly programmed. For the issue of CETI this does not matter in some instances. We can take as a given, that the AI may be a reflection, however many times removed and refined, of the attitudes and world-view of its creators. Communicating with an AI may well be communicating with an ETI via proxy, but it is communication nonetheless.

In contrast, a more complex view is offered by so-called "Strong AI". This is, essentially, the argument that to be labeled "intelligent' with all that implies, an AI would have to have some form of cognition, a recognition of itself as an ego, distinct from others as it is distinct from its sensorium. In such a case, however likely or unlikely, we would indeed be communicating with ETI: an artificial one, to be sure, but one that, if created by intelligent beings, would reflect its original 'culture of birth.'

5.4 AI as ETI

One of the early assumptions about interstellar communication and colonization was made by Bracewell (1974). He suggested that interstellar communities may well maintain communication by the use of physical 'probes' that would travel at high velocity between stars, carrying information. Moreover, he suggested, such probes may well be able to replicate themselves, and thus expand in a faster wavefront than would biological organisms, a concept originating with von Neumann (von Neumann 1951; von Neumann and Burks 1966). Rood and Trefil (1981) go even further, suggesting that any ETI we actually come into contact with are likely to be AI, rather than biological organisms.

Supporting Rood and Trefil, Stonier (1988) argues that computer programs are inherently more efficient than are biological programs in several ways. First, and most obviously, they compute faster, using larger and more accurate (though not necessarily more comprehensive) data bases to work from. Second, computer programs evolve (albeit, the evolution is *currently* directed by humans) at a much quicker rate than do biological organisms. Third, computer programs are migration-capable. That is, they are not dependent on a specific physical body. They can be cloned ("copied") and transferred from one machine to another with no impairment of function.

If we assume an artificial intelligence to be a program that adds *consciousness* to the rest of its qualities, then we have an interesting entity indeed. First and foremost, like any ILF, an AI will, presumably be able to create and operate computer programs. Since the AI itself is a computer program, we have the interesting possibility of an entity able to improve itself *consciously*. Moreover, given the speeds at which computers work, this will be done at much faster speeds than any human programmer could do the same task. In essence, an AI could "improve" itself (adapt to new circumstances) in a minimal time frame.

Another characteristic is no less interesting. AI are *programs*. They may be machine dependent to some degree, able to run not on any computer, but only on a specialized one, but that does not matter. Since they are programs, they could be transferred from one machine to another with no loss of consciousness. AI would be able to both replicate themselves (clone) *and* have what amounts to an indefinite life span—provided the appropriate machinery were available.

For some this is a frightening prospect. Stonier, obviously, considers these abilities, and the degree to which they are more efficient then human ones, to be a threat to the very existence of biological beings. It is just as likely, however, that in due course (and not necessarily any later than the problem of creating AI is solved, if at all) humanity will be able to understand the neurological function of the human mind. The step from there to complete melding of human and artificial mind is an obvious one. Human minds, enhanced by artificial programming, housed in either or both biological and electronic media, may well become the norm. Stonier (1988) has argued that while human intelligence is, in all likelihood finite, the same cannot be said of machine intelligence. Therefore, Stonier argues, AI may well be able to evolve themselves, effectively bootstrapping their own intelligence in a directed way. There are several implications of this. One, the most important from our point of view, is that species evolution may well include its own demise as a biological organism, as it develops, and is superseded or joined by, AI. This is of course particularly true if we accept the Darwinian paradigm as the reigning and only one, and if we accept the argument made in Chap. 4 about cultural evolution as a likely pathway for all ILFs. This does not presuppose a hostility of AI towards their creators (Science Fiction films notwithstanding) but simply the realities of evolution. Whether the biological organism will continue to exist along with its self-evolving creations is a separate matter.

Rood and Trefil's suggestion that ETIs are actually likely to be AIs is interesting. Enhancing intelligence, by sophisticated, self-programming computer systems (that is, software plus hardware) may be the next logical evolutionary step for any organic life form: a natural development of cultural evolution. There is already some discussion today of "wetware", hard-software packages that could be plugged directly into human neural systems to increase knowledge, or efficiency (Johnston 2008; Liu and Shimohara 2007; Teuscher and Adamatzky 2005), which we already do to some extent (Kalnikaité and Whittaker 2007). Additionally, if the method by which the brain does its coding is ever uncovered (not too unlikely), then it should be possible to copy a human mind to an artificial electronic matrix just as computer programs are copied from one machine to another (or to a storage medium) today.

Given what we said above about the apparent inevitability of the emergence of computer science among technologically-inclined ETI, we may well find three different classes of ETI in the universe. One class will be those who have avoided the technology/computer route, and perform computational functions without artificial aids. These would seem to be either technologically simple, or perhaps completely incomprehensible to us. A second class would be those cultures that have been overtaken by Stonier's nightmare: "The emergence of machine intelligence presages the demise of the human species as we know it" (Stonier 1988: 133): AIs that have somehow superseded their biological originators. Finally a third class might be cultures that have blended the biological and the electronic in some way, enabling biological-origin individuals to enjoy the benefits (and probably troubles) of both modes of living.

What forms would ET-AIs be likely to take? This depends, on the initial demands made by the designers. Conceptually, there are, two separate paths. The "culture-clone" path is one in which the constructors of the AI are intent, in effect, on creating an AI in their own image. This implies limits to the AI, but, may also mean that many of the ETIs biological traits must be incorporated as well. ETIs who chose the clone path and let them loose on the universe will be increasing certain factors in their ability to expand, at the cost of limiting efficiency.

The second possible path is the "super AI" path. By intent or accident, it should presumably be possible (and, in engineering terms, desirable) to design a self-improving AI. Given "self-improvement" as this type of AIs basic programming, it is likely to evolve away from its creators. It is thus likely that, even though intellectually and perhaps physically superior to organic ILFs, it will share very little of its "master"'s (and by now, the quote marks are necessary) interests.

Given that Self Improved AIs are possible, humanity is likely to be in a very precarious position indeed should those be the common form of ETI in the universe. Not because such AI's would be antagonistic, but because, given the time scale we are concerned with, and the possibility of self-improvement in that time, the relationship of the AI(s) to humanity would be that of a human to ants: not antagonistic, basically ignorant of their motives, and not concerned if they, or their nest, are stepped on.

5.4.1 Summary: Terrestrial Candidates for Non-Human Intelligence

To return to the discussion of identifying intelligence and its problems, we must remember that what has been said here is a sword that cuts two ways. Just as approximations of intelligent behaviors do not necessarily imply intelligence, so too, intelligence does not necessarily imply 'intelligent behavior'. The responses may be incomprehensible, the communication outside our ken, or the ETI so indifferent that communication may well never be established.

The final point to bear in mind is a simple one with some very complex implications: the definition and identification of intelligence are a crucial and critical step in the xenological process. As the examples of the three candidates for non-human intelligence on Earth demonstrate, identifying non-human intelligence is not a trivial thing. Human intelligence is the product of many factors, mostly evolutionary. Its expression as a product called civilization or culture (we will distinguish between the two terms later), the form in which we (self-) know it may be a purely human phenomenon, and doesn't necessarily, help in identifying intelligence.

Moreover, as the issue of UFOs demonstrates, we are inclined to assign intelligent origin to phenomena *that in our experience* are artificial. Under other suns, natural processes may have created natural phenomena that look like our artificial objects. The obverse is also true: developments in technology may well produce effects that we think of as purely natural. To quote Arthur C. Clarke "Any sufficiently advanced technology is indistinguishable from magic." (Clarke 1973). Transcendent intelligences may not only be out of our league, but not even identifiable as intelligent. Being able to *identify* intelligence is, however, only the first step: we also need to *accept* that intelligence, and what it does, something we tackle next.

5.4.2 Intelligence: A Unique Trait or a Continuum?

Perhaps one useful way of approaching the issue of ETI intelligence is to suggest that the measure of 'intelligence' can be portrayed as an N-dimensional space whose dimension *include* but are not limited to, the characteristics we humans identify as intelligence. Tool using? Yes, very likely, ranging from no tool use to the construction of tools to make tools to make tools to do functions. Information management? Yes, from little information management/retention to extensive memory use, indexing, and recall. Developing and communicating abstract ideas? Yes, from low or nonexistent, to complex and frequent. Goal setting and future modeling? Yes, from no anticipation or understanding of the future, to the ability to model and compare complex imagined futures.

This formulation has the advantage that it is cumulative: we may, and probably will, find ETI who have additional qualities we have not thought of, one or two missing qualities, and share with us (and other intelligent beings) certain of the continua that characterize intelligence. It is very likely, and easier for the xenological process, to think of intelligence as a composite. As a result, different individuals have strengths and weaknesses in different fields, creating a "profile" of intelligence, rather than a single scalar type. Among humans, one individual may score high in perception but low in verbal, and contrariwise. There is an argument that there are thirty different cognitive, and about 25 different affective components of intelligence, and they vary independently (Royce 1976: 186). A composite of different components may well be true of ETI as species, not only of individuals. Different ETI will have variations in a dimension or a subset of dimensions. They will also have additional dimensions (even in humans different dimensions can and do emerge as the result of organic change over time). This means that differences in modes of perception, or in environment can give rise to other components of intelligence.

Some obvious connections need be made between the idea of intelligence, as a concept, and the physical matrix in which the intelligence is carried. Four such connections have been considered by Sloman (1984: 178) as they apply to humans, but we can extend his concepts with some modifications to any ETI.

- The nature of the environment imposes constraints on perceptual systems. This implies that there will be differences (because different limits will apply) to the inputs and outputs of a particular ETI. How that will affect the operation of intelligence, the results the intelligence obtains, and the ways in which it reaches its conclusions is a question to be answered empirically.
- A second factor is the need to control the physical mechanism in which the computational system resides (i.e. the body) with its weaknesses, strengths, and limitations. Just for example, a body that is prone, for some reason, to a particular weakness—difference in atmospheric composition—will force the brain to expend time and ingenuity ensuring atmospheric stability, whether in the body or technologically.
- The ability to acquire new forms of knowledge, including social knowledge, is particularly important. A genetically-dependent species would have great difficulties in acquiring and processing new forms of knowledge, however brilliantly they are able to acquit themselves at the forms of knowledge they are familiar with.
- Finally, intelligences must be considered by their ability to cope with changing goals, not all commensurable, at different levels of time, immediacy, satisfaction, etc. This implies the need for strategies for comparing alternatives and outcomes, distinction between short and long term effects of decisions, etc. Presumably some ETI would be very good at that sort of 'multiple objectivizing' whereas others, perhaps as consequence of the environment from which they derive, would be very poor at the same process.

The foregoing analysis presents us with some hints about different forms that ETI intelligence might take. Each of the four variables above (and there may well be others) is a continuum on which any particular intelligence can be measured, and which collectively constitute a *profile* of intelligence. We can conceive of humans as being midway along each of these continua, not because this is the case *compared to other intelligent life forms*, but because, in the absence of empirical data, we can apply assumptions of mediocrity, to be proved as data become available. When we meet (first hand or via some medium) another ILF, we must be conscious of the fact that the species will likely differ in at least these variables. It is not until we have experience with several such encounters that we will be able to make global assumptions about whether these continua are causally connected, and if so, how.

5.4.3 *Communication Resistors*

Let us assume for both UFOs and Cetacea that they are intelligent in some fashion as yet undefined, or that may be identified using the above method. This is not the only problem. The Cetacea may be trying to communicate with us, and we may be too stupid to understand them. UFOs on the other hand may be purposely or casually ignoring us. How can one prove the intelligence of something that refuses to answer to any of the available tests? As it happens, we have cultural examples of similar behavior. Cultures as wide ranging as the Amazonian Jivaro headhunters and the very civilized Chinese have chosen historically to simply ignore other cultures as much as possible. Throughout most of the history of the Chinese Empire —some three thousand years—Chinese culture has chosen to largely ignore the doings of the barbarians beyond their borders. Though Chinese culture has been in contact with barbarians, and these have traded with and lived in China, the Chinese attitude has largely been one of "if they want anything to do with us, they had be better learn to be like us" In other words, the Chinese were simply not interested. This approach was markedly unlike that of other cultures we know: European travelers were constantly bringing back curios and reports from distant parts, since the sixth century BC (Woolley 1982) Western collectors have maintained collections and dictionaries of overseas material, as did the Chinese at some times in their history. The Japanese maintained a lively interest in foreign things even when the country was closed by its rulers: the Orandagakka ("Dutch learning") school of Japanese arts and sciences in the seventeenth through nineteenth centuries CE is one such example.

If we run into an ETI with the same isolationist attitude—let's assume the UFO are a single phenomenon of this sort—how do we prove its intelligence? There are some very tricky problems which might arise if we don't do so. Morally, we might be exploiting an intelligent species (cf. White 2007). We might also be facing retaliation? Exploitation? Missing cultural opportunities?

The answer to this problem is of course related to the problem of intelligence in general. If we can define intelligence in some non-anthropocentric manner, then we may be able to respond to that intelligence in the same way. If we can detect signs of the results of intelligent processes, then perhaps we have encountered intelligence.

5.5 Exotic Intelligences

There is of course the possibility that an ETI will be so outside our range of expectations that we will not be able to identify it at all. Two such suggestions have been made for appropriate cases. Insect societies have displayed what, in a single organism, would have been called intelligent activity. While it can be (and has been here) argued that this is an exhibition of genetic programming, or instinct, rather than intelligence, still some interesting puzzles keep the possibility that insect *nests* are intelligent alive.

To return for a moment to the equation of complexity with intelligence, it is well to remember that one of the most complex system for holding intelligence is coming into being right now, since the final decade of the twentieth century. This is the interconnected system of communication networks—the Internet—which, since the eighties, has become a major telecommunications effort. If such a system were to be intelligent, it would have no reference for the outside world, and thus, would have no motivation nor reason to contact us. In such a case, it would exist as intelligence, but be inaccessible to us. Such an intelligence may develop a purely discrete view of reality. We would have great difficulty in communicating with such an intelligence (DeVito 1991).

References

Alper, G. (1990). A psychoanalyst takes the turing test. *Psychoanalytic Review, 77*(1), 59–68.

Angeles, P. (1976). ETIs and the problem of intelligence. In J. Christian, (Ed.), *Extra Terrestrial Intelligence*: *The First Encounter* (pp. 161–175). Buffalo, NY: Prometheus Books.

Billingham, J. A. (Ed.). (1981). *Life in the Universe*. Boston: MIT Press.

Billingham, J., Michaud, M., & Tarter, J. (1991). The declaration of principles for activities following the detection of extraterrestrial intelligence. In J. Heidmann, & M. J. Klein (Eds.), *Bioastronomy the search for extraterrestial life—the exploration broadens* Lecture notes in physics (pp. 379–86), Vol. 390. Berlin, Heidelberg: Springer.

Blake, J. A. (1979). Ufology: The intellectual development and social context of the study of unidentified flying objects. *Sociological Review Monograph*, *27*, 315–337.

Brace, C. L. (1976). Ridiculed, rejected, but still our ancestor Neanderthal. In Poggie & P. Pelto (Ed.). *The evolution of human adaptations* (pp. 67–74). New York: MacMillan Publishing Co., Inc.,.

Bracewell, R. L. (1974). *The galactic club: Intelligent life in outer space*. Stanford: Stanford Alumni Association.

Carter, B. (2010). *Hominid evolution: Genetics versus memetics*. arXiv:1011.3393

Clarke, A. C. (1973). *Profiles of the future: An inquiry into the limits of the possible* (Rev. Ed.). New York: Harper & Row.

Coffey, E. J. (1980). Computers and intelligence. *Journal of the British Interplanetary Society*, *33*, 271–278.

Coffey, E. J. (1981). Visions of spaceflight. *Spaceflight*, *23*, 181–184.

Coker, M. (1984). Learning to communicate with an 'intelligent' computer. In M. Yazdani, & A. Narayanan (Eds.), *Artificial intelligence: Human effects* (pp. 128–142). Chichester: Ellis Hornwood.

Condon, E. U. (1968). UFO trouble in science. *Science, 3844*, 844.

Condon, E. U. (Ed.). (1969). *Scientific study of unidentified flying objects*. New York: Bantam.

Connor, R. C. (2007). Dolphin social intelligence: Complex alliance relationships in bottlenose dolphins and a consideration of selective environments for extreme brain size evolution in mammals. *Philosophical Transactions of the Royal Society of London B: Biological Sciences*, *362*(1480), 587–602.

Corballis, M. C. (1991). *The lopsided ape: Evolution of the generative mind*. New York: Oxford University Press.

Corballis, M. C. (2003). From mouth to hand: Gesture, speech, and the evolution of right-handedness. *Behavioral and Brain Sciences*, *2*, 199–208.

DeVito, C. L. (1991). Languages, science, and the search for extraterrestrial intelligence. *Interdisciplinary Science Reviews, 16*(2), 156–160.
Dick, S. J. (2004). The new universe, destiny of life, and the cultural implications. In J. Seckbach, J. Chela-Flores, T. Owen, & F. Raulin (Eds.), *Life in the universe: From the Miller experiment to the search for life on other worlds* (pp. 319-326). Berlin and Heidelberg: Springer.
Edmondson, W. (2012). The intelligence in ETI: What can we know? *Acta Astronautica, 78*, 37–42.
Falk, D. (1980). Language, handedness, and primate brains: Did the Australopithecines sign? *American Anthropologist, 1*, 72–78.
Flynn, J. R. (2006). Tethering the elephant: Capital cases, IQ, and the Flynn effect. *Psychology, Public Policy, and Law, 12*(2), 170–189.
Flynn, J. R. (2007). *What is intelligence?*. Cambridge: Cambridge University Press.
Goodall, J. (1970). Tool-using in primates and other vertebrates. In D. Lehrman, R. Hinde, & E. Shaw (Eds.), *Advances in the study of behavior* Vol. 3. New York: Academic Press Inc.
Gould, S. J. (1983). The wisdom of Casey Stengel. *Discover* (pp. 62–65), March 1983.
Gould, S. J. (1991). *Bully for Brontosaurus*. Boston: W. W. Norton.
Hall, R. H. (2001). *The UFO evidence: A thirty-year report* Vol. 2. Lanham, MD: Scarecrow Press.
Mann, J., Connor, R. C., & Tyack, P. L. (2000). *Field studies of dolphins and whales*. Chicago: The University of Chicago Press, 2000.
McClelland, J. L. (2009). Is a machine realization of truly human-like intelligence achievable? *Cognitive Computation. 1*, 17–21.
Hendry, A. (1979). *The UFO handbook*. New York: Doubleday.
Herman, L. M. (2006). Intelligence and rational behaviour in the Bottlenosed Dolphin. In S. Hurley & M. Nudds (Eds.), *Rational animals?* (pp. 439–467). New York: Oxford University Press.
Hovelmann, G. H. (1989). Animal 'language' research: The perpetuation of some old mistakes. *Semiotica, 73*(3–4), 199–217.
Howell, F. C. (1978). Hominidae. In V. J. Maglio & H. B. S. Cooke (Eds.), *Evolution of African mammals*. (pp. 154–248). Cambridge MA: Harvard University Press.
Hynek, J., & Allen, (1977). *The Hynek UFO report*. New York: Dell.
Johnston, J. (2008). *The allure of machinic life: Cybernetics, artificial life, and the new AI*. Cambridge MA: MIT Press.
Kalnikaité, V., & Whittaker S. (2007). Software or wetware?: Discovering when and why people use digital prosthetic memory. *ACM*, 71–80.
Karelis, C. (1986). Reflections on the turing test. *Journal for the Theory of Social Behaviour, 16* (2), 161–172.
Klass, P. J. (1975). *UFOs explained*. New York: Random House Inc.
Klass, P. J. (1968). *UFOs–identified*. New York: Random House.
Legg, S., & Hutter, M. (2007). A collection of definitions of intelligence. *Frontiers in Artificial Intelligence and Applications, 157*, 17–19.
Lewis, J. R. (1995). *The Gods have landed: New religions from other worlds*. New York: SUNY Press.
Lilly, J. C. (1975). *Lilly on dolphins*. New York: Anchor Books.
Liu, J.-Q., & Shimohara, K. (2007). Molecular computation and evolutionary wetware: A cutting-edge technology for artificial life and nanobiotechnologies. *IEEE Transactions on Systems, Man, and Cybernetics, Part C: Applications and Reviews, 37*(3), 325–336.
Lore, G. (1969). *Strange effects from UFOs - A NICAP special report*. Washington: NICAP.
Lovejoy, C. O. (1981) Evolution of man and its implications for general principles of the evolution of intelligent life. In J. Billingham (Ed.), *Life in the universe* (pp. 317–329). Boston: MIT Press.
Mann, J., & Patterson, E. M. (2013). Tool use by aquatic animals. *Philosophical Transactions of the Royal Society of London B: Biological Sciences, 368*(1630), 20120424.

Mason, R. L. (2008). The UFO experience reconsidered: Science and speculation. Lulu.com, http://www.amazon.com/The-UFO-Experience-Reconsidered-Speculation/dp/0615190456%3FSubscriptionId%3D0NM5T5X751JWT17C4GG2%26tag%3Dsonnysoftware%26linkCode%3Dxm2%26camp%3D2025%26creative%3D165953%26creativeASIN%3D0615190456

Minsky, M. (1985). Why intelligent aliens will be intelligible. In E. Regis (Ed.), *Extraterrestrials: Science and alien intelligence* (pp. 117–128). Cambridge: Cambridge University Press.

Partridge, C. (2004). Alien demonology: The Christian roots of the malevolent extraterrestrial in UFO religions and abduction spiritualities. *Religion, 3*, 163–189.

Preston, J. M., & Bishop J. M. (Eds.) (2002). *Views into the Chinese room: New essays on Searle and artificial intelligence*. New York: Oxford University Press.

Raup, D. M. (1985). ETI without intelligence. In E. Regis (Ed.), *Extraterrestrials: Science and alien intelligence*. Cambridge: Cambridge University Press.

Reiss, D., McCowan, B., & Marino, L. (1997). Communicative and other cognitive characteristics of bottlenose dolphins. *Trends in Cognitive Sciences, 1*(4), 140–145.

Royce, J. (1976). Consciousness and the cosmos. In J. Christian, (Ed.), *Extra Terrestrial Intelligence*: *The First Encounter*, (pp. 177–195). Buffalo, NY: Prometheus Books.

Rood, R. T. & Trefil, J. S. (1981). *Are we alone?*: *The possibility of extraterrestrial civilizations*. New York: Charles Scribners sons.

Sagan, C. Night walkers and mystery mongers: Sense and nonsense at the edge of science. *Skeptical Inquirer, 3*, 218–219.

Sheaffer, R. (1981). *The UFO verdict*. Buffalo, NY: Prometheus.

Sloman, A. (1984). In (Eds.), M. Yazdani, & A. Narayanan. Towards a computational theory of mind. *Artificial Intelligence*: *Human effects* (pp. 173–182). Chichester: Ellis Hornwood.

Stonier, T. (1988). Machine intelligence and the long-term future of the human species. *AI & Society, 2*, 133–139.

Strait, D. S., Grine, F. E., & Moniz, M. A. (1997). A reappraisal of early hominid phylogeny. *Journal of Human Evolution, 32*(1), 17–82.

Stringer, C. B. (1987). A numerical cladistic analysis for the genus homo. *Journal of Human Evolution, 16*(1), 135–146.

Swords, M. D. (1993). A guide to UFO research. *Journal of Scientific Exploration, 1*, 65–87.

Szalay, F. S., & Delson, E. (2013). *Evolutionary history of the primates*. New York: Academic Press.

Tattersall, I. (1986). Species recognition in human paleontology. *Journal of Human Evolution, 15* (3), 165–175.

Tattersall, I. (1992). Species concepts and species identification in human evolution. *Journal of Human Evolution, 22*(4), 341–349.

Teuscher, C., & Adamatzky A. (2005). *Proceedings of the 2005 workshop on unconventional computing: From cellular automata to wetware*. London: Luniver Press.

Vallee, J. (Ed.). (1993). *Dimensions*: *A casebook of alien contact* 2nd Ed. Berkeley: Contemporary Books.

von Neumann, J. (1951). The general and logical theory of automata. In L. A. Jeffress (Ed.), *Cerebral mechanisms in behavior – the Hixon symposium* (pp. 1–31). New York: Wiley.

von Neumann, J., & Burks A. W. (1966). *Theory of self-reproducing automata*. Urbana: University of Illinois Press.

White, T. I. (2007). *In defense of dolphins: The new moral frontier*. Oxford, UK: Blackwell Publishing Ltd.

Wood, Bernard. (1992). Early hominid species and speciation. *Journal of Human Evolution, 22*(4), 351–365.

Wood, B., & Collard, M. (1999). The human genus. *Science, 284*(5411), 65–71.

Woolley, L. (1982). *"Ur of the Chaldees"*: *The final account, excavations at Ur*. London: Herbert Press.

Zimmer, T. A. (1985). Belief in UFOs as alternative reality, cultural rejection or disturbed psyche. *Deviant Behavior, 6*(4), 405–419.

Chapter 6
Language and Symbols

In some form or fashion, intelligence relies on a coded method of transferring information between members of the intelligent species. We have seen that among terrestrial animals, K selection implies greater nurturance and training of the young, and thus a *need* for communication, which among most K selection species is accomplished by a mix of imitation and positive/negative reinforcement. Perhaps the greatest human "invention" is the development of the faculty of language, which made the whole process of educating the young a much more complex, but much more complete affair.

We can conceive, of course, of an ET species that consists of one undifferentiated 'intelligence.' Such an organic intelligence would find great difficulty in relating to other ILFs. It will not likely evolve into one interested in contact in any form. Communicating *within* a species is an important feature of intelligence. Not only is communication necessary in educating the young, but it is also an important feature of cultural adaptation, allowing the transmission over time and space of survival information. The method used by humans to communicate ideas that go well beyond immediate experience is language: audible vocal exchanges which can also be codified and transmitted through the use of specified signs. Language can be used in many ways, and it also can be codified in a series of ways that are important for further discussion.

The issue of language raises an important split in the possibility of communicating with ETI. Noam Chomsky, one of the world's foremost linguists and his students have argued forcefully that language is innate (Chomsky 1965, 1986, 2002, 2006). This in itself is unsurprising. What is still moot, is the argument made with varied degrees of forcefulness that human grammar—the essential structure of language—is generative. What this means is that all humans are born with a single language mechanism: a mental book of rules, if you will. This book of rules generates specific language emissions, which, though they differ from one culture (language) to another, have a fundamental, very basic unity. Specific grammars in different languages are *generated* by users from the fundamental grammar using standard rules inherent in the basic language mechanism. A human infant initially

M. Ashkenazi, *What We Know About Extraterrestrial Intelligence*,
Space and Society, DOI 10.1007/978-3-319-44456-7_6

has an instinctive (one presumes) knowledge of the rules of the ur-human language, from which mechanisms evolve along predictable pathways. This approach seems to tie human language very deeply into human physiology and psychology. One of the many objections to Chomsky's theories, however, is that these pathways have not been found anywhere else than in the logical analyses by Chomsky and his students.

Chomsky's approach, whatever it means for humans (and transformative grammar has been one of the bloodiest scholarly battlefields there are) has important implications for communication with ETI. It means effectively that humans will *never* be able to communicate with ETI, given our different physiologies and psychologies. If language is, in essence hardwired into brains at some elementary level, how could we possibly develop the necessary flexibility to converse with non-humans? As an analogy, I type these words using a Macintosh computer, whose physical substrate (the CPU and its associated ROM) are radically different from those of a PC. Indeed, given these differences, these two types of computers find it very difficult to communicate with one another. Would the same be true of communicating with ETI, when our language is hardwired into our biological ROM, and an essential part of our CPU? Before answering this question, let us look at an alternate possibility.

In contrast to Chomsky and his school. B.F. Skinner, a major figure in behavior studies has posited a simpler mechanism for language than Chomsky's 'deep grammar' (see e.g. Skinner 1957). All biological organisms respond to stimuli: complex structures of rewards (which increase or decrease the frequency of an emitted behavior) and can also explain the evolution of language in a species and in the individual. In this view, language is explained through an evolutionary process of reinforcement acting on the genetic nature of the organism concerned. As a species we have been rewarded by the presence of language (we can transfer information leading to positive rewards, and away from negative ones), and as individuals we are rewarded from infancy for getting the language of our nurturers right. For our purposes, Skinner's approach is both simpler, and less problematic. If language is learned rather than innate or inherent in biological 'deep' structures, than ILFs will be able to learn one another's language with some facility.

Beyond the arguments between the behavioral psychologist and the psychological linguist, however, lies a much greater chasm we must at least pay attention to. "…[D]o we believe that language is one of the ways in which we use the basic rules of cognition in order to communicate the workings of intelligence? Or do we believe that language is the basic tool with which our intelligences work?" (Coker 1984). The difference between the two positions is crucial. If we adopt the first, than we are essentially adopting a transcendental approach that says that we believe in some form of pure 'intelligence' as an entity. This comes very close to a position of saying there is some immanent difference between 'intelligent' and 'non-intelligent'. The other position is purely Darwinian, in that it ascribes the evolution of intelligence and the evolution of language as being intrinsically linked and thus we are likely to have greater difficulty in bridging the worlds of two different languages. If our cognitive world is entirely dependent upon our language, then different languages will produce different cognitive worlds: so much more so if we are talking about ETI. In fact, the latter position is supported by the Sapir-Whorf hypothesis (Sapir 1956), more correctly termed 'linguistic relativity' which argues that language does affect the choice of cultural categories and their relationships (Lucy 1992).

To summarize briefly, language is clearly essential for inter-ILF communication. It is likely that the structure and semantics of languages used by different ILFs, and the degree the language structures their conceptual world, will make any inter-ILF communication and minefield. Nonetheless, given the essential nature of this communicative form, we do need to assess both the pitfalls and potential meeting points. In this chapter, we shall examine some of those.

Coding, Transmitting, and Retaining Language as a Form of Information
We can probably accept as a given that any ETI will use some form of language: a structured system of self-generated signs used to convey information beyond the biological. If they were not able to do so, they could not, almost by definition, be an intelligent species. The use of language for an *intelligent* species goes further: into the temporal dimension. Language as used by humans differs considerably from the simpler "languages" used by terrestrial animals: it encodes and communicates *past* events, and is capable of communicating *future* events and even imaginative *possibilities* such as novels and counterfactuals (AKA 'lies').

On the other hand, I would argue that the mere *use* of language is insufficient to identify a species as intelligent for the purposes of inter-species communication. A language that encodes only experiences that are immediate is quite different from one that encodes past, future, and potential experience. From the perspective of communication with ETI, one major reason should be obvious: We expect a communicative ETI to be technologically savvy, which requires planning, the ability to conceive of non-material artifacts (design), social allocation, and probably social differentiation. None of these can be carried out successfully without the use of signs—communicative items that stand for other, often more complex 'things—and symbols, which are particularly complex types of signs. In sum, the unlikely possibility of a single intelligence making up a species aside, we could expect, as a universal, that any ETI would have a grasp of language. Given any minimal social and cultural sophistication, we could argue for the likelihood of *any* ETI possessing language, that language encoding not only immediate experience, but also past and potential experiences. You will of course be saying "This part is obvious, but how do we understand such a language, or even identify it *as* a language? Surely not all ETI communicate by modulating sounds?"

Modes of Communication
Among humans, as among most primates, communication emerges from a set of vocalizations that are dependent, physically, on our breathing apparatus. We have discussed before why this is a likely arrangement: sound propagates fairly fast, and requires little expenditure of energy, so many animals rely on it heavily. While this is true in theory, in practice, humans accompany and elide their communication by

non-vocal signs. Non-Verbal Language (NVL) has been an interesting element of study for social scientists for several decades. Hall and Hall (1990) and Morris (1956, 1979, 1982), have shown repeatedly how humans, through use of body signaling, enhance, change, and even contradict their spoken words. Now, aside from telling us that humans can lie (something we shall deal with subsequently) it also indicates that language can be expressed through a mix of different modes: sound modulations, bodily posture, manipulation of objects. Among *humans* primacy is given to voice. However, ignoring the possibility of exotic senses for communication (e.g. modulated electromagnetic broadcasting), even humans do not restrict their communication to voice alone.

What about ETI? Assuming some analogues of our 'mundane' five senses, they would be using these to good effect. But would that necessarily be sound? Simple signs—'this is my territory, stay out'; 'I'm looking for a mate'; 'I'm bigger than you'—can be conveyed by very different senses: making a particular sound, as birds do, flashing a visual signal, as many deer do, or scent marking like dogs, bears, and others.

On Earth, there are other forms of complex communication. Many squids have the ability to manipulate minute chromatophores in their skin (Hanlon et al. 1990). This means they can, very rapidly, and with great accuracy, change their color in whole or in part. This has advantages in the natural world including mimicry as both predator and prey, sexual signaling, and group dynamics, all of which are important to the squids' existence. Given the speed and accuracy with which squids can change their color, it would not be surprising to find color coded communication among non-humans too.

The generation of signs—through color, sound, scent, or any other somatically relevant system—is in a sense puerile. It is useful to keep in mind, however, that the value of a language lies not in its ability to transmit (or even retain: e.g. scent) *signs*. We need to be able to transmit and store concepts and ideas: non-material, non-immediate, complex data, that is 'symbols'. That is, roughly speaking, we need (1) a finite list of signs representing concepts which (2) can be out together in some systematic way to generate additional signs for more complex concepts, and (3) a set of rules by which we present those signs so the input and output meanings are the same (or as close to the same as feasible): in other words, a grammar.

This is where the previously cited argument between grammaticists and behaviorists comes to a head: is grammar somehow inherent in a species through evolutionary processes, or is it something that develops as a cultural artifact? We *do* know that Darwinian processes are capable of developing extremely complex mechanisms (as evidence, the human species). Whatever the case among humans, we would expect of ETIs that, since they need to be able to express complex thoughts to one another, and to their descendants and correspondents, their language would have a grammar. Most importantly from the perspective of inter-ILF communication, grammars are structures of elements, that is, codes. Once one recognizes the *medium, even* if the meaning of the individual signs is unknown, one can recognize the existence of intelligent communication. This has two implications: anyone using a grammar is, almost by definition, intelligent, and the existence of many grammatical instances teaches a great deal about the user, even if one is more or less ignorant of the dictionary.

6.1 Language as Memory Storage

At the most minimal level, language is conceived of a method for passing on information between members of a species. We know that many non-human species communicate as well. Dogs bark and growl, primates howl and grunt and fluff their hair, squids send complex patterns from their skins. Insofar as oral sounds alone are concerned, mynah birds, dolphins, and several other species are perfectly capable of producing the full range of spoken words used by the human species. One could claim that these sounds produced by say, a parrot, bear no meaning to the bird. On the other hand, anyone could teach an American four year old to recite the Ramayana, and the sounds would be no more comprehensible to her. The point is that communication can be achieved by numerous devices, and *in that* aspect, human speech is no different qualitatively from feather fluffing.

There is one difference, a critical one, between human language and non-human signaling. Language *encodes* experience, separating the experience from its immediate occurrence. Removed from its immediate ken, a baby can nonetheless articulate a desire for a particular item. The four year old who has learned to recite the Ramayana might, in the fullness of time, also write a learned dissertation on the poem. Language can, in fact, store information. Unlike feather fluffing, which essentially says "you there, go away," language enables an organism to store the information and recount it: "Today I met another person and told him to go away." This makes language's role in intelligence and its emergence critical. Without it, neither the concept of time, nor the evolution of non-immediate thinking are at all possible.

Having said that, it is necessary to retreat a bit. Language consists of the ability to store, in some commonly accepted coded form, bits of retrievable information about experience (later in a species development it does other things too). Given our interest here, *any other naturally occurring storage system could do as well.* Obviously, in language as in all other developmental processes, there are better and worse ways of going about things. Theoretically, the storage of data could be carried on using colored and size-sorted pebbles. Some penguins do indeed use this method. Unfortunately, once your "vocabulary" passes ten words, you are required to carry several kilos of rock: not efficient for storage. Words—immaterial sequences of agreed upon sounds—weight little, are easily transmissible, and can be stored and multiplied without risk of hernia in various formats are obviously an excellent evolutionary choice.

Are there any other possibilities? Arrangements of body posture or organs suffer from the problems inherent in all sight systems: strictly point to point. Communication by smell is a definite possibility and should not be discounted. Given the propensity of smell molecules to disperse rapidly, it seems rather strange that our evolutionary ancestors chose sound over smell. Perhaps because smell is not so easily controlled by organic methods, and tends to give one away rather quickly. Or perhaps because scent molecules leave long-lasting traces, as any dog owner does (or should) realize.

6.1.1 Natural Languages: Mistake and Ambiguity

One of the more interesting, and probably important aspects of natural languages is that they are *not* accurate. They are not accurate representations of fact, of intention, or of ideas. Many people, particularly physical scientists and all married people, lament this phenomenon, often at great length. Some language groups—Japanese is an example—take great pride in their language's ambiguity, lack of clarity, and contextuality (Passin 1980). Here I take the side of the ambiguous minded. Ambiguity in a language, however problematic it might be for dyadic (between two speakers) communication, is nevertheless a great invention. To understand how, let us look at ambiguity's triplet sibling: honesty.

6.1.2 The Social Disadvantages of Honesty

Honesty is an interesting concept. It implies that internal states (what the person perceives and stores) and external realities expressed in language, must somehow match. Given the inherent flexibility of mind needed to be a successful intelligent species, the likelihood of this being adopted as a survival mechanism is doubtful. In a previous chapter we discussed Darwinian versus cooperative evolution as concepts. Let us here consider only Darwin-evolved species. Is there an evolutionary advantage to an intelligent species in complete 'honesty' as a strategy? If one were completely honest, how would one fool either one's predators or one's prey? I cannot conceive of any evolutionary mechanism that would give an advantage to complete honesty. Situational honesty? Now that, as Don Corleone might say, is a different kettle of fish. Under certain circumstances and playing certain roles, there is an advantage to an individual making their word their bond. As a power play, provided one has staked out a particular turf, this is of course advantageous. Indeed, it seems possible at least, that as a culture, some ETI would adopt that course of action in communicating with other ILFs. But that is a cultural choice, one that would depend to a great degree on the advantages to accrue from such a position.

6.1.3 A Single Message and Honesty

An interesting case for dishonesty is made in the ongoing SETI debate about the 'unified message'. In 1991, Billingham, Tarter and Michaud suggested that a message from Earth in answer to one from an ETI should be a unitary message; politicians, lawyers and scientists will sit down and compose a single message from humankind (Billingham et al. 1991). There are two technical issues with this point of view. First, if the ETI's technology is superior to ours, they will likely have radiation evidence of the fact that humanity is *not* unified. Though some, such as Korbitz (2014) and Goodman, (1990) have argued the signals from e.g. TV and radio would be too faint, the giant X-band US and Russian anti-missile radars are *definitely* not indications of peaceful coexistence, nor are a decade of flashes from nuclear explosions, which pump out far more electromagnetic power than a radio broadcast. A second issue is far more *critical.* Who is going to develop such a message? As in most international activity, "International agreement" usually means that those with power and cash—the US, Europe, Russia, the PRC—will determine (presumably after lengthy debate and political horse-trading) what the message will be. The rest will be shut out; no dissent allowed. So right off the bat, this unified message is problematic. Add to this the fact that once a message is detected, multiple responses messages will be initiated, and we have a recipe for a proper Clochemerle (that is, an epitome of miscommunication and crossed intentions, all of them inherently good, together, a cacophony. See Chevallier 1951). In contrast (intentional or not) to the single-message position (which has been adopted by international astronomical bodies) is Harrison's firm support of [linguistic] honesty: "Perhaps, honesty is the highest value in interstellar communication since there would be so few ways of verification." (Harrison 2011). An ersatz message from the power-that-be purporting to represent a unified humanity is clearly not 'honest' in any sense of the word. There is a major contradiction between honesty and a single message: honesty would compel us to admit and allow many voices speaking for humanity, rather than one faked singular message.

Put bluntly, the more varied the messages received, the more inclined *I* would be to accept them. If I saw only one type of message emerging from a particular culture, I would most likely be very suspicious (and so should you): it means, basically, that someone is massaging the message, trying to manage impressions, in short, lying.

I am very fond of a cartoon I saw many years ago. Two psychologists pass in the street, and each says "Good morning" to the other. After a few steps, one of them says to himself "What did he mean by that?" We, a fractious, politically and otherwise diverse species, receive a comprehensible single message from another species. How are we to view this message? There are two possibilities. One, this ETI species is a united species, with no or little internal dissent. Two, this species does have internal dissent, but it is cloaked over by prevarication. In both cases, there is the potential in a simple message for something fishy. Bluntly, what are they hiding? Has all dissent been suppressed (pleasantly or otherwise)? Or have they truly reached the Latter Days of the lion and lamb resting together? In other words, while dishonesty is to be expected (subject to ethical issues we shall discuss subsequently), even the SETI process does, for some occasions, advocate providing messages that are inherently dishonest.

6.1.4 Errors as a Form of Testing the Waters

Erroneous statements are excellent ways of testing the waters in discourse. Does alter accept ego's mistakes and move on? Does alter try to correct the errors? React to them? The answer can always be "Sorry, just a mistake" as a low-level way of smoothing things over, during which ego has learned a great deal about alter. Thus ambiguity is tactically advantageous.

Crucially, errors can also be used to display weakness. Deliberately displaying a weakness goes against Western cultural thought. One is expected to display strength at all times. The same is not true for Asian thinking. Many of the Chinese Seven Martial Classics, including the better known *Bing Fa*, or Art of War (Sawyer, 1993), as well as the more populist '33 sayings' provide much space to the question of *when* to display weakness to gain a tactical or strategic advantage.

Japanese culture, which contains many refinements of ideas of Chinese origin, has even a specific term for managing superiors and powerful opponent by displaying weakness. *Amaeru* as it is called, may not be as pervasive as some (e.g. Doi 1973) believe, but it is nevertheless an effective and well-used tactic. Other cultures and sub-cultures use this tactic with more-or-less consciousness and forethought. The *primary* means of expressing weakness is often linguistic. Some languages such as Asian ones have a vast repertoire of terms to express weakness grammatically. Others such as French and German retain these in an attenuated form.

In the communicative environment we are discussing, which would likely be primarily linguistic, and possibly visual, *images* of weakness (such as e.g. the human habit of making oneself smaller and more harmless by bowing and exposing the neck to superiors) may not have the same meaning. Verbal exchanges, which are once removed from pictorial representation may well do so. And being taken in by such a display, whatever its motivation could be extremely costly to the fooled side.

6.1.5 Ambiguity and Social Progress

Ambiguity is an important communicative concept. For non-Darwinian species that have evolved cooperatively, one can suggest that ambiguity is an important 'oil' to pour on delicate social relationships. Many human languages use ambiguity in such a way. Japanese, Javanese, and other East and Southeast Asian languages use ambiguity creatively. The object is to ensure smooth *social* interaction. Data can always be sent later, as one Japanese friend told me.

This does, however, pose a problem. Ambiguity works fine within and among members of a cultural group. Japanese are convinced they have clear unspoken understanding among themselves that no interlopers from other cultures share (Passin 1980). The ability to understand the unspoken derives largely from shared experience, as well as knowledge of cultural cues. Very often the knowledge of how these cues operate is unconscious: one learns them through a process of trial and error as one grows up in the culture concerned. One can even learn them, albeit incompletely, if one immerses oneself in the culture concerned. What happens when we communicate with an ETI society that is highly contextual [that is, as Nakane (1970) says, one where ambiguity and implication are important parts of the communicative process]? After all, the essence of the problem of communicating with ETI is that ILFs are by definition, alien to one another.

This implies that at least when communicating with ETI, 'managed honesty' may be the best strategy. Those elements of our statements that are paralleled by factual data we have no control over (e.g., pictures of warfare) we should be manifestly honest about. Implications of such features of humanity? Well, those we need to think about. A clever and experienced ETI, notably one with several contact experiences under their belt, may well be expected to anticipate such communicative chaff, and indeed, to be able to account for prevarication and protective dishonesty in newly contacted races.

6.1.6 Constructed Languages

Natural languages evolve (at least in the human species) as the twin of intelligence. As we have seen, languages are likely have a great deal of ambiguity, and offer opportunities for mis-communication. This is fine for humans, and probably for many ETI as well, whatever their genetic history. However, dissatisfaction with language problems and the difficulties of inter-cultural communication, and the desire to construct artificial languages to overcome communications barriers, have been a human hobby for some centuries. The best-known example is Esperanto ("Hope"). L.L. Zamenhof published the first book detailing the rules of Esperanto in 1887. Zamenhof's somewhat naive assumption was that inter-ethnic quarrels (he was born in Bialystok, a mixed Yiddish, Russian, Polish speaking town) were the result of linguistic misunderstandings, and he set out to develop a language along purely logical lines. Of course Zamenhof was attempting to solve the human problem of misunderstanding with a human solution. While there is no evidence that Esperanto speakers (there are currently several million, and the language provides amusement and leisure-time interest to many more) are less (or more) inclined to quarrel, there is also no evidence that speaking a common language leads to less bloodshed, as recent evidence from Syria, Rwanda, and elsewhere show. The Esperanto enterprise raises two questions, one somewhat puerile to my mind, the other more important. The puerile question is whether a common language would ensure greater amity, or at least security, in inter-cultural (and in this instance we can conceive of inter-ILF communication primarily in its cultural dimension) discourse. The second more fundamental question is whether, given two distinct languages, what role could a third, artificial language, fill?

6.1.7 Some Artificial Languages

In an attempt to develop a means of communicating with non-human intelligences, Dr Hans Freudenthal, a mathematician, published a book suggesting a mathematics-based language he called LINCOS as a universal medium of exchange with ETI (Freudenthal 1960, 1974). DeVito and Oehrle (1990) suggested another variant which would allow for the exchange and comparison of technological fundamentals. Lincos allows for the gradual development by two correspondents of a means for the exchange of information free of human grammatical and syntactical quirks. Lincos can be developed gradually provided one of the correspondents knows the theory of Lincos, and both correspondents are mathematically literate. Lincos embodies, once it has been evolved during a series of exchanges, some very sophisticated ways of exchanging information, and can be readily learned with by parties willing to do so.

The only problem with Lincos is that, as is often the case in grand endeavors, Freudenthal was barking up the wrong tree. Let us consider the situation of message exchanges suggested by Freudenthal's approach. Two parties, *ego* and *alter* both unfamiliar with one another try to communicate over interstellar distances. In order to do so, some prerequisites need to be in place: they must be mathematically sophisticated (else they would not be able to understand the Lincos mathematical principles), technically capable (else they could not have a mode of interstellar communication) and linguistically experienced (how else could they communicate among themselves?). Given these three things, they should have no problem communicating using Lincos (or any other artificial language ranging from computer instructions to pictograms).

In the 1990s, as a member of NASA's CASETI program, I spent a hugely interesting and intensely frustrating day with a sub-committee of the meeting trying to unravel what theologians, engineers, and social scientists meant by the term 'model': members of the three disciplinary areas had completely different ideas of what constitutes a model, and the implications of the different positions were quite radically different. The moral of the story is simple: engineers (or mathematicians) can always talk to others of the same profession, because the issues that interest them are context-free, or almost so. As can social scientists and theologians *within their disciplines*. The same is manifestly *not* true for communication *across* disciplines.

In terms of Lincos and other artificial languages, this teaches an important lesson. The basics of technical communication—table of elements, mathematical concepts, astronomy—would be figured out very quickly by any two intelligent species, provided each side is willing, using any mutually achieved system. We/they could easily transit information about physical states, locations, etc. We could also, possibly transmit concepts such as ‘don’t go there, it is mine’. However, ‘don’t go there’ is also problematic: consider the simple lexical meaning (‘the coordinates of your physical body should not be x, y) and contrast it to the American colloquial meaning (‘let us not discuss the issue you have just raised’).

Thus Lincos and other artificial languages, while they will make initial communication much easier, are unlikely to lead to better understanding. Why is that area of space ‘mine’? What does ‘mine’ mean? And how do we deal with situations in which alter’s ‘mine’ and ego’s ‘mine’ are, or appear to be in conflict? Lincos provides no answers precisely because it does *not* deal with the ambiguities which are inherent in social communication.

6.1.8 Summary: Language and Lies

Common thinking including among the SETI community, has it that language exists to enable communication between two disparate species, separated by time and space, wishing deeply to communicate. To be blunt, humans don’t act that way. Nor is it likely that non-humans will act in that way either. We need to recognize that any ETI we come into contact with will have an agenda and interests of their own. This is not a bad thing by any means. It certainly does not mean that we need to push the panic button and stop trying to find, and when found, communicate with ETI, as the British Astronomer Royal once proposed. It does mean that linguistically we should expect business as usual: they will lie to us, and we will lie to them *at some points*. They and we will be completely honest *at some points*. A much less frightening proposition than an ILF that tells the truth all the time, for then we need to ask the question “What are they hiding? Why? And how are they hiding it?”

6.2 Symbols to Make up a Language

An important element in human communication is symbolizing. Humans, apparently unlike most animals, are able to symbolize objects and concepts in various ways. That is, they are able to posit objects, words, and other transferable items in such a way, that they represent in some fashion other items and concepts, some of them quite complex. These ‘symbols’ can then be manipulated in various ways, some of them themselves symbolic. For example, the cross generally symbolizes a very complicated and complex concept called Christianity. Doing things with a cross: kissing it, burning it, touching it with other objects, addressing it with words, implies, to the humans concerned, that they are doing this to, or with Christianity.

Are symbols and symbolization necessary? To answer that theory, we needs discuss some elements of communication theory. The quality of messages between a transmitter and a recipient can be enhanced by the utilization of two important principles, *redundancy* and *compressibility*. Redundancy implies that elements in the message are repeated a number of times, to ensure that the full, complete, and authentic message is actually transmitted and received. Human communication is redundant in a number of ways. For example, we often use facial gestures and body language to repeat a message. This is one reason communication on the telephone may well be ambiguous. Compressibility, in which messages are compressed to smaller physical units, ensures that less time is taken for the message and among other things, also ensures that the odds against message interruption and malfunction are lessened. Symbols, among other things, offer a great degree of compressibility. Showing a cross which under many contexts implies a host of things to do with Christianity is one example. The importance of contextuality in the use of symbols is illustrated by one historical example. When Christian missionaries first arrived in Japan, they were heartened by the use of a cross as the symbol of the Shimazu clan of southern Japan. Unfortunately for the missionaries, the cross in this case signified a horse's bridle, and the powerful Shimazu were among the strongest opponents of Christianity in late medieval Japan.

We can assume that symbolization is a likely component of intelligence, simply because it is such an efficient mechanism for ensuring message transmission. Moreover, as we shall see below, culture is partly a function of symbolic message transmission.

It seems logical therefore, that any ETI, even a non-technological ETI, will need some way of ensuring redundancy and compressibility of its messages. For humans, symbolization is an appropriate and useful way. After all, we do this all the time. This may be a function of our communication physiology. We produce immediate personal messages on a single band (sound modulation) with a few auxiliary and strictly limited channels: facial gestures and poses (i.e. vision), tactile messages, and to a much more limited degree among humans than among other primates, olfactory messages. It thus makes sense for us to use the communication channel as efficiently as possible, that is, compressing, ensuring redundancy, and symbolizing our messages.

6.2.1 Symbols and the Construction of Culture

We have not, so far, dealt in any coherent way with the concept of culture. Previously we asked whether symbols and symbolizing were a prerequisite, or even *the* prerequisite of intelligence. Now we must ask the same thing about culture: a product of intelligence which exists through a given time and space. Symbols, as we have seen, allow individuals, through a poorly understood but often argued-about process, to mask their actual meanings and intentions while representing them in some form that is acceptable to a collective. This can be seen in two rather opposite ways. In the first, as the foundation of a collective delusion, and in the second as a kind of structuring principle which allows people to act in a directed manner.

Cultures consist, at one level, of arrays of symbols which people hold to collectively, and which motivate them to collective action. The sociologist's "norms" and "values" can also be represented in the form of symbols which imbue in themselves a "call" to action: consider the use of national flags. The fact that the symbols obscure the actual individual meaning, also means that individuals will interpret this "call" in different ways, and may be motivated to act in different intensities and under different circumstances, as indeed in human society they do.

The symbols in a culture themselves derive from common experience, and actually depend on this commonality. Cultures are therefore quintessentially "local" phenomena. It is not that someone outside the culture will be completely unable to understand the culture—though it unlikely that anyone *outside* the culture will experience it in precisely the same way, or in the same intensity—as that a 'culture' will be modified quite severely when it tries to exist in a different setting.

6.2.2 Conversations with ETI

The mere possession of language is insufficient for two representatives of different cultures, speaking different languages, to converse with one another. By definition, conversation is not a monolog, but a dialog (at least). Crucial to this view is the fact that one party's emissions are going to affect the other party's and vice versa.

We need, therefore, to look carefully at conversations (even delayed over centuries) to understand some of the complexities inherent in communicating with ETI. Given our own languages' ambiguity, it will be reason to suggest that conversations with ETI will be full of ambiguities. This does make interpreting messages difficult. On the other hand, it also allows for a great deal of smoothing the interaction, as both sides puzzle out the actual meaning. This does have a major implication: *all* messages from ETI should (and presumably will) be subject to intense examination. Meanings need to be accepted pro-tem, until further data supports any analytic conclusion, and even those should not be firmed up until and unless great reliability is established.

Unfortunately, the scientific establishment is often biased against innovative and unorthodox interpretations (Grinker 2008; Kuhn 1970). Thus the real hidden danger for interpreting ETI messages may not lie in the message itself. Once an orthodox opinion about the meaning is established, any contrary interpretation risks being a voice in the wilderness. If there is one trap science can fall into quite readily, it is the trap of orthodox wisdom. This would be bolstered by the elders of the tribe, whose position would be threatened by new interpretations.

References

Billingham, J., Michaud, M., & Tarter, J. (1991). The declaration of principles for activities following the detection of extraterrestrial intelligence. In J. Heidmann & M. J. Klein (Eds.), *Bioastronomy the search for extraterrestial life—The exploration broadens* (Vol. 390, pp. 379–386). Lecture Notes in Physics. Berlin, Heidelberg: Springer.

Chevallier, G. (1951). *Clochemerle* (J. Godefroi, Trans.). Harmondsworth: Penguin Books.

Chomsky, N. (1965). *Aspects of the theory of syntax*. Boston: MIT Press.

Chomsky, N. (1986). *Knowledge of language: Its nature, origin, and use*. Westport CT: Greenwood Publishing Group.

Chomsky, N. (2002). *Syntactic structures*. Berlin & New York: Walter de Gruyter.

Chomsky, N. (2006). *Language and mind*. Cambridge and New York: Cambridge University Press.

Coker, M. (1984). Learning to communicate with an 'intelligent' computer. In M. Yazdani & A. Narayanan (Eds.), *Artificial intelligence: Human effects* (pp. 128–142). Chichester: Ellis Hornwood.

DeVito, C. L., & Oehrle, R. T. (1990). A language based on the fundamental facts of science. *Journal of the British Interplanetary Society, 43*(12), 561–568.

Doi, L. T. (1973). *The anatomy of dependence [Amae No Kozo]*. Tokyo & New York: Kodansha International.

Freudenthal, H. (1960). *Lincos: Design of a language for cosmic intercourse*. Amsterdam: North Holland Publishing Co.

Freudenthal, H. (1974). Cosmic language. *Current Trends in Linguistics, 12*(2), 1019–1042.

Goodman, A. E. (1990). Diplomatic and political problems affecting the formulation and implementation of an international protocol for activities following the detection of a signal from extraterrestrial intelligence. *Acta Astronautica, 21*(2), 103–108.

Grinker, R. R. (2008). The politics of knowledge: Julian Steward, Leslie White, Melville Herskovits, and L. Luca Cavalli-Sforza. *Reviews in Anthropology, 37*(2), 259–276.

Hall, E. T., & Hall, M. R. (1990). *Understanding cultural differences*. Maine: Intercultural Press Inc.

Hanlon, R. T., Cooper, K. M., & Budelmann, B. U. (1990). Physiological color change in squid iridophores. *Cell and Tissue Research, 259*(1), 3–14.

Harrison, A. A. (2011). The search for extraterrestrial intelligence: Astrosociology and cultural aspects. *Astropolitics, 9*(1), 63–83.

Korbitz, A. (2014). Toward understanding the active SETI debate: Insights from risk communication and perception. *Acta Astronautica, 105*(2), 517–520.

Kuhn, T. S. (1970). *The structure of scientific revolutions* (2nd Enl. Ed.). Chicago: University of Chicago Press.

Lucy, J. A. (1992). *Language diversity and thought: A reformulation of the linguistic relativity hypothesis*. Cambridge: Cambridge University Press.

Morris, D. (1956). The feather postures of birds and the problem of the origin of social signals. *Behaviour, 9*(2/3), 75–113.

Morris, D. (1979). *Gestures, Their Origins and Distribution*. New York: Stein and Day.

Morris, D. (1982). *The pocket guide to manwatching*. London: Triad/Granada.

Nakane, C. (1970). *Japanese society*. Berkeley: University of California Press.

Passin, H. (1980). *Japanese and the Japanese: Language and culture change*. Tokyo: Kinseido.

Sapir, E. (1956). *Culture, language and personality*. Berkeley: University of California Press.

Sawyer, R. D. (1993). *The seven military classics of ancient China*. Boulder: Westview Press.

Skinner, B. F. (1957). *Verbal behavior*. New York: Appleton-Century Crofts.

Chapter 7
Collective Action

Humans do not live alone. This statement is crucially important when dealing with ETI (as it is with humans). Not only that, it is impossible to *make* a human without other humans (biology aside). Human young must be *socialized* into being humans: the price we pay for not relying on instinct in our reproductive strategy. We have already noted that it is highly unlikely that ETI will consist of a species of one (though, we should be prepared, if only as a mental exercise, to accept the possibility of an ETI species that acts as a single gestalt organism). If there are two of a species, somehow or other they will need to find interpersonal accommodation, in what we call roles (see below). In this chapter, we shall explore some of the implications of the fact that intelligent species need to organize collective action, to get things done in larger groups than dyads or nurturance/socialization groups (AKA 'families'). This is not to say that families cannot also be larger groups, but as, we shall see, the same principles will apply to large families as they do to large groups.

This chapter also provides the obverse of the following one. It is extremely difficult to separate 'society' from 'culture'. So, as a matter of convenience and convention, I shall be largely addressing 'society' in this chapter, and 'culture' in the following one. This is partly for historical reasons [the history of British anthropology, which favored studying 'society' (Kuper 1973), versus the American school which favored studying 'culture' (Watson 1995)], and largely in an attempt to frame the issues more simply. However, one must always be aware that 'society' and 'culture' are little more than two sides of the same coin, or perhaps more correctly, two different views of the same general phenomenon.

7.1 A Tripartite Division of Activity

Once again, it is useful to start from a human template. So let us for a moment consider what humans *do* in some schematic way. The results of all human activity can be fitted into three categories, first articulated by a biologist, Huxley (1955).

M. Ashkenazi, *What We Know About Extraterrestrial Intelligence*,
Space and Society, DOI 10.1007/978-3-319-44456-7_7

Artifacts are material objects used, manufactured, or otherwise physically manipulated. *Sociofacts* are the ways in which human beings arrange their social lives, whether at the simple dyadic level, or in more complex domains such as nations. *Mentifacts* are the product of thought; the images, mental maps, notions and other verities people hold in their heads, including valuation of artifacts and sociofacts.[1] This rather simple template is useful when we try to identify what people are doing. Obviously, there are problems with the scheme. "Artifacts" for instance do not exist in a person's consciousness until and unless there is also some mental image, a symbol of the artifact that can be manipulated, thus making the artifact meaningful. For years I would teach an undergraduate course called 'Material Culture', in which, as a demonstration, I would bring some simple objects—a woven sago starch strainer, a dried fish *katsuobushi* stick from Japan, and other objects. Very rarely would a student be able to identify these, except for the lubricious-minded who immediately assumed the worse of the phallic-shaped *katsuobushi* comestible. A further problem is that mentifacts are unknowable except by second-order derivatives (what people do, or say they do, or say they believe or think) or through artifacts (books, statuary) or sociofacts (plays, actions, speech acts). Nevertheless, it enables us to ask a more directed question about any ETI society: what artifacts are in use? How are sociofacts structured? And what mentifacts are expressed? It directs our attention to certain persistent questions, and, significantly, it is not biologically determined. *How* ETI think is, in this instance, less important that what they think about (or express what they think about).

7.1.1 Analytical Levels

Another set of analytical concepts will help. It is a useful, if not necessarily valid at all times, convention to consider human beings as living within a hierarchy of levels. Every human being is at once a singular entity, a part of a network of social relations peculiar to her or himself; a functioning element in a conceptual and behavioral set of rules; an atom in several definable social cohorts; a node in social networks; and the consequence of a particular set of historical circumstances. At the level of an individual, human actions are constrained by physical requirements (we all need light, food, air, water); learned behaviors (we all learn from our environment); social norms (people around us influence our behavior); and even international contexts (personal space, for example, varies from nation/culture to another).

If we take the upper level in the hierarchy, the so-called 'world system', composed of nations, international organizations, powerful international and national corporations, diplomatic and political arrangements, we find similar complex

[1] I apologize in advance to my anthropological colleagues, notably my daughter, who find this approach simplistic and contrary to the current interest in mental images and interpretations as the supreme *locus* of human activity.

embeddedness. A large state such as the US is simultaneously directed by its own interests (the distillation in some fashion of some or all of its citizens' interests) trade and diplomatic practices, temporary and permanent alliances, and particular individuals from lowly clerks to presidents who may, or may not, act in foreseeable ways.

Crucially we need to recognize that this discussion of levels of organization is an analyst's convention. It is codified, in the social sciences, by professional identification. A 'scholar of international relations' deals at the upper level of the world system. A 'psychologist' at the lowest level of individual behaviors and motivations. In rough fashion, anthropologists, sociologists, and political scientists deal at ever higher levels of organization. They do so for the simple reason that it is humanly impossible to encompass all levels at the same time and with the same clarity and detail. Using the convention of analytical levels makes it easier to see how such concepts might play out among ETI.

7.1.2 Individual Level

Consider a conversation between any number of individuals, even a conversation in which several people are speaking at once. 'A' might be speaking to a group of excited hangers-on 'B', 'C', and 'D'. D might turn to 'Z' while A is speaking, and say something, while 'X' is happily prattling on to an audience composed of 'Y' and Z. It is nevertheless possible in theory (and given modern technology, in practice) to separate these complex interactions into a finite set of *dyadic* interactions between two people. A can be described as having a conversation with B and B with A (even if B's sole contribution to the loquacious A is "Yeah man, cool, right on" or "Indeed, sir."); Simultaneously a conversation between A-C and C-A is also taking place, and a third conversation A-D/D-A, D-Z/Z-D, etc.

For convenience sake, and following sociological convention, we call one of the conversants in such a dyadic interaction "Ego" and the other "Alter". In this form of analysis there are four elements we need to keep track of: *Ego*, *Alter*, *emissions from Ego* (which Alter receives), *emissions from Alter* (which Ego receives). Ego and Alter have particular roles vis-à-vis one another. In practice, both Ego and Alter have, as do all humans, multiple roles *but for any given dyadic exchange one role normally predominates*. Ego emits a message. This message may be simple or very complex (even simple emissions can have hidden complexity, something we know from the nature of symbols). Ego's emission may stimulate Alter to an emission of his/her own, which may provide a stimulus for Ego, and repeat. Thus in any dyadic interaction, we need keep track of Ego and Alter and of their respective emissions in order to understand the interaction.

A short example is in order

Store clerk: "Good morning sir, how may I help you?"
Customer: "I'd like to buy a pair of Testoni shoes, please."
Store clerk: "Certainly, sir. Would sir take a seat? What size and color would sir care for?"
Customer: "…."

Now, most of us cannot afford Testoni shoes, but that aside, this is a perfectly normal exchange. For our analysis, it is important that it sets each of the dyad into a role (customer:clerk) and even refines that (very *rich* customer:clerk). The actual words of the exchange are a function of this role identification, and can be, and are, updated throughout the exchange, triggering realignments of role and of emissions.

What we need to understand here is that exchanges between humans are *flexible* and context dependent. They are also composed of substantive content ('buy a pair of shoes'; 'what size and color') and a sort of 'wrap' of emissions whose major function is one of displaying emotions, including respect (clerk to wealthy customer, wealthy customer to clerk), acknowledgement of status ('sir') and so on. These wrapping meta statements are heavily dependent on context: the set of acknowledged rules which the two actors adhere to. These contexts are often unsaid, may be class dependent, affected by the gender of the speakers, time of day, mood, and so on.

Would ETI have the same paradigm? Two questions frame this issue. The first, is the question of cognition/consciousness. As we have seen, *some* form of consciousness is needed for an organism to be considered 'intelligent'. Thus the odds are that indeed, where the boundaries of one ETI's 'self' and that of another's are tangential, it would be possible to map an ego-alter exchange. Second, suppose a given species does not communicate verbally? Telepathic communication aside (possible, admittedly, but far fetched), the need for consciousness of self would mean that one could not exchange gestalts (unless our understanding of reality is so provincial, that we will *never* be able to understand that particular ETI). So, once again, we fall back on the assumption of mediocrity: we need to assume that (a) ETI individuals have consciousness of self, and that this is self-bounded (there is a boundary between "I" and "not I"). This implies (b) that there must be some form of ego to alter exchange (whether conducted by sound, chromatophore, or waving tentacles does not alter the principle).

7.1.3 Social Level

Humans find it incredibly difficult to function without a group. Why this is so should be obvious: phylogenetically, most primate species to which the hominids belong operate as small face to face (usually thirty individuals or less) groups. That allows for long nurturance, and may well be a prerequisite in all species for lengthy, multi-year nurturance of offspring. The social level of analysis is where most person-to-person interaction is visible.

Among humans, groups can be arranged in an infinitude of ways. Sociologists generally define groups as an assemblage of individuals with a shared mutual purpose (which may be expressed overtly, or may be unspoken). Within that definition, however, there are varied forms of purpose, organization, size, and complexity, with some groups being relatively simple (e.g. a nuclear family) and some being incredibly complex (e.g. a publishing house). Individuals are very often members of several groups, some of which are presumptively perpetual such as a financial corporation, others being quite temporary, such as a pickup football game. Moreover, groups can nest within groups, also in an infinite array of complex nested units.

We can assume with almost certainty that ETI will have groups. Without them, individual energy and abilities cannot be leveraged to create emergent features of technology or culture. Moreover, we also know that they are likely to have complex groups: groups which contain specialists who deal with particular group activities. It is *conceivable* that every member of a large ETI group handles the books, designs the project, allocates personnel, and coordinates everyone else's activities. As can be seen from the preceding sentence, it seems unlikely. So, whether pro-tem, or perpetual, we can expect ETI groups to include specialists.

7.1.4 Categorical Level

At a more abstract level of analysis, it is possible to identify the outcomes of collective behavior of categories of individuals. This, in modern society, is the basis for much commercial activity. Identifying what category of individuals uses a particular product, for instance, is the basis of most successful advertising campaigns. Generalizations about large categories—eighteen year-olds, drivers of sports cars, users of perfume—are also possible, and we often make them, including or alluding to their supposed qualities, which, when replicated over time and between speakers, bring rise to stereotypes, negative and positive.

It is also a tool in cultural analysis. "Among the x" is a stock phrase used conventionally by anthropologists when they are trying to identify cultural rules in a particular group of people, the 'x'.[2]

The professional role of the anthropologist is to attempt to tease out these regularities, and provide an interpretation of them and their consequences. Not all anthropology (or anthropologists) is able to do so, less a criticism of anthropologists

---

[2]For a humorous, but essentially accurate picture of the nature of the "Among the X" statement, it is worth reading the classic "Body Ritual Among the Nacirema" (Miner 1972).

(though that should be included) as much as a testimony to the difficulty of extracting scientific, prediction-useful regularities from a mass of ever-changing data.

These issues aside, we would expect to find the following in any society/culture:

1. A set of regularities, that would allow, with some defined variance and confidence, to predict behaviors at the categorical level.
2. Covariance of categorical behavioral rules with ecological surroundings.
3. Covariance of categorical behavioral rules with economic/social/status/political classes.

In other words, what we would expect to see in *any* ETI society is "variance within regularity" where the variances are associated with intra-societal groupings.

7.1.5 Culture and Social Relationships: Human and Other

What we call society—the set of perduring relationships between humans based on the rules we call culture—may well be a purely human, or perhaps terrestrial construct. Whether or not ETI will have similar bases for their society remains to be seen. What we must do here is separate the purely human aspects of society from the generic ones. For instance, human society is intimately tied to human sexuality and reproduction (that is, it is species specific) and to the acquisition of food (which is ecologically [natural or artificial] specific). All variations on human society encompass various modalities of these two fundamental elements. It is therefore possible to claim that "society" is a purely human construct. Yet at another level human sexuality is but a special case of reproduction which we have seen to be a universal necessity for any species, even an ET one, in some form or other. We should, therefore, be able to tease out those social functions that are necessary for the existence of a number of ETI existing in concert.

7.2 Small Groups

Groups, that is interconnected assemblages of humans, come in many shapes, and forms, but there is a major difference between small, simple, undifferentiated face-to-face groups and large, complex, differentiated groups. Both of these provide support—emotional, material, and otherwise—to their members, but inevitably there are differences. Primate societies, notably those of our closest relatives, the pongids consist of small groups. These rarely exceed a few dozen, and generally number, as among humans at about 30 people (children included).

In practice, the important quality of such groups is not so much their size, as the fact that all member are familiar with one another by face, name, and characteristics (hence, 'face-to-face' groups). Groups such as this are tied together by many collective forces: affection, kinship, security and safety, production and consumption. In the modern world, most normal people have small face-to-face groups of intimates. People one goes to for emotional, sometimes economic, political, or other support. Two points are crucial:

- Among humans, face-to-face groups are an essential social feature, and
- Small groups are mutable, flexible, and their memberships interpenetrate each other.

We need therefore to ask a number of questions, all of which have implications for ability to deal with ETI.

- First, is the need for small groups inherent in human phylogeny or is it a universal need? That is, do we need small groups simply because our genotypical ancestors did?
- Second, if small groups occur among ETI, do they have the same qualities human small groups have, that is, are they essential, and mutable, flexible and interpenetrative?
- Finally, on a more global note, if a given ETI race does not have small groups, how are small group activities, including co-operative tasks, emotional and intimate support, and so on accomplished?

7.2.1 Are Small Groups Inherent in Human Phylogeny or Is It a Universal Need?

Consider the number of connections between five points. Each point (AKA 'node') has four lines (AKA 'edges') that connect it to each other points. If we arranged the points symmetrically, they will form a pentangle, with five nodes and ten edges, which is easy to visualize. Now, if one tries to do the same thing with fifty nodes, the figure obtained is much more complicated, since each node must be connected to 49 others. Now let us provide one node, call it 'One' with an item—a message—we want passed on to all the other points. It will take a finite amount of time for One to pass the message on serially to all other nodes. If done simultaneously, the time will of course be reduced. However, what happens if messages have different weights, sizes, and priorities? Each message would then have to be read, analyzed for each of the different characteristics, then queued before being sent on.

Now let us complicate the picture: let us assume that a finite number of messages originate from all nodes One, Two, Three …, Fifty. Each node can originate any number of messages at any given time, each message has a weight, a length, and a priority. Prioritizing with fifty nodes, is a much more complex task by an order of magnitude. If the messages also have qualities, the problem becomes almost insuperable.

This exercise illustrates the advantage of smaller groups: fewer messages, better prioritization, and whether the messages are simultaneous or sequential, there is a chance of dealing with them (assuming a point or person needs to process the message, not merely repeat it) by each individual, or by someone assigned to coordinate.

There is however an inherent *disadvantage* to small groups which has to do with prioritization as well: some nodes in the pentangle will receive more messages, some less. If the messages need processing, which is likely in a real-world scenario, 'One' might be overwhelmed, whereas 'Three' might be loafing: a waste of processing power for the collective. Large groups, thus have an advantage dealing with multiple messages (or issues, or problems, or anything that requires processing) if there is some message switching mechanism that adjusts and directs the messages to idle members of the 'network'. In the real world, if we look at collectives of people, small groups can function perfectly well without leadership or status differentiation among members *up to a point*. They are excellent solutions for low-cost tackling of single problems and problems that have few priority options. Large groups tend, among humans, to fissure into a lot of small face-to-face groups, each dealing with a specific issue. Efficient use of this phenomenon requires someone who can serve as a switch, directing effort, switching messages to individuals and/or groups, in other words, some form of leadership. Under those circumstances, the small groups formed can deal with specific tasks or chunks of the overall task—they specialize—while the large collective as a whole can take on multiple or complex tasks. The problem is that very large groups—in the thousands and millions—are likely to develop into smaller groups, and those even smaller, with different ideas, focus, interests and abilities: in other words, the world as we know it.

Alternatives to small groups The purely communicative advantages of small groups as coherent, relatively focused units seem obvious. However, this 'obvious' might be no more than a human quirk. So, we need to ask, under what circumstances could small group tasks be accomplished better by a larger group?

Let us return to our pentagon. Assume that 'One' is specially qualified as a processor of information. Two … Five, are specialized as operators. It would make sense, if the group were given a task, for One to be the coordinator, receiving and transmitting messages to each of the others on the basis of processing the messages received, assessing their importance, giving priority to certain messages, etc.

Let us make a contrary assumption. Take our fifty node assemblage. If *each* node can function as a low-level processor, or if messages are clearly coded as to their qualities, the entire collective can act as a small group, saving energy for each individual. Here is how it could work: One, Thirteen, and Forty-nine each receive simultaneously a different message. Each of the messages is coded for three qualities: priority low or high, criticality low or high, length long or short. One and Thirteen's messages are coded high priority. One's message is coded highly critical for the task, as is Thirteen's. Thirteen's message is coded short, whereas One's message is coded long. Each one of the three nodes sends a message to all other nodes. Node Twenty-five receives all three messages simultaneously. They are processed. Forty-nine's message is low priority, and is kept in a buffer for later treatment. One and Thirteen's messages are equal in weight for priority and criticality. But Thirteen's message is shorter, thus quicker to process, and, as an added benefit, the information in the message might be useful in processing the other two messages. In addition, if the collective has previously agreed on a rule that either longer or shorter messages are to be addressed first, then prioritization would be much faster. Assuming all members of the assemblage are receiving the same messages, a consensus quickly emerges as to how to proceed.

The example is of course a sandbox. In reality, many messages will emerge from many sources, require lengthier or shorter times for processing, etc. However, we can see how a large group *could* function with the effectiveness of a small one provided:

(A) A large fraction, or the total population of the collective are equally qualified to make decisions on data received;
(B) The messages are clearly coded (whether socially or intrinsically) and the coding is largely agreed upon and not fakeable;
(C) There are agreed upon rules as to how to prioritize messages based on the messages' unambiguous coding.

We could conceive of an ETI species that has the physiology and social organization to function in a group in that way. However, certain realities intrude. First, external stimuli are rarely unambiguous and easily coded. Second, where a group becomes extremely large—almost a requirement for a technological society for reasons discussed earlier—*some* sort of specialization must occur, if only because different sub-groups may be detached from the main group, and must evolve their own communicative patterns.

This implies that we should expect ETI to have two kinds of groups, as do humans. They are likely to have large groups that are able to act collectively, but 'pay' for it somehow by the presence of inefficiencies, internal competition, and mutual misunderstandings. These problems and discontinuities may not be as large, nor as pernicious or pervasive as those among humans, since our example ETI might have better communication protocols (language) and media (electronics instead of sound) but they will exist. ETI will also have small face-to-face like groups with specialized functions and abilities. These may or may not have the emotional weight, and importance attached to them as among humans. They may not have the same functions or abilities as small groups have among humans, but they will exist.

7.3 Complex Groups and Corporate Entities

In contrast to small groups, which have an apparent *physiologically-determined* upper limit among humans, complex groups can be of any size. There is, however a cost for greater size. As a rule of thumb, the greater the size of the group, the more likely it is to be heterogeneous, with officers, internal divisions, and sub-groups embedded within itself. Moreover, the larger the group, the weaker its ability to command loyalty, and the more heterogeneous the ideas, concepts and activities within it. Parenthetically, that is where symbols are most effectively deployed, since they allow two people with different, even opposed ideas to hide their differences under a shared symbol: think national flag.

Among humans, we find two types of large groups: temporary and presumptively perpetual. These two types *also* correspond to the degree of internal organization of the group, something that may well be universal.

7.3.1 Ephemeral Groups

In the late nineteenth century, Gustave Le Bon, a French psychologist, suggested that the seemingly unstructured nature and behavior of a crowd of people actually had underlying structure to it (Le Bon 1897). This insight is the foundation for more detailed studies of people in mass, including riots and mobs, festivals and parties, as well as mass consumer behavior. The characterization also fits smaller face-to-face groups, including the family (which lasts only during the adolescence of the second generation, then gradually dissipates). What all of these assemblages have in common is that they are time bounded. That is, they are groups by definition—assemblages of people with a common purpose—and yet, as soon as the purpose is achieved, they disperse. Temporary groups *may* (indeed, Le Bon argues, often) have a coordinating or leading component, but these components too are ephemeral, and their ability to control the crowd disappears with the achievement of the goal. Pity the leaders: all dressed up, but no one to direct. Enter the corporate group.

7.3.2 Perpetual Corporate Groups

We are all familiar with what M.G. Smith calls "presumptively perpetual groups" (Smith 1974). Groups that have no temporal association, in effect, supposed to last forever, are a feature of all complex-developed societies. Governments and their myriad offices, agencies and departments immediately spring to mind. Official football/basketball/ teams, commercial corporations such as IBM and Google, your local needlepoint club, are all presumptively perpetual: they are established to go on essentially forever. Even when such a group achieves its goals, the organization reorients itself, and carries on with a new goal in mind. This is a "March of Dimes" phenomenon, named for a well-known charity whose goal was the eradication and treatment of child polio in the US: once that goal was achieved with the invention of the Salk and Sabine inoculations, the March of Dimes reoriented it's goals to what they are today is common in presumptively perpetual corporate organizations.

A successful corporate organization can be visualized as a biological cell. It has a list of members (the 'membrane' of the cell), internal specialized parts, a central coordinating element, and some sort of template, or idea which directs its actions: an ideology, 'corporate vision', or objective (consider the cell's genetic load). Crucially, however, the corporation, like the cell, is a *living* organism. It may swallow up other groups, it's membrane is permeable and new individuals are incorporated or dropped, material is taken up from the surroundings as well as from the specialized parts of the cell. The corporation may incorporate a huge number of subordinate, encapsulated corporations. Corporations, moreover, have specialized *offices* (individuals with a responsibility and a scope of matters they are to see to), as well as *colleges* (groups of people responsible collectively for some matter). This entire structure is ostensibly dedicated to a single goal or objective. In practice, sub-goals, some of them detrimental (overtly or substantively) for the corporation develop over time, because control, even very restricted control, is never perfect.

One thing we can say with almost certainty about ETI: they will have large groups, with some form of internal organization. The reason initially is a concept derived from management called 'span of control'. Simply put, a single individual would find it virtually impossible to accomplish a complex task (which by definition requires parallel inputs from many individuals) effectively without delegating (as we've seen in the discussion of small groups). In fact, it has been suggested that the more technologically complex the task assigned a complex group, the smaller the span of control, that is, the fewer people a supervisor can control (Bell 1967).

For humans, span of control is a complex subject, but basically boils down to the fact that any given individual can effectively control at the most about 30 other individuals, give or take. Anything larger than that (and in most real-life cases, even controlling ten subordinates can be a problem), and the group will not function. Does the number sound familiar? Hierarchy is necessary not only because of our ancestors, but simply because it is impossible to accept inputs, process them, and provide outputs above a certain volume. This is true of machines (and therefore the need for parallel processing) as for biological persons. To take a lengthy and very well documented historical group, a Roman century of soldiers had a formal complement of 64 men, and two commanders (supervisors); as Marius, the great Roman military innovator found, any smaller ratio of commanders turned the well-disciplined century into a mob (Keppie 1998). Any larger ratio would, of course, be a waste of resources.

ETI may, as we saw earlier, may have a better physiological or psychological basis to control larger numbers of individuals and combine them into a single coherent collective (not necessarily a pleasant thought under all circumstances), but at some point in the process, subordinates and complex organization will have evolved, notably if the species is a technological, world-spanning one.

7.4 Networks

One very useful way of describing societies was initially conceptualized by an early anthropologist, A.R. Radcliffe-Brown. Radcliffe-Brown suggested that societies could be seen as a network, whose nodes were members of the society, and whose edges were the social connections between those individuals (Radcliffe-Brown 1965). Though Radcliffe-Brown did not pursue this idea systematically, it has since become a major tool for social analysts as diverse as sociologists (Bidart and Charbonneau 2011; Feld et al. 2007), marketing specialist (Imai 1987), Internet developers, and spy-agency analysts (Xu and Chen 2005; McIllwain 1999). There are two very important issues for us in network analysis. First, identifying and characterizing a network is data based, as we saw earlier in this chapter. That is, looking at a network as a system of message transmissions and receptions yields information *even if one does not know the contents*. Second, based on message traffic, we can identify subdivisions of networks and networks of networks. Both concepts are important for xenology.

To illustrate, here are two hypothetical cases. Take three individuals/nodes A, B, C. There is traffic of high volume, say 10 transmissions either way per time period, between A and B. Occasionally, at a very low frequency, there are individual messages between A and C, but never between C and B. As a result of lengthy observation, we can assume that A and B are relatively close, perhaps members of the same group, with some interest or interests in common. C, on the other hand, may be completely unknown to B, and in any case, is only peripherally important to A, or to the interaction A-B.

Now let us change the rules. A and B have high traffic volume. Rarely, there is unidirectional communication A to C. On even rarer occasions, there is a single unidirectional communication C to A. Immediately after this happens, traffic A-C doubles, or else ceases completely for several time periods, then resumes. What can we make of that? First, the normal A-B traffic, whatever it consists of, is less important and deterministic than traffic C to A. That A-C traffic overrides normal traffic implies one of two things, in human terms. Either C is a leader or superior of A (and possibly B as well), or C is a functional node who provides some kind of warning or specification that A (and possibly B) require. In the human world, C might be a parent ("Stop that racket" or "It's too quiet up there, what are you up to?"), a leader or manager issuing orders ("Get to work") or maybe simply a sentry ("Someone is coming." Or "There is a storm on the horizon").

Now, we know nothing about the content of the transmissions, but even so, we have a great deal of useful data, which can lead us to other useful data. And for those readers like myself who are paranoid about government prying, yes indeed, the same principles can be, and probably are, used on anyone. For instance, does C have similar relations with others such as X, Y, or Z? That will tell us that there are two subgroups A, B, C and C, X, Y, Z. Another possibility is that we can relate C's emissions to some environmental feature: every time it is about to rain, C broadcasts a message, and traffic ceases between the others; most likely, we might say at least until more data comes in, C is an observer, rather than a leader. Whether we identify C as a decisive member of the network (its manager, if you will) or as a simple mechanical drone can, in most cases, lead to a better understanding of the social underpinnings, such as e.g. membership in a corporate group. In fact, with a sufficiently long baseline and point data, we can even track an individual's whole career and activities (see previous footnote!).

To get the most out of our understanding of networks, three factors are needed.

- A good mapping of the network itself: who the nodes are, what is the volume and direction of the traffic, etc.
- A good mapping of the environment, and the correlation between environmental variables (changes in the weather, or in the international environment, or in the price of goods in the market) and network traffic.
- An understanding of the contents of the messages, in whole or in part.

Crucially, however, an interested party does not have to have all three datasets. One alone will yield a great deal of information. Two will make the understanding stronger, and all three is a winner. This has obvious implications for SETI (both active and passive) and for CETI and xenology. If we can map the transmissions of a putative galactic communication network, we are in a better position to develop and 'who is who' guide to ETI.

7.4.1 Networks of Networks

We can also look at networks in another way. Networks can be considered as single-data points. In this case, we are not interested in the doings of a particular network, but in the relationship *between* different networks. Consider a network composed of a number N of nodes, and a second network consisting of M nodes. The frequency of traffic *within* both N and M is much higher than that *between* N and M. For example, internal message traffic *within* any country, is much higher than *between* countries (except perhaps for Andorra, Monaco and a few other tiny nominal states). Thus the object of interest shifts from any individual network, to the communication between networks: a network of networks. This is particularly important when we consider the oft repeated claims that (a) ETI are likely to be ahead of us technologically, and (b) they are likely to be organized into an interstellar civilization (McConnell 2001; Oliver 1986). So tracking and identifying networks of networks, ignoring for the moment the technical problems associated with this issue. tells us a great deal about the nature of the ETI we are in contact with. Are they a single culture? A complex, multi-cultural civilization? How did either of these states come about? A sufficiently long baseline of looking at the traffic would provide at least initial answers (and concerns) for a sufficiently astute observer.

7.4.2 Politeness and Courtesy

Every (human) culture we know about, adds social markers to communicative utterances. These *meta statements* provide a variety of empty-sounding, but socially significant information. Consider

A: "Good morning, how are you? Lovely weather, isn't it."
B: "Morning, fine, and you?"

There are two things worth noticing in this very common exchange. First, there are two conventional statements 'good morning' (or its contraction 'morning') and a weather statement. Second, B feels no need to respond to respond in detail to the personal question, nor to the weather question. So, what is going on here?

All human societies tack on *meta-information* to social exchanges. A and B in the example may not even stop for the exchange. In fact, stopping what they are dong (walking, working) for the exchange, would tack on additional meta-information. What this (and other, similar exchanges) are about, are what Goffman (1959, 1967) calls 'presentation of self' and 'interaction ritual'. What A and B are doing is to present themselves (as a social being, as a person with some social standing comparable to their alter) and a public announcement of their social relationship. We know from this brief exchange that:

- A and B are roughly socially compatible: the exchange is brief, they don't stop moving, and there is no data exchange beyond the meta statements. Had there been a social gulf between them, the one with lower social status would have stopped, perhaps made some conventional movement of respect (bowing, touching their hat, added the equivalent of 'Sir/Ma'am').
- A and B are not intimates, inasmuch as they did not exchange *any* data, nor did they conventionally express social closeness (e.g. 'Sorry mate, gotta rush,' or 'See you later') which would exhibit some ongoing social expectations.

Meta-statements are *conventional* (that is, they are culturally determined and structured by usage); may be verbal, proxemic (body movements), aural or visual; their proper usage indicates membership or potential membership in a society. They have no 'meaning' except to express conventional statements of *deference* (ego's supposed attitude towards alter) and demeanor (ego's estimation of *ego's* social position and worth) in a social exchange (Goffman 1956). Of course things get interesting (that is touchy, and sometimes deadly) when deference or demeanor clash with alter's expectations. Truly deadly: in Edo-period (1600–1868) Japan, *bushi* (warrior/administrator class) had the right of *kiritsu-gomen* (cut-and-pardon): to kill any lower class individual who did not display appropriate deference without suffering any penalty. In other societies, such meta-statements may become a sort of debased coinage: statements about someone else's statement, or even about the seriousness of one's own: consider the widespread use of smileys, such as :) ;(:p.

Meta-statements, because they are conventional, are very useful as a gauge of many things, such as the social structure of a society, its tolerance for deviance, as well as a map of who is who in terms of social relevance. As such, they are very useful for users and observers alike. People in a society do not have to constantly negotiate their status—like all symbols, meta statements are often vague and polysemic—but can state what they think is their, or another's worth, without too many challenges. In other words, meta-statements smooth social interaction considerably, allowing people to make statements about their social others with a large amount of leeway: most people are susceptible to flattery, and talk is cheap, as long as it does not make real sentiments overt.

Interaction rituals, where social positions are overtly exhibited and presented, and where each individual in the exchange comes away with a feeling that their worth has been acknowledged publicly, can be defined as *polite* exchanges. A polite exchange does not have to be between social equals, so long as each side follows conventional acknowledgement of the other's supposed status: "Hi, Queen" as one American diplomat was reputed to use when greeting the queen of England, however friendly, is not polite. In our example above, if B had responded with a grunt and moved off, A would probably have said some uncomplimentary things to himself. And possibly have curtailed his exchanges with B in the future. That is to say that politeness can be crucial in human relations, and misunderstandings over polite exchanges can, and have, escalated to warfare among humans.

7.4.3 Politeness and ETI

Expressions of politeness are highly dependent on culture (different human cultures have different expectations of politeness) and partially on physiology [head and body positions and movements, eye contact, touching, spatial distance and orientation (Morris 1982)]. The topics of politeness are also culturally determined. Nonetheless, politeness and presentations of self with all it implies are so ubiquitous, that we must as whether this will be true among ETI as well. If the answer is yes, what are the implications?

Will ETI be polite? Politeness is not merely a tiresome production imposed by the majority on a minority of free spirits, as some tend to think. It plays a very important role in ensuring that societies—at all levels, and in all types of social interaction and groups—can run smoothly. Consider the difficulty of running a meeting where no one knows anything about anyone else. It necessitates long introductions, perhaps not only of oneself ("I am a deputy chief Poobah, which means my opinions are more important than those of anyone who is merely an assistant deputy chief") but also of one's group ("I work for the department of widget production, that is more important than the department of cleaning, but less important than the department of sales, in a mid-level company"). Business organizations solved that problem by the by-now universal practice of carrying business cards, which state who and what, to avoid confusion and set up a 'status map' everyone can adhere to.

The presence of meta-statements, and their polite expression works in many ways. They allow participants in an interaction to make meta-comments about both other participants (present and absent) and about the contents of the exchange. Politeness can also be used as to rein in impudence, suppress subordinates, and demonstrate allegiance and rebellion. As authors from Jane Austen to Haruki Murakami and Alexei Panshin demonstrate, in the hands of a master, the use of politeness is a multi-functional tool for ensuring one's goals are met and authority acknowledged. On that account, at least, I would suggest strongly that ETI messages, whether internal (within the race, culture, or civilization) or external (to peers, superiors, or primitives), would contain meta-statements, whether, as in Chinese communications, to test others for compatibility (AKA 'are they civilized?' Cf. Friese 1963; Ho 1964; Huang 2011) or to demonstrate to them that they are not: all according to intention and circumstance.

If they are, what are the implications Meta-data yields useful hints about relative statuses, and the relationships among human communicants. We can assert that the same thing would be true among ETI. For example, communications which regularly include elements which are identified as ritual indications of politeness, would likely indicate societies in which hierarchical relations are central. At which point it is worth asking how we (humans) would fit into such a discourse. Our human experience (see for example British, Chinese, Japanese, and traditional Spanish societies) is that foreigners are relegated to the lower orders, with all that implies in terms of access to resources and power. Fewer meta-statements which display deference might not indicate an *equal* (all have the same access to resources and power) society, but the lack of deference presentations would indicate that the society might be more *egalitarian* (professes, and perhaps allows, a rise in status and access), which is visible, for instance, in changes in the vocabulary of greeting in personal communications (Murphy and Levy 2006; Waldvogel 2007).

Understanding the structural relations and hierarchies within a society one has newly come into contact with is crucial in two areas. First, in terms of presentation of self, it helps to present oneself as someone who is part of the culture, and able to function within it. In other words, demonstrating that one will act as a proper member of society is the single most important aspect of being able to function *within* that society (even if one is only an electronic pen-pal). Second, both in terms of exchange—diplomatic, economic, artistic, or of knowledge—it is important to know who one is dealing with, if one is to get (at least) a fair bargain for one's wares (something we address later in the book). Among humans, for instance, virtually all societies that exhibit courteous (that is, expressions of deference towards a person who is a stranger) behaviors, such as e.g. a man opening a door for a woman in European cultures, or bowing in apology in Asian ones, are highly sexually dimorphic (men are expected to open doors only to women, women bow more deeply than men of equivalent status). That is, "…act[s] of common courtesy may be viewed … as a political statement designed to perpetuate traditional views of masculinity and femininity." (Walum 1974, as cited in Harris 1992: 1400).

Finally, it is always an advantage to know what 'they', whoever 'they' are, expect. Do they have a strong competition for dominance? In such a case, should we pretend to strength, or be humble? Or do they practice, as in many Asian cultures, self-deprecation, which does not reflect real power relations? If so, what are the consequences of being humble, or, to the contrary, demonstrating power? In any case, knowing what is expected allows a broad spectrum of choices for ego, in this case, the human species.

7.5 Gossip: Informal Information Flows in a Society

I am, I admit, very fond of gossip. Gossip is partly what an anthropologist does: listen to, and sort through, gossip. Indeed, gossip, for humans is a crucial form of inter-personal exchange, a sort of personal intelligence network. Like all intelligence networks, it is hugely unreliable, and sorting the wheat from the chaff is a major job, notably since wheat and chaff change places with amazing rapidity. Of course, gossip has a bad reputation as destroyer of trust and "a poisonous vine" (Gluckman 1966). However, it does have an important positive side for human society, and understanding gossip may have a crucial importance in both SETI and xenology.

7.5.1 What Is Gossip?

Like all organisms, humans exist in an environment of indeterminacy: not knowing whether the next minute will be good for the self, bad, or indifferent. Indeterminacy affects all organisms, whether prey, predator, or neither. Some mollusks develop strong shells to lower indeterminacy and increase the odds of survival, whereas predators such as the many species of sabertooth cats, develop extended canines to ensure a quick, determinate kill. Human culture—material and social aspects alike—ensure regularity. Having a store of grain ensures food will be available for long periods, and having friends and allies ensures someone will feed and possibly protect me in times of trouble.

However, humans being humans, are not really 100 % reliable (nor are insurance companies). And so, humans engage, to a greater or lesser degree, in swapping information about others. This information may not be, in itself reliable, because we offer, and receive, three types of information: factual information (what we think we know), affective information (our opinions and emotions concerning the subject of gossip) and speculation, which we often present as factual. The strategy of increasing reliability of the information ("Should I trust this individual with a loan of my lawnmower?" or "Should I choose this insurance company over that one?") is to gather as much information/gossip from as many sources as possible. To do that, we need some information/gossip of our own to trade. And of course, we may use gossip as a tool: to increase or decrease the subjects reputation, in a way that is not formally attributable (Besnier 1989). Gossip appears to fulfill three primary functions (Fine and Rosnow 1978). It provides information about socially relevant subjects; it influences the behavior of subjects and emitters, and it serves as an entertainment which brings people together.

7.5.2 And the Internet

There is another aspect to gossip which is relevant to ETI communication. "Gossip" is a strategy sometimes useful for network communication between computers. Like human communication, computer gossip protocols are opportunistic, and rely on the availability of a number of nodes, none of which have priority in sending their messages on. However, by passing messages on to another computer node in a network, whose address is slightly closer to the target of the message than the sender, a message will, eventually, reach its destination (Boyd et al. 2005; Jelasity et al. 2005).

Given the realities of a star-spanning civilization (of which more anon) a gossiping protocol would help ensure communication. The same message would reach any given intended recipient via several different sources, ensuring redundancy and therefore a lower error rating. From an external observers perspective (the human race), this means that any message will likely be repeated several times, increasing the ability to listen-in, and thus gain information. And gossip that has been loaded with additional comment and opinion (such as human gossip invariably is) would yield additional useful information.

7.5.3 Would ETI Gossip? And If They Don't What Would that Teach Us?

It is conceivable that an ETI civilization does not gossip socially. In other words, they do not repeat and broadcast the mix of hearsay, supposition, affect, and outright invention that characterizes human gossip. We can learn from that as well. Gossip, as noted, serves as a means for social control among humans, by allowing non-formal critiques of members' behavior measured against community norms. Where such is not expressed, we may contemplate the existence of a culture that has very strong internal controls over behaviors (including, presumably, broadcast behavior) which presupposes a type of soviet in which individual behavior is highly restricted, and in which control by the collective over the individual goes very deep indeed. This would have obvious effects on our ability to communicate with them, and, more to the point, their acceptance of humanity's more freewheeling ways.

7.6 Trust

Humans, (and some animals, cf. Bekoff 2001) have expectations of fair play in their social exchanges. Trust is learned early in life during the socialization period, for pack and group animals. As Bekoff (2001) notes, morality consists of behaving fairly during play, which helps in developing trust among members of the group.

Trust is also generated by repetitive sets of dyadic acts in which one's positive expectations are fulfilled. Consider for example, a case of a simple purchase of some item. The shopkeeper hands over the item. The customer takes hold of the item. The customer pays the money. The shopkeeper hands over the change. Consider how improbable the scenario is. The customer could walk out after receiving the goods. The shopkeeper could keep the entire sum without rendering change. Of course these things do actually happen, and of course there are external variables such as police and the courts. However, essentially, the transaction relies on the trust, of both partners, that the exchange will be carried through. Trust can therefore be defined as *the degree to which ego maintains belief that a social expectation be fulfilled by alter*. The very fact that sometimes these expectations are *not* fulfilled, strengthens those that do. And, trust inheres in *social personae*, be they individuals, corporate groups, or social systems. In most developed societies, we 'trust' the judicial system to be fair and balanced, or else the value of our society drops in our own eyes.

We need to keep in mind, however, that trust is evaluative, and thus flexible. If, in the dyadic exchange above, the customer does *not* pay, after years of successful exchanges, the shopkeeper will no doubt chalk it down to the customer's forgetfulness or other circumstances, and perhaps mention the problem at a later date. Where there is no trust, on the other hand (the two parties are strangers) the shopkeeper is likely to chase after the customer and demand payment. And if the reliable customer repeats the nonpayment, eventually the shopkeeper will stop trusting him. Not only that: the shopkeeper will likely gossip about the customer, both as a warning to others, and as a form of punishment (damaging the trust *other* people have in this particular individual).

Now, suppose you've ordered a frost-free refrigerator via the internet. You put your money (or credit card) down. The company sends a refrigerator that does not meet the specs. Both sides must trust one another, but there is a remove of weeks, sometimes months, before negotiation can establish who and what is to blame, and more often then not, one has to call in a mediator or arbitrator of some sort. If the gap between payment and receipt, or between what has been promised and what is actually received is on the order of years, or even centuries, serious problems can emerge. Establishing a reputation for trustworthiness is crucial for interstellar/inter-ILF exchanges. We can reverse the issue as well. Suppose party A wants to warn party B off for some reason. How is party B to be convinced? Again, this depends on many variables, but one of the most important is A's trustworthiness: Will A's threat actually materialize, and is the threat credible enough to motivate B to actions A desires? As we have seen from recent cases of diplomatic brinkmanship, politicians, even presidents, who fail to make a case for their trustworthiness invite further deleterious actions.

7.6.1 Encyclopedia Galactica

One of the staples of SETI has been the benevolent advanced ETI race/culture/civilization (Ćirković and Bradbury 2006; Deardorff 1987; Dick 2010; Kent 2011; Webb 2002). Some scholars have even gone further, suggesting that benevolent advanced races might put out a data base of knowledge: nicknamed Encyclopedia Galactica in the age before Wikipedia (Drake and Sobel 1992), and sometimes Galactowiki subsequently. The Encyclopedia Galactica *could* be a freebie broadcast of information useful to younger developing ILF races. That is tenable. However, just as likely, it will be a paid for service. In such a case, how does a developing race—us humans—pay for the service, and what would that payment consist of?

Beyond the issue of fee or free service, lies another problem. Can we trust Encyclopedia Galactica? Wikipedia, a similar service in that is a free broadcast, has faced a number of issues of reliability. These include problems of cultural bias and variation (Callahan and Herring 2011; Hara et al. 2010; Nemoto and Gloor 2011), content-creation issues (Feldstein 2011), and has been accused of causing a lowering of technical standards (Keen 2007). Clearly, the crowd-sourced process of providing information (which we are supplying, gratis, free and for nothing to any external observer with sufficiently good electronic means at their disposal) has benefits and risks. Wikipedia, which is more reliable than most due to its internal self-regulating processes (Reagle 2010) is more trustworthy than Joe Blogg's blog. But, to reverse the positions, how are we to know that if we receive Encyclopedia Galactica, the information is reliable? Interestingly enough, Wikipedia does provide an answer that established encyclopedias do not: the meta-discussion and editing process are open to all to examine, and test, at least to some degree *and for humans* the trustworthiness of the process. Anyone can open Wikipedia's editorial pages and examine the discussions, emendations and editing process. It is thus inherently more trustworthy than any commercial encyclopedia, whose trustworthiness relies on two points: its ongoing reputation (Larousse and Britannica have both been published for over a century), and the personal reputations of its writers.

Encyclopedia Galactica may provide internal bases for trustworthiness, or it may rely on some obscure (to kibitzing humans) indicators of trustworthiness. And, of course, the entire thing might be a giant con, either to lure us into the jaws of hungry aliens, or for some other nefarious purpose, as Langston so charmingly puts it (Langston 2013).

Trust is developed through a long process of exchange between two parties. It can also be engendered through the good services of a trusted third party who can guarantee each transaction, and who might serve as an unbiased go-between. Truly trustworthy go-betweens are very hard to find at either the personal (cf. Bailey 1971) or national level (cf. Melin et al. 2013), so trust through a third party is hard to develop. To complicate things, trust requires constant reinforcement: something difficult to achieve in communication exchanges that could last for centuries rather than minutes.

7.7 All Work and No Play Makes ILF Dull

Collective action among humans is not all survival oriented. From childhood on, humans engage in activities that have no expected productive value; we engage in play. The young of many species of mammals and some adult mammals and birds engage in play as well. Non-purposive behavior of that sort does of course have reproductive and productive benefits. Young cats of all species practice their pouncing skills, and sporting animals display their fitness to mate in a non-threatening environment. So much so that some ethologists have argued that play is largely training for 'real' life (Byers 1998). However, as all of us who have engaged in play as adults know, there is more to the activity than that.

7.7.1 Homo Ludens and the Hedonists

Play is, indeed, at least partly a test bed for 'serious' activities of production and reproduction. It allows individuals to practice and enhance their physiological and mental capacities. However, given that play has artificial boundary rules, and that it can be broken off at any time, we have to recognize that play goes beyond the immediately instrumental (Huizinga 1955). Play, and particularly sport, also bring about states of happiness and joy, even transcendence (Macaloon and Csikszentmihalyi 1983).

Pleasure and joy are partially wired-in for terrestrial organisms. The sensation (or spectrum of sensations) can be triggered by satisfying essential needs, as well as by satisfying those needs higher up the Maslow pyramid (to be discussed later), such as self-realization. In effect, the generation of pleasure in an organism is a crucial mechanism for ensuring species survival: positive rewards reinforce: rewarding behaviors, which will then be repeated, until a new, more rewarding positive reinforcement arrives. We can therefore predict that—notably given the simplicity of the mechanism of conditioned behavior—ETI will also have some sensation of pleasure *and* that mechanisms will evolve to increase this sensation, when time and location permit. In other words, we would predict that ETI are likely to indulge in non-immediately purposive behaviors of some sort.

We can suggest another element in non-purposive behavior as well. We've already noted that ETI are likely to engage in the creation of both face-to-face and larger organized groups. We can thus predict that some of those groups will be for the purpose of generating pleasure, whether solely as participants, or, as among humans, as a mix of participants and spectators. This is not to say that they will have football teams and cheerleaders, but that some of the behaviors of any ETI we come into contact with, would be equivalent to our play.

7.7.2 If I Bring My Football, Will You Play with Me?

The important thing about *most* play is that it is voluntary. Humans, notably children, devote a significant amount of time and energy to playing together, and social play is fun. It is more than that: for children (and many other species' young) play is crucial for the development of behavioral flexibility, peer-ranking, and practicing roles for later life. For adults, playing provides synesthetic pleasures, but it also helps in displaying and ordering social ranking, social pretensions, and provides and arena for (generally) harmless competition.

Thus two aspects of play may be particularly important: first, does the ETI culture in question play or not? And second, what kind of play is it? Those who do not play represent a serious enigma to humans, because a great many social issues —from open displays of sociability to presentations of humor and irony— are associated with play. How can we deal with a culture that is serious all the time and in which every expression has instrumental value? Assuming the ETI culture in question *does* play, the form and meta-expressions related to play tell us a great deal about how this culture deals with other issues. Do they play in groups or only as individuals? Is play competitive or cooperative? It is frequent or restricted to specific trigger occasions? Is it structured by qualities such as age, gender, sex, or some other quality, innate or achieved? Since play more than many other activities is likely to be public and overt, this allows the opportunity for an observer to learn more about hidden currents and values within the observed society.

7.8 The Socio-cultural

Social scientists use the term 'socio-culture' as a sort of place-holder: a term that allows identification of a specific group of people by their behaviors, language, artifacts, etc., which, as a sort of rough whole, seem to hold together. There are many deficiencies in this term, but as a loose, non-obligatory label, it is quite useful. In this section, we shall parse this term a bit, to make it useful for the xenological enterprise.

7.8.1 Society, Consumption, and Socialization Groups

We have spoken earlier about the close relationship between the family as a socializing unit (and its potential analogs among ETI) and the need to produce stuff. Here we need to add the other two elements of the economic enterprise: exchange and consumption.

Exchange Exchanges among humans take on some creatively varied forms. Exchanges under capitalism—market, or direct exchanges—are not as universal as some scholars and believers would like (us) to think. In practice, three types of exchange are common everywhere in the world: *direct/market* exchange (ego provides goods to alter, alter reciprocates with some good, immediately or at some agreed-upon time); *redistributive* exchange (ego provides goods to an alter, who then distributes the goods as alter sees fit, to other individuals); and *delayed* exchange (ego provides alter with goods, alter later provides those goods to an agreed-upon third party). As with all exchanges, we must recognize that things do not always proceed smoothly. Ego may or may not provide all the goods honestly and fully, alter may not do with the goods as expected. In other words, trust plays a major part in the process. However, ignoring the reality of improper exchanges, chicanery, and general human dysfunction, let us concentrate first upon the theory.

Direct exchange is of course the simplest. In modern societies, we run across such exchanges on a regular basis. We go to a shop, pull out a wallet, hand over money for goods: a stick of gum, a kilo of tomatoes, or a sports car. The principle is the same. In *theory*, Economics 101 would tell us, there is no social element involved: purely a financial transaction. In reality, social ties factor in (one might walk and extra twenty paces to a shopkeeper who is pleasant, rather than shop at an impersonal automatic vendor), and there is always an element of trust, between passing on the money and receiving the good (and the reverse, of course). Thus even within a purely financial exchange, social issues intrude. Billions of units of currency change hands every day using this system, generating not only wealth (sometimes through very complex transactions) but also trust, sometimes very abstract trust (consider the effects on trust of the Volkswagen emission standards scandal in 2015).

Most of us (few manage to avoid, and if caught, are likely to suffer penalties) participate in redistributive schemes. Even in the most stridently capitalist societies such as e.g. the US, individuals provide wealth to a corporate individual (AKA 'Uncle Sam') who then redistributes the accumulated resources as 'he' sees fit (not of course without much complaining and gnashing of teeth by both donors and recipients). Even in simple/primitive societies, redistribution is a common feature, whether a successful San hunter distributing his catch as a form of social insurance (Lee 1979), or a Kwakiutl potlatch feast (Benedict 1977).

Delayed exchange is the most abstract of the three forms of exchange. The raising of children is one such example. Ego has children, with the expectation that at some due date, these children will pass on some of the wealth to ego's grandchildren, and some of it to support ego in the infirmity of old age. In some cases, the demand for delivery may be explicit (as in Chinese society, for instance) and in others implicit in the process of raising offspring. In modern society, where pressures of child-rearing and old age are distributed into several social agencies (family, school, insurance, government) delayed exchanges can also be viewed more clearly in for example, charities and various trust arrangements.

Consumption Consumption is obviously a very broad subject, since humans in a modern setting consume a wide range of goods, some material and essential (food and drink) and some abstract, yet nonetheless considered important (entertainment, knowledge).

What one consumes, *when* one consumes, *with whom* one consumes, and *how* one consumes give evidence about the *meaning* of consumption, as an emic (internally relevant to the individual) process, and also ties in consumption to broader issues in the society concerned. Since consumption is such a broad issue, we'll only deal with two aspects towards either ends of the spectrum of consumption possibilities. We all, ETI included, must consume "food" in order to live: some set of substances that produce energy for body use, as well as necessary chemicals. In the case of ETI these different inputs may be separate.

Human consumption, has, and has had, an influence on the environment far beyond sheer numbers. In prehistory, humans in Australia altered the environment substantially by starting hunting fires that, besides killing off local megafauna, substantially altered and desertified parts of the continent. In North America, Clovis culture did something similar; and there is reason to suppose that the Amazon forest was created partly by humans through the manufacture of *terra prieta*. As humans entered the industrial and post-industrial ages, these environmental changes have become massive and potentially catastrophic (Turchin and Ćirković 2008). What is more important, they create changes in the atmosphere which are visible from space, even across interstellar distances (New Scientist 2009): changes in atmosphere (e.g. more than natural presence of certain gases from industrial emissions) would indicate an industrial society on the planet. The same would likely be true of a culture which exploits it's entire solar system: depending on the materials used, ionization and electromagnetic emissions would give clear indications of non-natural activities. Obviously, this is particularly important for SETI. Nonetheless, such emissions and changes in atmosphere (planetary or solar) would also tell human scientists something about the technological achievements of the ETI concerned, and possibly something about the types of consumption they engage in. Consumption (that is, use) of weapons would become obvious sooner rather than later, as would the degree of extra-planetary settlement, population numbers and distribution, and even something about their social and political arrangements.

7.8.2 Artifacts, Sociofacts and Mentifacts Among ETI

For convenience sake (though not good methodological practice) we can arbitrarily assign the ways we think about sociofacts, mentifacts, and artifacts and manipulate them in our heads to a box called 'culture'; and the ways we organize ourselves to physically make our mentifacts 'real' to a box called 'society'. This is a very rough-and-ready division, and most anthropologists reading this are already straining at the leash of professional outrage; bear with me, however.

Among humans, there is a very strong correlation (probably even a causative connection) between items in the 'culture' and 'society' boxes. That is because due to evolutionary pressures that shaped the ways we think (see Chap. 5), to our phylogenetic social history (consider ape troops), and to the process of socialization in our societies, these two boxes can only be fictions: artifacts are not created, social rules not manipulated, nor thoughts emerge except that they are embedded, somehow, perhaps in individual histories, in social and mental matrixes. Among humans, at least, society and culture are inseparably interlinked (not in a simple one-to-one way, I hasten to add). The same is true of the other three boxes we discussed earlier: mentifacts, artifacts, and sociofacts.

We always need to keep in mind that the tripartite division into socio/menti/artifacts is an analytical convention. Nevertheless, it is a useful one for looking at ETI as well as at humans. Roughly speaking, we should be asking four questions: what do/can we know about each of these three analytical boxes, and, perhaps most importantly, how do these boxes tie together? How is a particular artifact manipulated socially? What kind of meaning is attached to it, that is, what role does it play as a mentifact? Of course the answers to these questions can only become apparent, if at all, after collecting many data points and a great deal of information. Nonetheless, we need to be aware that artifacts in use (the most obvious of the boxes) have social uses as well as purely instrumental ones, and that social acts and knowledge may have physical and mental elements as well. And mentifacts, of course, may be recorded, expressed, or symbolized by particular practices, relationships between individuals, and the use, creation, and manipulation of artifacts. Contra Menken, a woman is never only a woman, and a cigar is never only a smoke.

7.9 Primitive, Developed, Advanced, and Transcendental

The term 'primitive' has received a justifiably bum rap. It has largely been used by anthropologists, philosophers, historians and others to denigrate non-European societies, notably during the nineteenth and early twentieth centuries. Used as an epithet for African, Asian, and Native American societies, it basically meant "someone who is not up to our standards, and whom we may, and should, treat with (at best) paternalistic disdain, if not outright contempt as a target for exploitation."

However, as we shall see, it is an extremely useful term, if used with precision and without bias. Being so, we shall see it usefully also connects us in precise ways to ETI.

Let us consider a band of stone age people about twenty thousand years ago. They are very likely a face-to-face group, that is, twenty-thirty people at most. There is a reason for that size: land carrying capacity. Thirty people, let us say 2/3 of them adults, require around 3500 k/cal a day per person for survival practicing manual labor. Children about half that. By foraging—collecting vegetable material and some grubs—and hunting, they need a collective quantity of some 105,000 k/cal/day. They probably spend about four days of every week doing that, but they *must* have available a minimal magic number of about 2200 k/cal day *every day*. If they don't do that on a regular basis, they weaken, and eventually die off unless and until the magic number can be restored. Now let us consider you, the reader. Get up, go the wall, and switch on the light. If you're using incandescent bulbs, that's it: you have your 105,000 k/cal. That's about 140 W/h, about two incandescent bulbs. The difference between that simple stone-age band and you is extremely important for your (and everyone else's life) as well as for xenology.

'Primitive' and 'Developed' can thus be easily and neutrally compared. Of two societies, the one in which the average individual has immediate access to less energy, is more primitive. If you as an individual have available for consumption in 1 h more energy than an entire band can in a day, by definition, you are a member of a developed society.

7.9.1 Simple Societies

Now, if this were merely a matter of energy, the above would be a relatively puerile fact, xenologically-speaking. After all, *most* SETI literature assumes that any ETI we might contact will be far in advance of us (Bracewell 1982; Vaidya 2007). But the fact is that the minor issue of caloric consumption has huge implications on the social side as well. So, let us return to our (by now, we can label it without quote marks) primitive band. Suppose they know winter is coming, and some bright spark says "OK, in winter there is less food to forage and hunt, let us store up enough to last the winter." We can assume the thinker was a woman, and she would likely be greeted by her spouse with "Duh, how?" which is unusually perspicacious for rather dull males. How indeed? In good times, hunter-gatherers such as our stone-age band ate well when the foraging was good, and suffered when the foraging was bad. Technology to preserve food over the long term was simply nonexistent. The foraging lifestyle requires *two* prerequisites, one social, the other technological. First, socially, foragers can exist only in small bands. Even so, they eventually exhaust the carrying capacity of their foraging area, and *must* eventually move elsewhere, or perish (Steward 1968). Second, the only technologies they can develop are technologies that allow easy movement, since otherwise they would need to abandon it as they moved, and the effort invested would go to waste.

Eventually, humans figured out that surpluses, rather than being consumed in mass feasts, *could* be stored. Storage comes in many forms. Technologically, storage runs the gamut from salting/smoking/drying through building granaries and agriculture. Socially, one can use a surplus in one season to trade for energy or energy-producing tools, or create systems running from the Northwest Native American *potlatch* to modern insurance schemes.[3]

Characteristically, simple societies can afford few free-loaders. No full time artists, politicians, engineers, or judges: everyone, or almost everyone, must work to acquire resources. Thus, simple societies are simple not in the intellectual sense, but in the politico-economic organizational sense: they *cannot* afford non-producers, and can therefore not develop complex, extensive non-productive institutions.

7.9.2 "Developed" Complex Societies

How we got from there (stone-age band) to here (you) is relevant as well. Once one has an enduring surplus (by increased yield or storage technologies) there is one human social factor that immediately raises its head: some strong-arm robber appears, and creams off the surplus. The robbers have many names: kings, chiefs, tax collectors, priests, bandits.[4]

How does this happen as a general process? Remember our stone-age band? One thing that characterizes *all* such subsistence societies is the fact that they cannot afford specialists. *Everyone* must be able and willing to do every task the band needs. Of course some proto-specialists might emerge: one woman is excellent at weaving baskets, another always knows where the most succulent mushrooms are to be found. Nevertheless, each one, in a pinch, could do what the other is better at. What a regular surplus allows, is the allocation of surplus to specialists. The band can support a shaman (a religious specialist), a maker of implements, a painter. The more surplus is available (sometimes *because* there are specialists such as smiths and potters) the more specializations can emerge, and as said in the previous

---

[3]A system that in it eponymous form, was used by Northwest Coast Native Americans to ensure long-term survival. The Northwest Coast (Washington, US to Alaska, US) is an area of great diversity and wealth in wildlife and food sources. However, these are not fixed, but vary from year to year. In a flush year, a village might gather huge surpluses, in a poor year, it might starve. To overcome this, a flush village would have a potlatch, a major feast where people from other villages were entertained royally, and given gifts of food and valuables. In a poor year, the same village would feast with its more successful neighbors. In effect, this was a simple insurance system built on reciprocity, pride, and ritual. See Drucker (1966).

[4]I am being only mildly sarcastic here. Nonetheless, and however they are termed, they emerge from surpluses. What is significant is that *because there is a surplus these robbers can come into being*.

paragraph, more often than not, those specialists specialize in organizing others (and, not incidentally, creaming off the best for themselves). This type of society, to which ours belongs, many forms removed, is called a 'complex' society. In point of fact, modern civilization is partly characterized *universally* by fewer primary producers (farmers, hunter gatherers such as fishermen) and more specialists by a factor of about 1:25 or even 1:50. That is, 2-4 primary (food) producers in 100 people. We shall use the term 'developed' not as an encomium, but to identify societies that are characterized by (a) the availability of surplus energy that exceeds the caloric needs of the society (minimum caloric needs/person x number of persons) by an exponent; (b) a size greater than a face-to-face group; (c) consequently by the complexity of relationships this implies, and finally (d) the presence and influence of non-primary production specialists.

7.9.3 "Advanced" Societies

By now one major question should have popped up, and once again, it has social implications. Suppose the society does indeed have an energetic surplus, and specialists as well, *but the energy is not distributed evenly in the society so that everyone has the same access?* This is indeed the case in virtually every society any reader of this book lives in. In fact, for complex developed societies, while famine is unusual, it is not unknown. See for instance the Great Chinese famine in the fifties (the direct result of specialists not knowing what they were doing), the Ethiopian famine in the eighties (lack of proper social storage) and so on.

This means we can add another operative characteristic of types of societies. An *advanced* society is one in which *every* individual has access at will to sufficient energy and knowledge (at its most basic as an artifact, a form of energy) to fulfill all individual needs and wishes. Note that we use the terms 'needs' and 'wishes' quite precisely. Humans need many things. Maslow, a psychologist has suggested that these needs must be seen as a set of layers or a pyramid. Once one set of needs is satisfied, humans naturally try to satisfy the next in line (Maslow 1987). Maslow's posited a hierarchy of 5 needs: physiological needs (eating, sleep, breathing, reproduction), safety (from harm), belonging (to a group, a family, some reference group), esteem (both self-esteem and esteem by relevant others), and finally self-actualization (the ability to accomplish everything one wishes to do and can be). "Wishes" broadens the concept of self-actualization to include those elements of a person's being that demand other-than-personal needs. The desire for change, for particular aesthetic surroundings, for individuality, which I believe are insufficiently covered explicitly by Maslow's needs hierarchy, belong in the 'wishes' category. From our perspective, an advanced society not only allocates energy in such a way that every individual has sufficient energy for their needs, but also to indulge their wishes.

7.9.4 "Transcendental" Societies

Now, There Ain't No Such Thing As A Free Lunch, or TANSTAAFL, as Larry Niven taught us. Somehow, some way, advanced societies must pay for all that energy. They may have specialists, as we do, who do nothing but provide energy, guide it to its users, and monitor systems that ensure energy provision. They may have other ways of balancing the books. But, somehow, the energy is likely to be finite. So, we needs define another level of society. *Transcendental* societies are those which, for one reason or another, are free from the basic energetic equations of the universe. How they would do that, and in what form, and, perhaps more importantly, what the social implications would be, are hard to imagine (We shall discuss and critique one such possibility, suggested by Kardashev, below in this chapter). It is certainly conceivable that transcendental societies have dispensed with specialists, and have returned to being complete generalists, notably if they have made massive use of communication, AI technologies, and the blending of biological and physical sciences. However, given that one basic premise here is that we accept the assumption of the mediocre position, we simply cannot know what Transcendentals are like without more data, and even that data may well be incomprehensible to us.

7.10 ET Societies in the Mirror of Xenology

Now that we have some tools to work with, let us consider ETI societies. Let us start with the issue of energy as a whole. One of the first *technological* characterizations of ETI civilization was a crude division into three classes of ET civilizations based solely on their ability to exploit energy sources (Kardashev 1964). Class I are civilizations that use the entire output of their planet, Class II use the entire energy output of their sun, and Class III the entire output of their galaxy.[5] There are many possible criticisms of Kardashev's classification, including the temporal assumptions (which are straight line extrapolations). A recent publication argues that Kardashev stage II and III would be identifiable by their infrared signatures (Carrigan 2010). The implication is that inasmuch as no such signatures have been shown, Kardashev II/III civilizations do not exist. This is not, to me, a real argument, since they may have come into being at the other end of the galaxy a mere few thousand years ago, or our observations may be faulty.

The critical problem with Kardashev classification, however, is that it lacks any social explanation: what *kind* of society would need such major works, and how would the society be organized? As we have seen above, maintaining social networks and social groups over time is, at least for humans, an insuperable task.

[5]There ought to be a Class IV—those that use the entire energy output of their universe—but that would bring them into the realm of transcendentals.

The oldest continuing organization among humans is the Catholic church, and that has lasted less than two thousand years, has undergone many changes, and has lost membership and sub-organizations over the years.

Consider an ETI civilization A. It has corralled *all* the energy of its sun into useful forms. This supposes the presence of some technology—a material shell? A superconductor guide?—that brings all energy available into use. There must be X members of the society concerned. X may be a very large number, notably if the mechanism involves a Dyson sphere (Dyson 1966) which would allow for a very large number indeed, even if most of the available interior surface is dedicated to natural reserves or manufacture. Whatever the number, it is highly unlikely to be a face-to-face group in human terms. That means we are talking about a developed society at the very least.

It is highly unlikely that the A have a face-to-face society. If they do have one, what possible need could they have for the Kardashev shell? Whether they use a Dyson sphere, which is astronomically unstable, or some other form of engineering, *someone* has to ensure that, over the eons, the structure is maintained and if necessary repaired.[6] This means that we know at least one thing about civilization A: it is a complex society. Which in turn means that the *social* complexity of this society tends towards the infinite, including the growing possibility of social breakdown, conflict, and even violence.

The major criticism of Kardashev's scheme is thus not the technological nor physical one. What is more important is the issue of how is such a society organized? Given the difficulties humans face in organization (which may be extreme or minimal among ILFs in the universe) it would seem that L (= the length of existence of an intelligent species. See the Drake/Green Bank Equation) would be too short for any civilization to become a Kardashev II civilization, let alone a Class III *unless* they can first solve the problems inherent in organization. My own suspicion is that at that point, they are more likely than not to simply not need a Kardashev shell, since the simplest solution to the need for energy for a species is to become more efficient about using it, along with slimming down the demographic.

7.10.1 Social Mechanisms Among ETI

So, rather than using a simple Kardashev measure, let us see what the implications of the discussion of social mechanisms means for understanding ETI. As usual, a warning is in order: human models *may* be local expressions of some universal principles but until more data emerges, we need to be cautious in accepting them wholesale.

---

[6]This line of argument deliberately ignores the possibility of something like a super-AI maintaining all aspects of the 'civilization' on its own. Among other reasons, this would mean we are not discussing a civilization, of which more anon, but a single individual of some sort, however transcendent.

Small groups Partly, at least, our preference as a species for the intimacy and support of small groups is phylogenetically determined. If modern apes are a good model, our phylogenetic groups were composed of troops of females and their young with dominant, competing, hierarchical-minded males ruling the troop (Parker and McKinney 2012; Sahlins 1968). The same is not true of other species on Earth. Lions have small groups of sisters who accept a dominant male. Hyenas have the same structure, except the females are dominant, and share in raising cubs and providing direction for the group. Meerkats and canids have groups dominated by a single breeding pair who may kill other offspring, and who determine the group's activities.

While it is likely that *some* sort of socialization group (AKA 'family') exists among ETI, small groups may be little more than functional arrangements, which have no phylogenetic basis, but are, for the species concerned a 'modern' managerial development which has no deep emotional content. In other words, while humans derive a great deal from small groups in terms of emotional support, feeling of belongingness and other immaterial benefits, this may not be true of (all or any) ETI. If this is the case, what benefits are there, beside functional ones ('We need three people to build a bridge Volunteers, take one step forward')? To manage the building of a complex bridge (or of it's prerequisites, say a computerized factory) one needs some sort of organization and direction, that is, management, not merely spontaneity. This in turn implies some sort of hierarchy, fixed or malleable. It also implies that an ETI society will consist of numerous specialists, able to handle the complex and varied tasks inherent in maintaining a society, particularly, and by definition, a developed society.

Perhaps a species may overcome that need; they may all be genius level polymaths, able to handle any and all technical and scientific challenges. How likely this is, each reader is invited to consider. A safer bet is to go with the doctrine of mediocrity: they may be smarter and better developed, even better organized than humans are, but transcendentals make any kind of calculation difficult, and a society that is composed of a few super-geniuses is unlikely to be encountered.

Complex groups Complex groups of some sort are vitally necessary. Of course, ETI who have some sort of gestalt mind, or arise from a social arrangement like starling flocks or schools of fish (Pearce et al. 2014) may exist. Nevertheless, the same principles that apply to the potential of humans to develop technologically apply here as well: the need for specialization and specialists, which implies, at the energetic level, an amount of energy that is an exponential multiple of basic energetic needs, and specialists (robbers or not) to manage the system. On the other hand, we can know very little about *how* these groups are organized or function. Our own group types, like our intelligence, are inherently bound into our phylogenetic evolution. We have 'leaders' and 'followers' competition for status and position, prevarication and the use of false information and lying, and devote a great deal to advancing our own individual causes and views (and doing down our perceived opponents). None of this is necessary for running a complex society: it happens to be the human way, but cannot by any means be considered a universal set of processes.

Networks Given a neutral and unbiased approach, network analysis (provided we have access to data about the network) suggests a useful approach to understanding any ETI society. Even the simplest of face-to-face groups has many thousands of interactions a day, all of which can be mapped if someone is interested in making the effort (Bidart and Charbonneau 2011; Diesner et al. 2012; Feld et al. 2007; Hafner-Burton et al. 2009). Such mapping clarifies how a society is organized by tracking messages between different individuals, or between networks of networks, that is, between groups. No matter how complex (or simple, if one can conceive of an ETI society that consists of low number of individuals) a society is we would be able, given the technology, to track the structure of the society and understand how it is organized, where the power is located; in other words, who does what to whom.

Crucially, as we shall see, the concept of networks of networks also allows us to say something about interstellar civilizations (Bracewell 1960; Learned et al. 2008; Loeb and Zaldarriaga 2007), of which more later. Principally, as we shall see in a following chapter, civilizations are best conceived of as cluster of social objects and forms—networks of networks, networks of cultures, nested social collectives—to which the network of network concept is admirably suited as a descriptive frame.

7.10.2 When Society ≠ Culture

Among humans, the phenomena characterized by the terms 'society' and 'culture' are analytical conventions that describe a complex phenomenon from two different perspectives. 'Society' implies that the analyst is concentrating on the description and understanding of the learned behaviors having to do with social relations within some sort of bounded group. 'Culture' in contrast implies concentration on all learned behaviors within a matrix of the social relations of some bounded group. In practice, of course, neither the understanding of any society/culture, nor it's actual nature/working can be captured by looking at culture alone or society alone: the two are inextricably intertwined.

The above raises a crucial question: could an ETI species *not* have this intertwining? What would the implications be? Clearly, some form of accepted social relations are necessary to keep any collective of individuals working together. This is more crucial for a technologically advanced, or at least developed society. Clearly, too, keeping the technology, as well as the social relations going, should be a learned thing. But do they have to go together?

Consider two alternative examples. A species' social relations might be hard-wired, perhaps dependent on a sensed cue (visual, oral, electrical, or olfactory). Presumably, they would still be able to learn and teach other topics—electronics, physics, art—while having very little awareness of their social triggers and an inability to modify these. This is not unknown among humans: years ago I had a conversation with a well-educated Sudanese Muslim, who was completely unable to accept the possibility that just as a man could have many wives (as is common in Sudan), so too could a woman have many husbands, and that some societies accepted polyandry as a desirable form of marriage. He was completely unequipped to accept even a cited, scientifically observed reality. With proper acculturation, and some experience, my Sudanese colleague may have learned differently, but in our ETI example, that would be impossible for biological reasons. Scientific, as well as popular scientific analysis of the social system would be hampered by lack of example, as well as by the evident reality all members of that society experienced: social systems are immutable.

An ETI species with a flexible and fluid social system such as humans have, but with no culture is harder to conceive of. It would mean that individual would have to operate based solely on genetic clues, that is on instinct. We *could* conceive of the instinct to build radio telescopes, even space travel, as the result of local Darwinian processes,[7] Humans could, in theory, become members of such an arrangement, but it would be the same arrangement we have with other non-sentient species (though most likely, if they can build spaceships as result of Darwinian pressures, they could be very well adapted to dealing with species such as ours that seek to exploit them). We *might* be able to tap into their communications, for instance, even learn from them, but the odds are against any form of exchange of ideas and culture.

The main point of the discussion above is to show how carefully we need to apply human understandings of our collective existence and knowledge to ETI. They *may* have some substitute for either society or culture, which, because of *our own* biases and blindness do not occur to us. So long as we recognize two points— First, there will be both a behavioral component that deals with the induction of new members and ordering of cooperative activities, and a component that deals with the production and induction of new knowledge. Second, both these aspects are amenable to analysis and understanding by an outside observer—we will at least be taking the first step towards being able to communicate fruitfully.

7.10.3 Societal Types Among ETI

It is useful to ask what *types* (points in a common typology) of ETI societies could be identified. The primitive to Transcendent continuum helps to define the issues.

Simple/primitive Could we have simple primitive ETI societies? Clearly we could, but, given Einsteinian realities, we would be unlikely to know anything about them, and even their existence would not register on our radio or other telescopes. They

[7]And such have been discussed in science fiction literature, by among others Mack Reynolds and Larry Niven. But they would most likely not be intelligent in any sense that we would understand.

could not create a sufficiently artificial effect on their atmosphere (which would be observable cf. Flores 2015) because large populations of sentient beings would, automatically, have to develop some form of social organization beyond the simple, which would mean, by definition, that they were complex societies. By SETI terms, unless they were sufficiently developed to build radio telescopes, they would be beneath notice in any case.

The interest in this book, however, does extend to non-radio-telescope ETI. Notably, given the growing sophistication of astronomical devices, it may well be possible in the future to be more discriminatory than we are at present. Crucially, knowing that there is a sentient, albeit non-technological, 'out there' would have four important effects.

1. We may have identified a technology that is radically different from our own, some of whose side effects nevertheless, have been identified as non-natural. Such a species may be very advanced in, for example, biological knowledge, with no, or little interest in industrial development. This, we may agree is unlikely, but nevertheless needs to be considered.
2. We (the human species) should be *deeply* concerned that the catastrophic actions of Europeans in the Americas and Australia not be repeated on the indigenous people of other planets. While this possibility is admittedly remote, given Einsteinian limitations, we should be very wary of doing anything that could potentially be catastrophically disruptive (which goes for worlds that do not harbor intelligent life, but may harbor life nonetheless: consider Io, Ganymede and Europa, and the disastrous potentialities of industrial exploitation even in our own solar system).
3. At the most egoistical level, even the most primitive societies have things they can teach the most advanced, whether it survival in an extreme environment (consider water-management among the Moche and Inca) or a socially-beneficial practice such as conflict mediation (consider conflict reduction among the San). Ruining a potential source of new knowledge is inherently stupid and wasteful.
4. Finally, from a scientific point of view, the clear evidence of ETI, even if at a remove, would finally serve as proof that $N > 1$, and settle a major argument that this book is part of. If that argument can be settled in favor of $N > 1$, it is likely to stimulate an interest in astronomy and in the space sciences (Astronautics, Cosmology, etc.) in general and specifically reinvigorating the search for other ETI.

Developed A developed society is possible, notwithstanding the common SETI argument that ETI we become aware of are going to be technologically and scientifically more advanced than the human species is. If the human species is an example of others, developed societies tend to make themselves known to others simply out of lack of care for their environment and the type of technology they use. They pollute their atmospheres as a side effect of industrialization. They change the surface of their planet. They radiate wildly in the infra-red and radio spectra (and possibly others). If they have developed space capabilities, the same effects are likely to be seen on other locations (planets and non-planetary orbits) in their solar system.

We do need to keep in mind that by the time we have identified the presence of a developed ETI society, they may well have changed. They may equally have reached some sort of stasis point, regressed (through war or mismanagement) or developed, perhaps even become an advanced society. Development is *not* a straight line process, and there is no reason to suppose that the kinds of technological development on Earth—a sort of generalized Moore's law effect we have seen since the end of the nineteenth century—will continue forever. Any data we derive from distant observation will be largely out of date. How out-of-date depends on the distance, as well as the dynamics of that particular race.

Advanced Advanced societies—those in which all members enjoy equal access to energetic and information sources—are *one* potential pathway for an ILF species. Whether the kind of galactic civilization envisaged by e.g. Bracewell (1960); Sagan (1966); Tough (1986) and others is likely, we may well expect at least an advanced civilization to exist, if only, as Larry Niven says, because "the perversity of the universe tends towards a maximum".

Paradoxically, advanced ETI species may be more difficult to detect than developed civilizations. If only based on the laws of conservation of energy, they may be of a Kardashev-like type, in that they waste as little energy as possible, and thus ensure that stores of energy are kept up. For example, since 1897 we have been broadcasting promiscuously into space. First in the form of radio, since then, television and powerful radar. All of these are clearly observable as artificial.[8] Weak as these signals are, a sophisticated listener (and we are talking here about an advanced society, whose technology may well be better than ours by an order of magnitude, and limited only by the realities of the velocity of radiation) at a range of 119 light years from Earth (in 2016) and growing, in any direction, might be aware of our presence. A more sophisticated, aware, and technologically developed society is unlikely to waste energy on broadcast, however. Some kind of information and energy transfer (laser, maser, or graser or ??: take your pick), proper recycling of waste heat, and so on, are more likely. This of course has implications for SETI as well.

Whatever the cause, and in the absence of intentional broadcasts such as galactic beacons (Benford et al. 2010; Learned et al. 2008) or other intentional indications, advanced ETI species are likely to be hard to detect accidentally. The results of such

---

[8]So much so that certain eminent scientists, including Stephen Hawking have raised the alarm about potentially arousing the wrath of powerful aggressive ETI.

detection are not necessarily beneficial, but as in the detection of *any* ETI, it would settle the N = 1/N > 1 controversy once and for all.

Transcendental The issue of transcendentals keeps on popping up, however much we might want to stay within the zone of the assumption of mediocrity. By their very nature, it is almost impossible for someone from a non-transcendent society to say anything useful about transcendent species. At this point in the discussion we need to concentrate on the issues of society and culture: would transcendentals have either of these? We can, I believe, assume that they have, *if* we delete a critical component of human scientific understanding: the presence of identifiable patterns of social intercourse, as of cultural learning. While it is almost certain that a transcendent species will have some form of social association and of cultural learning, the patterning of these, if there is one, would be so mediated by whatever form their transcendence takes, that non-transcendentals would be very unlikely to be able to identify or understand. One of the risks of sensing transcendentals, their works, or their artifacts, is that, given our appreciation of the infinite variety (and perversity) of the universe, we might not be able to discriminate between natural phenomena and transcendental-artificial ones. The best we can say is that given a transcendent species, insofar as mediocre species such as us are concerned, they might as well *be* natural phenomena. In practical terms, it is likely that transcendentals would be the cause for scientific argument, but unlikely to be a factor in CETI.

Implications of social types The simple typology set in this chapter is a tool intended to make some sense of potential communicative scenarios. At either end are societies whose culture and society are such that they are unlikely to be communicative. We also ignore outriders—species with culture but no identifiable social structure, or with a social structure but no culture—with which we would not be able to communicate at all (though we could, perhaps, interact with). Developed and advanced species, with all the difficulties entailed, we could understand if only because they share with us certain meta-principles: society and culture.

Another issue is relevant notably to SETI, and to what I have called 'the space impulse' (Ashkenazi, 1995). *Should* we identify the signs of social or cultural processes, our own civilization's interest in space would get a boost. This would also bootstrap SETI because, to paraphrase Isaac Asimov, we can argue that $N \neq 0$ (as some of us doubt human intelligence); we know N = 1, but N = 2 is a ridiculous number: if there are more than one intelligent species in the universe, it is impossible that there be only two (Asimov 1972).

In the next chapter, we shall go more deeply into the issue of culture and its implications for what we know about ETI. From this point, While based on known science, informed speculation will make more of an appearance.

References

Ashkenazi, M. (1995). Will ETI be space explorers? Some cultural considerations. In G. S. Shostak (Ed.), *Progress in the search for extraterrestrial life* (pp. 507–520). San Francisco: Astronomical Society of the Pacific Conference series, 74.

Asimov, I. (1972). *The gods themselves*. New York, NY: Doubleday.

Bailey, F. G. (Ed.). (1971). *Gifts and poison*. New York, NY: Schocken Books.

Benford, G., Benford, J., & Benford, D. (2010). Searching for cost-optimized interstellar beacons. *Astrobiology, 10*(5), 491–498.

Bekoff, M. (2001). Social play behaviour. Cooperation, fairness, trust, and the evolution of morality. *Journal of Consciousness Studies, 8*(2), 81–90.

Bell, G. D. (1967). Determinants of span of control. *American Journal of Sociology, 73*(1), 100–109.

Benedict, R. (1977). Central elements in Kwakiutl culture. In W. Goldschmidt (Ed.), *Exploring the ways of mankind* (pp. 15–20). New York, NY: Holt, Rinehart, and Winston.

Besnier, N. (1989). Information withholding as a manipulative and collusive strategy in Nukulaelae gossip. *Language in Society, 18*(03), 315–341.

Bidart, C., & Charbonneau, J. (2011). How to generate personal networks: Issues and tools for a sociological perspective. *Field Methods, 23*(3), 266–286.

Boyd, S., Ghosh, A., et al. (2005). Gossip algorithms: Design, analysis and applications. In *IEEE* (Vol. 3, pp. 1653–1664).

Bracewell, R. L. (1960). Communications from superior galactic communities. *Nature, 186*(4726), 670–671.

Bracewell, R. L. (1982). Preemption of the galaxy by the first advanced civilization. In M. H. Hart & B. Zuckerman (Eds.), *Extraterrestrials where are they?* (pp. 34–40). New York, NY: Pergamon.

Byers, J. A. (1998). *Animal play: Evolutionary, comparative and ecological perspectives*. Cambridge, UK: Cambridge University Press.

Callahan, E. S., & Herring, S. (2011). Cultural bias in Wikipedia content on famous persons. *Journal of the American Society for Information Science and Technology, 62*(10), 1899–1915.

Carrigan, R. A. (2010). *Starry messages: Searching for signatures of interstellar archaeology*. F—arXiv:1001.5455

Ćirković, M. M., & Bradbury, R. J. (2006). Galactic gradients, postbiological evolution and the apparent failure of SETI. *New Astronomy, 11*(8), 628–639.

Deardorff, J. W. (1987). Examination of the embargo hypothesis as an explanation for the great silence. *Journal of the British Interplanetary Society, 40*, 373–379.

Dick, S. J. (2010). The postbiological universe. *Acta Astronautica, 62*(8-9), 499–504.

Diesner, J., Carley, K. M., & Tambayong, L. (2012). Extracting socio-cultural networks of the Sudan from open-source, large-scale text data. *Computational and Mathematical Organization Theory, 18*(3), 328–339.

Drake, F. D., & Sobel, Dava. (1992). *Is anyone out there? The scientific search for extraterrestrial intelligence*. New York: Delacorte Press.

Drucker, P. (1966). Rank, wealth, and kinship in northwest coast society. In J. Jennings & E. A. Hoebel (Eds.), *Readings in anthropology* (pp. 211–216). New York, NY: McGraw Hill.

Dyson, F. (1966). The search for extraterrestrial technology. In R. E. Marshak (Ed.), *Perspectives in modern physics* (pp. 641–656). New York, NY: Wiley.

Feld, S. L., Jill Suitor, J., & Hoegh, J. G. (2007). Describing changes in personal networks over time. *Field Methods, 19*(2), 218–236.

Feldstein, Andrew. (2011). Deconstructing Wikipedia: Collaborative content creation in an open process platform. *Procedia—Social and Behavioral Sciences, 26*, 76–84.

Fine, G. A., & Rosnow, R. L. (1978). Gossip, gossipers, gossiping. *Personality and Social Psychology Bulletin, 4*(1), 161–168.

Flores, M. C. L. (2015). Convergent evolution and the search for biosignatures within the solar system and beyond. *Acta Astronautica*. http://www.sciencedirect.com/science/article/pii/S0094576514004457. Accessed 12 March 2016.

Friese, H. (1963). Qualification for office: Expertise or character. In J. M. Menzel (Ed.), *The Chinese civil service* (pp. 77–83). Lexington, MA: D.C. Heath and Co.

Gluckman, M. (1966). Gossip and Scandal. In J. Jennings & E. Adamson Hoebel (Eds.), *Readings in anthropology* (pp. 332–338). New York, NY: McGraw-Hill Book Co.

Goffman, E. (1956). The nature of deference and demeanor. *American Anthropologist, 58*, 473–503.

Goffman, E. (1959). *The presentation of self in everyday life*. New York: Anchor Books.

Goffman, E. (1967). *Interaction ritual*. New York: Anchor/Doubleday.

Hafner-Burton, E. M., Kahler, M., & Montgomery, A. H. (2009). Network analysis for international relations. *International Organization, 63*(03), 559–592.

Hara, N., Pnina, S., & Khe Foon, H. (2010). Cross-cultural analysis of the Wikipedia community. *Journal of the American Society for Information Science and Technology, 61*(10), 2097–2108.

Harris, M. B. (1992). When courtesy fails: Gender roles and polite behaviors. *Journal of Applied Social Psychology, 22*(18), 1399–1416.

Ho, P-t. (1964). *The ladder of success in imperial China*. New York: Wiley.

Huang, X. (2011). The formation of modern concepts of 'civilization' and 'culture' and their application during the late Qing and early Republican times. *Journal of Modern Chinese History, 5*(1), 1–26.

Huizinga, J. (1955). *Homo Ludens*. Boston: Beacon Press.

Huxley, J. S. (1955). Guest editorial: Evolution, cultural and biological. *Yearbook of Anthropology*, 2–25.

Imai, K. (1987). The corporate network in Japan. *Japan Economic Studies, 16*(2), 3–37.

Jelasity, M., Montresor, A., & Babaoglu, O. (2005). Gossip-based aggregation in large dynamic networks. *ACM Transactions on Computer Systems (TOCS), 23*(3), 219–252.

Kardashev, N. S. (1964). Transmission of information by extraterrestrial civilizations. *Soviet Astronomy, 8*(2), 217–221.

Keen, A. (2007). *The Cult of the Amateur: How today's Internet is killing our culture*. New York: Doubleday/Currency.

Kent, A. (2011). *Too damned quiet?* F—arXiv:1104.0624

Keppie, L. (1998). *The making of the Roman army: From republic to empire*. London: Routledge.

Kuper, A. (1973). *Anthropologists and anthropology: The British School*. Harmondsworth: Penguin.

Langston, M. C. (2013). The accidental altruist: Inferring altruism from an extraterrestrial signal. In D. A. Vakoch (Ed.), *Extraterrestrial altruism: Evolution and ethics in the cosmos (the Frontiers collection)* (pp. 131–140). Berlin: Springer.

Learned, J. G., Kudritzki, R.-P., et al. (2008). *The Cepheid galactic internet*. F—arXiv:0809.0339

Le Bon, G. (1897). *The crowd: A study of the popular mind*. London: Fischer.

Lee, R. B. (1979). *The !Kung San: Men, women and work in a foraging society*. Cambridge, UK: Cambridge University Press.

Loeb, A., & Zaldarriaga, M. (2007). Eavesdropping on radio broadcasts from galactic civilizations with upcoming observatories for redshifted 21 Cm radiation. *Journal of Cosmology and Astroparticle Physics, 2007*, 020.

Macaloon, J., & M. Csikszentmihalyi. (1983). Deep play and the flow experience in rock climbing. In J. Harris & R. Park (Eds.), *Play, games and sports in cultural contexts* (pp. 361–384). Champaign Ill: Human Kinetics Publishers.

Maslow, A. H. (1987). *Motivation and personality* (3rd ed.). New York, NY: Longmans.

McConnell, B. (2001). *Beyond contact: A guide to SETI and communicating with alien civilizations*. O'Reilly Media, Inc.

McIllwain, J. S. (1999). Organized crime: A social network approach. *Crime, Law and Social Change, 32*(4), 301–323.

Melin, M. M., Gartner, S. S., & Bercovitch, J. (2013). Fear of rejection: The puzzle of unaccepted mediation offers in international conflict. *Conflict Management and Peace Science, 30*(4), 354–368.

Miner, H. (1972). Body ritual among the Nacirema. In J. Jennings & E. Adamson Hoebel (Eds.), *Readings in anthropology* (3rd ed., pp. 282–285). New York, NY: McGraw Hill.

Morris, D. (1982). *The pocket guide to manwatching*. London: Triad/Granada.

Murphy, M., & Levy, M. (2006). Politeness in intercultural email communication. *Journal of Intercultural Communication, 12*, 1–11.

Nemoto, K., & Gloor, P. A. (2011). Analyzing cultural differences in collaborative innovation networks by analyzing editing behavior in different-language Wikipedias. *Procedia—Social and Behavioral Sciences, 26*, 180–190.

New Scientist. (2009). Editorial: Pollution sends an interstellar message. *The New Scientist,* 204 (2730).

Oliver, B. M. (1986). Technological civilizations. *Journal of the British Interplanetary Society, 39* (7), 327.

Parker, S. T., & McKinney, M. L. (2012). *Origins of intelligence: The evolution of cognitive development in monkeys, apes, and humans*. Baltimore, MD: Johns Hopkins University Press.

Pearce, D. J. G, Millera, A. M., et al. (2014). Role of projection in the control of bird flocks. *Proceedings of the National Academy of Sciences, 111*(29), 10422–10426.

Reagle, J. M. (2010). *Good faith collaboration: The culture of Wikipedia*. Cambridge, MA: MIT Press.

Radcliffe-Brown, A. R. (1985). *Structure and function in primitive society*. New York: Free Press.

Sagan, C., & Shklovskii, I. S. (1966). *Intelligent life in the universe*. New York, NY: Dell Books.

Sahlins, M. D. (1968). The social life of monkeys, apes and primitive men. In M. H. Fried (Ed.), *Readings in Anthropology* (pp. 263–276). New York, NY: Thomas Crowell.

Smith, J. M. (1974). The theory of games and the evolution of animal conflicts. *Journal of Theoretical Biology, 47*, 209–221.

Steward, J. H. (1968). The concept and method of cultural ecology. M. H. Fried (Ed.), *Readings in anthropology* (pp. 159–171). New York, NY: Thomas Crowell.

Tough, A. (1986). What role will extraterrestrials play in humanity's future? *Journal of the British Interplanetary Society, 39*, 491–498.

Turchin, A., & Ćirković, M. M. (2008). *Structure of the global catastrophe. Risks of human extinction in the XXI century*. Moskow: Transhumanist Society.

Vaidya, P. G. (2007). *Cosmic contact censorship.* arXiv:physics/0702172v1 physics/0702172.

Waldvogel, J. (2007). Greetings and closings in workplace email. *Journal of Computer-Mediated Communication, 12*(2), 456–477.

Watson, P. J. (1995). Archaeology, anthropology, and the culture concept. *American Anthropologist, 97*(4), 683–694.

Webb, S. (2002). *If the universe is teeming with aliens—Where is everybody? Fifty solutions to the Fermi paradox and the problem of extraterrestrial life* (p. 2010). Berlin: Springer.

Xu, J. J., & Chen, H. (2005). Crimenet explorer: A framework for criminal network knowledge discovery. *ACM Transactions on Information Systems (TOIS), 23*(2), 201–226.

Chapter 8
The Cultural Parameters of ETI

Human beings function within a complicated relationship of rules of behavior that we call *culture*. The term is a sort of shorthand which means a number of different things. And, importantly, where it comes to culture, we have only one data point: ourselves.[1] Nevertheless, we can make a weak hypothesis that ETI posses culture (or something analogous). As a result, we need to understand, in principle, what this concept means: those aspects of collective living which are learned from others in the society the individual is embedded in.

It was the German philosopher Herder who first noted that we use the term ‘culture’ in somewhat different ways (Barnard 1969; Berlin 1976). On the one hand, culture is used as a term for the aesthetic products of a set of people. Thus, ‘high culture’—classical music, painting, sculpture, the theater, etc.—contrasting with ‘low’ or ‘pop culture’: the culture of the masses, rock n’ roll music, and comic books. On the other hand, a culture is used as the collection of learned behaviors that a set of people exhibit and practice. The first use is probably more common, among the self-styled elite and is also the basis for an entire sociological discipline “Culture Studies” (Crane 1994). The second use is more common in anthropology. Like the term ‘intelligence’, this is a contentious term. So much so that two early anthropologists devoted an entire book to the definitions of the term (Kroeber and Kluckhorn 1963). Roughly speaking, and for the purpose of this book, culture will refer to the learned, non-instinctive way of doing things that a group of ILFs use. Crucially, culture is *learned* by contact with, and intentional socialization by, other members of the group. Culture lends itself to doing things in new and novel ways, and is transmitted by some means, directly or indirectly, to other members of the group. Among humans, any given culture tends to covary with marriage and

[1]It is true that zoologists often speak of different cultures among other animal species, notably orca (killer whales; *Orcanis orca*) and chimpanzees (*Pan troglodytes* and *Pan paniscus*). However, since there is little indication that either these species, or others, use symbolization, which we have seen is a central concept in the kind of culture we are interested in, we can afford to ignore these examples as outliers.

M. Ashkenazi, *What We Know About Extraterrestrial Intelligence*,
Space and Society, DOI 10.1007/978-3-319-44456-7_8

kinship patterns (cf. Kroeber and Parsons 1970 and 'socio-cultural' above), though this covariance is not absolute.

The label "culture" is a linguistic element in a domain that practically everyone is familiar with. It includes behaviors, estimates of the quality and morality of behaviors, learned things, social practices, and so on. What "a" culture is, however, is something very different. Traditionally in human history, cultures were relatively isolated from one another. The term "relatively isolated" require explication however. Almost all cultures known to us came into *some* contact with other cultures. Isolated Australian tribes nonetheless possessed material artifacts originating hundreds of kilometers away. Politically isolated kingdoms like Bhutan, Japan, and Korea nevertheless received occasional strangers. What we mean by isolated, therefore, is usually that strangers did not have a major political or economic impact on the society concerned, and secondarily, that the impact of imported ideas and genes was vanishingly small.

When European anthropologists first started conceiving of and using the idea of "culture" they came into contact with societies which were indeed relatively isolated. It became useful, and indeed possible, to formulate a term called "culture" which objectified and summarized the things done in those societies. This was possible because in many (not all, and anthropologists were often great sinners in that they ignored examples to the contrary) such societies, it seemed possible to identify boundaries between one "culture" and the next. Those boundaries were permeable, more often than not, with people, goods, and ideas exchanged across boundaries, which existed if at all, primarily in the anthropologists' minds. However, because of these boundaries, which were useful heuristic devices, the idea of "a culture" could flourish.

The modern era brought about a major change. One aspect of modernization, perhaps one of its crucial aspects, is a change in modes of communication. Communication need not be physical (we don't have to meet a person) it is also stored, manipulated, manufactured, in diverse, flexible, and often overwhelming ways. Such a communicative flood obviously carries culture, or elements of cultures along with it. Indeed, when moving from one country to another in the modern world, many cultural factor are, or appear to be the same: drink signs, clothing, fast-foods. One might assume that this is a process of homogenization of the human species. However, there are strong signs that no matter the superficial resemblance between cultures in a modern era, fundamental differences remain *and* new cultures emerge as a consequence of the process of modernization.

Cultures, we must remember are *impositions*. Participation in a culture is sometimes forced, unwilling, and people participate in them for lack of choice, as much as because they want to. To take an extreme case, people of a particular sexual preference *must* in most societies, adhere to social and cultural norms of the majority. Given freedom of choice and the ability to express it, they may perhaps create a culture of their own choosing, emphasizing the values of their sexual choice: abstinence, homosexuality, heterosexuality, pachydermophilly, …. Modern communications allow for the creation of these atomized cultures because they allow for better communication, which means (a) that like-minded co-culturists can get together *and* anyone can move away from a disliked culture, and (b) that such cultures can isolate themselves while *still* enjoying the benefits of participation in the economic and political processes that constitute modernity.

There are some strong implications in this discussion of culture for the question of ETI. granting that any ETI we hear from through SETI is going to be technologically more advanced, it is also likely to be more modern. Accordingly, the cultural alternative offered by the contact with the human species is going to be a drop in the bucket. If we assume that ETI cultures are 'globalized', that is shared broadly through various communicative media, we might also find that the introduction of items from human culture will become consumables within a globalized (or more accurately, solarized) culture, or at worst, simply be added to the roster of potential cultural oddities with an expression of "so what?" by the ETI culture concerned.

Rules of Cultural Variation When discussing ETI it is necessary to attempt to derive a paradigm that not only considers differences deriving from chemistry, physiology and elementary thought patterns, but one that can deal with differences in social and cultural orientation.

Three general approaches to the study of culture are possible here. One is exemplified by Leslie White, who felt that the rules of culture were totally independent of the biological nature of man. That is, while biology in the broad sense structured the physiological and mental limitations and characteristics of the species *Homo sapiens*, the rules by which the species acted *collectively* were independent of biological nature. "Culture exerts a powerful and overriding influence upon the biological organisms of *Homo sapiens*, submerging the neurological, anatomical, sensory, glandular, muscular, etc., differences among them to the point of insignificance." And so it follows that culture "... changes and develops in accordance with laws of its own, not in obedience to man's desire or will" (White 1959). If this assertion is correct, the social sciences and humanities are independent of biology and the natural sciences, and must chart their own course. This has implications for xenology too. *If* culture is indeed independent of nature, then it may have universal rules of its own. Therefore, we may assume that these have truly universal validity, and that we may use them to understand ETI cultures as well as our own.

A view that is diametrically opposed to cultural autonomy is the view generally known as sociobiology. Exemplified and originally codified by Edmund Wilson, the sociobiologists feel that the origin of culture is, at its extreme, an expression of genetic imperatives, though Wilson seems to have retreated to this position from his original, more extreme formulations (Lumsden and Wilson 1983). Behaviors are explained almost purely by the generative impact of genetic rules. But it seems no more likely that the human mind is a slave of culture than that it is a slave of the genes. There is a powerful tendency for the brain to evolve into a perpetually growing system that combines cultural innovation with genetic influence. In the end, when any intelligent DNA-based species such as *Homo sapiens* emerges, the individual mind must be able to reflect upon problems and make choices, but its growth and development are biologically programmed to take certain directions in preference to others. Genes and culture are held together by an elastic but unbreakable leash. As culture surges forward by means of innovation and the introduction of new ideas and artifacts from the outside, it is constrained and directed to some extent by the genes. At the same time, the pressure of cultural innovation affects the survival of the genes and ultimately alters the strength and torque of the genetic leash. Against this conceptual background we can try to examine the permutations of ETI culture.

A third view, somewhat intermediate between the two previous ones, and one that is roughly adopted pro-tem here, is what is known in its various permutations as 'cultural ecology' (Steward 1973) or 'cultural materialism' (Harris 1979). This view suggests that cultural rules must be firmly based on nomothetic principles, that is, on the idea that material conditions do much to limit and control the ability of cultures to evolve and to regularize themselves. But intelligent choice, transmission of information between generations and between individuals, and chance decisions play an important part as well. In more or less coherent format this has been the position of many sociologists and anthropologists. The advantage of this approach is that 'material conditions' themselves are mutable and self-reflexive: intelligent beings are embedded in three ecologies: the natural, the social, and the technological. Ecologies, by their nature are active, not static, and, when talking about three ecologies functioning simultaneously, the end result becomes highly unpredictable and dynamic.

8.1 Culture and the Ecologies

It is useful to think of human culture as existing within the area bounded by the intersection of three ecologies. An ecology is a dynamic system in which objects at different resolutions (small, grain structures to large complex ones) interact, bringing about both apparent homeostasis, and real change.

The natural ecology is familiar to us all, at least by reputation. It is a dynamic system in which life is a major element supplying both dynamism and homeostasis. The homeostasis is more apparent than real—a in many ways—since each 'life' (organism, group, and species) is effectively pulling in a different direction. Two other ecologies need to be acknowledged too. Ecological rules apply to *social* behaviors as well. Just as any individual exists in, is acted upon, and acts upon its natural surroundings, every individual is also enmeshed in, is acted upon, and has effects upon its social surroundings. For all of our characterization of human exchanges as a 'network', the network analogy is weak in also representing the constant flux and change, that is the hallmark of a living ecology. For some occasions, it is easier to conceive of these relationships as a fixed set of nodes and edges. In reality, these are never fixed as a network map would suggest. Finally, we also have a *technical* ecology. Ideas (memes) and artifacts intersect with one another, creating whirlpools of interaction, dynamism and constraints, that can dart off into new directions without anyone meaning to direct them there, what is often called ‘emergent properties’.

Ecologies have a pervasive and subtle effect on the ways in which a culture can develop. They limit the development possible to a culture, though they do not necessarily encourage development in any particular way. Rice cultures, which emerged in Asia in prehistoric times, quickly learned that wet rice (i.e. rice grown in an artificial swamp) is far more productive than dry rice. Artificial swamps require great sophistication in water control, and thus require both stability and extensive organization. The history of most Asian wet-rice civilizations (e.g. particularly China) has been an attempt to organize as many people in as enduring a form as possible (Wittfogel 1956). Highly reliant on co-operative labor for rice cultivation, Sinitic societies developed principles of social behavior, which are still evident today in their modern societies. All Sinitic cultures have dealt *technically* with wet rice in similar ways, because the natural and technological ecologies provided strict limits on how one could deal with growing wetland rice. Nonetheless, notwithstanding cultural exchanges, their social ecologies constantly differed, and so the societies of China (and to some degree, it’s provinces), Korea, Japan, and Vietnam differ quite considerably. The same can be said about Pre-Colombian Meso-American cultures: similar technologies and natural ecologies, but different cultures overall. Culture, in these cases and others, has been the outcome of a complex interplay between particular ecological factors, organizational demands that derive from them, and technical ecologies of materials and tools: within the boundaries of impossible-to-do because nature or available materials were not available, many different choices can and are made.

Once these limitations are overcome—because some bright spark figures out a new way of doing things, or because some natural (in the form of a calamity, for example) or human (in the form of a new group encountered) change occurs—new boundaries are set, and a whole new set of possibilities occurs. Thus groups of hunters and gatherers are small because of ecological considerations (Steward 1973); once they figure out or learn agriculture from others, they can become larger, or they may simply ignore the new possibilities. Cultural development is neither a necessity, nor a natural progression, and there is no automatic pathway, as was once believed, from savage to barbarian to civilized.

8.1.1 The Effects of Distance and Demography

Demography is a critical variable for cultures. Absolute number are less important than are things such as the physical distribution (dispersed or clumped) of individuals in it, and their characteristics. To take but one example, Japanese society in the pre-feudal period (seventh to tenth centuries AD) consisted of two very distinct groups: the aristocracy, numbering a few tens of thousands living almost exclusively in the capital of Heian, and almost everyone else. The lifestyles, access to power, and worlds of these two populations were completely different. Among other things, this chasm (one court lady wondered quite seriously if the peasants were human at all) led, eventually, to the rise of a warrior class, the *bushi*, who not only deposed the aristocrats as the political masters of the country, but managed to forge social institutions which held their society together very well for centuries, at least on the vertical axis. It took several attempts, and the injection of new ideas from the West, to forge horizontal integration and the Japanese nation.

The effects of demography, whatever they may be, must be accepted as a purely human phenomenon, until such time as we can identify, if at all, universal features. For example, it has been seen in man *and some other animals* such as rats, that overcrowding (even when plentiful food is supplied) leads to aggressiveness and pathological behavior. What has to be kept in mind is that this is far less true of other animals such as terns and mole rats who actually enjoy great crowding.

8.1.2 The Effects of Environment

Distances between population centers can and do affect the opportunities a culture has to develop through interaction with other cultures. In practice, the greater the physical distance between these cultures, the more difficult, and thus infrequent the contact between them. As a result, such centers are almost bound to develop in different ways. In effect, they will cohere into different cultures relatively quickly, even if they originate from the same base.

The issue is of course complicated by a further variable: the means and modes of communication between those centers. Absolute distance is not as crucial as the degree in which these centers are integrated into a larger unit— a civilization— which has some commonalities. The distance between Los Angeles and New York is physically greater than between ancient Rome and any one of her colonies, and yet there seems to be greater integration between the two America cities than between Rome and Alexandria in the first century.

The availability of land is of course one major environmental factor. The availability of free land does not constrain a culture to develop in any particular direction. Cultures in which land is scarce usually force the development of institutions able to deal with this shortage, for good or ill. These institutions might be judicial (ultimately backed by the threat or actual employment of force), religious or social, backed by more subtle pressures. In order to understand what may be the issues, it is necessary to identify the resources in question.

We can also expect a technologically developed culture to exploit all available ecological niches, including outer space. This implies the likelihood of permanent space colonies, with virtually all their ecologies being artificial. There are likely to be great differences between Earth-based and space-based cultures: after all, they exist in different ecological circumstances, and the distance to their 'home' culture and its pervasive effects are enormous. What the effects are is difficult to predict, but assumptions that the social and cultural forms that will emerge will be copies of the source culture do not seem tenable (Ashkenazi 1992; Bertka 2009; Billings 2006;

Turner 1985; Zubrin 2000). This has an important implication for contact with ETI: whomever we contact, might be not a single society, but one of several. Given the likelihood that developed and advanced ETI will move many operations to space, and thus develop multiple expressions of culture, who will we be dealing with?

Resources For the purposes of this discussion, a *resource* is some good that is in limited supply, and that can be controlled by some actor. For us (the readers and myself) air is not a 'resource' because it is neither in short supply (quality of air notwithstanding) nor is it controlled by anyone. It would be different in a space colony, where air is both in limited supply, and controlled by some authority which stores, allocates, and perhaps produces air. Sunlight is another example. In cities with multi-storey buildings, sunlight is a resource which is allocated based on legal regulation in many cities. In near Earth orbit, sunlight is effectively free, given the size of the available space for expansion and spread of solar collectors. The collectors themselves are a resource, and, for example, are likely to be a primary target in case of conflict: deprive someone of solar panels in space, and sunlight energy is effectively a resource in short supply.

This implies that conflict among developed (presumably not advanced) ETI societies is potentially possible. Whether this descends into violence is another matter: conceivably some form of conflict reduction and management will develop as well. However from a purely theoretical viewpoint we should *not* assume that ETI are necessarily peaceful *even among themselves*. Indeed, we should at least be ready to acknowledge that for a given ETI actor in a contact situation, the contact itself is likely to be a resource, and therefore be sequestered. In other words, if we are dealing with A, A might make a point of ensuring that whatever benefits we provide will accrue to A, and other potential partners for contact *of the same or different species and cultures* are excluded from the situation.

Nature Nature is the concept we tend to think of when we use the word 'ecology'. In the nature that surrounds and permeates us there is constant patterned, sometimes repeated interaction from the chemical level all the way to the atmosphere and entire planet. These interactions are at one and the same time extremely robust and paradoxically very delicate (as we are learning to our cost).

Species of animals in an ecology have restrictions on their ability to expand demographically. These limits are primarily the availability of resources: food and water. Whether or not there are predators to control growth, or parasites (other animals, bacteria, viruses and so on) are also factors in species expansion. As a general rule, predators are limited to the availability of easily acquired prey, so predators tend to be fewer in any environment than herbivores the number of prey (and their reproductive capacities) are a limiting resource.

For an intelligent species, the natural environment has restrictions as well. Are there extractable resources such as metals to make complex durable tools? Are there resources to make fire? If not, cooking, which may be one cause of greater intelligence (among humans, at least), may not be possible, or may be restricted, and certainly will mean a huge increase in labour (as it is today in many developing countries). The types of resources available will also mean other resources will be prized, trade or warfare for these may emerge, and again, changes will occur in the natural ecology, as say, domesticated animals protected from predators by their herders will displace naturally occurring animals.

Planetary ecologies are present at all scales, so we need also to consider the total natural ecology of a planet when discussing the natural ecology with which intelligent species interact. The complexity of the natural ecology is so great, that the Gaia hypothesis suggests an almost mystic connection between a planetary ecology and its intelligent denizens (Gardner 2003; Lovelock and Margulis 1974; Margulis and Lovelock 1974).[2] Whether the Gaia hypothesis in either of its two forms is demonstrated or not, the complexity of planetary ecologies is beyond doubt; and the cultural necessity of intelligent species interacting with the limits the ecology places on their choices likewise.

At the very least, the planetary ecology sets the tone for any intelligent species phylogeny. Did they evolve from predators? Prey? Omnivores? How were manipulatory organs, vision, speech, affected by their surroundings, before they could modify them artificially? All of these issues are a consequence of the natural surroundings they emerged from.

Society Human beings, embedded in the smallest of groups or the largest, are constantly acting upon and being acted upon. Very few actions are not a derivative of other people, just as very few actions do not impact on others, even if removed in time. In other words, 'society' constitutes an ecology, which like the natural ecology, can be viewed in the micro—as role behaviors or face-to-face behaviors—or as a planetary whole: the entire human race.

While we may take the natural ecology for granted, the social ecology is in our face. Mess up, and we soon get negative feedback from other units in the ecology: friends, neighbors kin, the judicial system, the media. Whatever we do, we do as units embedded in this ecology. Moreover, we have rules by which these social relations are carried out. The rules are not absolute: they too are part of the social ecology. They are modified by individuals who, in their efforts to succeed in their goals take shortcuts, game the rules and those who use them, and, by repeatedly doing so, modify the rules on the fly. As Moore and Myerhoff (1977) argue, as norms are codified into law, individuals violate some of the norms on a regular basis (either because they wish to advance their interests or because the environment has changed) until eventually the codified rules fall into disuse and disappear.

While one can hypothesize a race so predictable and orderly that no social change occurs, such an ETI race is unlikely to be able to lift itself into any kind of technical level. The ability to change *socially* is a prerequisite for the adoption of new sciences and technologies that are at the basis of any technological culture. Of course technical change brings about social change, just as social change is fertile ground for technical change. So we should expect any technically advanced culture we come into contact with, to also be socially dynamic, probably (though not necessarily: after

[2] I am aware I am doing some injustice to the enormously complex Gaia hypothesis, but its proponents will have to excuse me.

all, their communication processes may be better than human ones) with a great deal of cultural variety and blended sub-cultures. In the case of a star-spanning culture, where communication would be particularly slow, we would expect manifold pockets of different social practices—that is, different cultures or sub-cultures—that will emerge at the different settlements points around different stars.

Intellect Science is just one of many intellectual pursuits that human beings as a culture are actively engaged in. Intellectual pursuits that demand a formal set of logical processes, formal statements and transformations have been used by human civilizations for millennia: theology, philosophy, logic, music, and mathematics have antecedents that go back three thousand and more years, in different human cultures, tackling different topics.

A relative newcomer—one that did not even have a name until a couple of hundred years ago—is the intellectual endeavor called 'science'. What differentiates science from other forms of intellectual endeavor, both those relying on logic and others (think of poetry, painting, chess), is a formal feedback loop of hypothesis<—>empirical testing. That is, Mrs. Ogg, tired of having to run up the volcano to light a twig to make tea every time Mr. Ogg forgets to keep the fire going, reasons after some years of cogitation that fire = warmth and rubbing sticks = warmth and that therefore, rubbing sticks = fire. After some experimentation, Mrs. Ogg discovers that rubbing a hard stick with sufficient velocity against a softer one will indeed cause fire to come into being. She can now *predict* that rubbing sticks together under specific conditions, will create fire. If she manages to record her findings and process, so that others will learn of her discovery, she has invented a process of thought: science, that is transmittable to others.

In the real world, the feedback loop, as well as precise recording of the results, and of the intellectual reasoning that leads to the results, took some centuries to get established. In the final word, though, all science that is recorded is no different from Mrs. Ogg's improvement. Mrs. Ogg's scientific progress, also has social and natural implications. Humans could now burn down forests, without waiting for lightning to start the fire for them. They could forage at night, able to light fires when convenient, instead of wandering about with a pot of coals all day.

Intellectual procedures, notably those that are used to provide material effects have widespread effects on the natural and social ecologies. Large human concentrations—cities and towns—would not be possible without strong, fundamental, scientific changes that brought about roads, motorization, building techniques, and many other technical devices upon which large concentrations of people who do not produce their own food, clothing, or shelter, rely.

The technical environment we live in, and have lived in since before recorded history, mutates and changes as well. The rate of change may be slower or faster. We tend to believe that in the world of the twenty-first century, technical change occurs faster than in earlier historical periods, and that belief may be correct. Nonetheless, the technical environment, whether rapidly changing or not, is fundamental to our culture as humans. Moreover, different technical environments have existed side by side for millennia—cultures based on working stone, iron, and non-muscle energy were all common throughout the nineteenth and twentieth centuries, and affected the ways in which people that supported this environment could live. Different technologies—in communication, medicine, food collection, mobility and many more—mean that different cultures evolved and functioned, some changing more slowly than others, and the more effective ones have been, and are, subsuming the less effective ones: ecologies are always Darwinian, and rightly or wrongly those that cannot compete go to the wall.

This too has implications for our understanding of ETI. For convenience, we can divide ETI cultures into those that cannot, and those that can build large-scale radio telescopes. The first category is of interest because in our current discussion, which ignores the realities of the Einsteinian universe to a degree, we are interested in all ETI. However, besides (possibly) being able to detect their presence in the future, they are unlikely to be communication partners. The second category is more interesting. To build a large radio-telescope an ETI race must have developed, probably through Darwinian means, intellectual processes under which ineffective technologies and scientific discoveries were superseded by more effective ones. While it is barely conceivable that this was a purely hit-or-miss process, it is far more likely that some system of organized intellectual scientific reasoning and recording was a driving force. Of course this process might have been longer or shorter than humanity's experience (let us say, 3000 years of unconscious trial and error, with the last 400 years to develop a shortest-path process we call 'science'). And, possibly, some other system will supersede science inheriting science's mantle. An ETI race may have a culture that supports clear thinking and communication, and perhaps the process took much less. Or, to the contrary, they may have been more scatterbrained or more deliberate, and the process of developing the idea of science and the technology of radio telescopes might have taken longer. Whatever the case, *some form of scientific thinking, and the development of tools and instruments to test that thinking* (ranging from microscopes to macroscopes, as well as the technologies to manufacture them) *would have come about.*

8.1.3 Intellectual Effects

It was noted earlier that the realm of intellectual pursuits includes a variety of endeavors that intelligent beings (however defined) engage in. Certainly among humans, there are a variety of intellectual pursuits that seem to provide sometimes contradictory effects. We need to keep two things in mind from our perspective. The first is that we know of all these pursuits only from the artifacts or sociofacts that they produce. The artifacts may be second-hand recordings of first hand thoughts, or they may be direct expressions of the intellectual process behind them: the plans for Gaudi's Sagrada Familia cathedral in Barcelona, the cathedral itself, and Gaudi's dreams of the cathedral represent this kind of effect. The second is that different intellectual pursuits may use quite different sets of formal systems.

Consider, for instance, the difference between science and mathematics. Both are intellectual pursuits, and both rely on one another, and are associated deeply together. Yet there is an inherent and deep difference between them: science is fully dependent on physical demonstration of its propositions. Such demonstration can take years between conception and demonstration, as was the case with the theory of relativity. Mathematics, on the other hand, requires proofs of its propositions *which cannot be physical, but must be logical.* A small, but nevertheless, crucial difference.

In order to avoid a too-lengthy discussion, we can examine some examples of intellectual processes, with a view to understanding whether these are purely human, or perhaps universal avocations.

Science: The realm of reason We have already discussed the nature of science: an intellectually rigorous, spiral process of hypothesis and empirical testing. The value of a scientific theory, to paraphrase Robert Heinlein, is its ability to predict. Where a scientific theory—the most famous case is the Ptolemaic theory—is unable to predict correctly, it is discarded (in this case, replaced by the Copernican theory). We reach the conclusion by a process of trial and error, or sometimes direct reasoning, but it always must be checked against empirical reality. Is this the only way to go about it? I admit I have no alternatives to offer. This does not mean, however, that ETI also do not have alternatives. This could be a physiological advantage (supreme rationality, perhaps?) or some elaborate process. On the one hand, if we can communicate with an ETI, they *must* have some form of teasing out the regularities of the universe. On the other, they may have other ways of doing so than we do.

There is another aspect of science that needs to be kept in mind. The substrate of science is a social form, an institution, which engages in science. This includes scholars, experimenters, administrators, governments, universities, laboratories and so on. In other words, there is a social ecology of science. This institution is regularized, and many of its components are of some antiquity. And not all of it engages in science. It also has distinct human tinges of emotion, jealousy, competition, suppressing heterodoxy and upholding orthodoxy: human behaviors we are familiar with from other institutions.

We cannot assume that ETI have evolved the same institutional substrate, even if they engage broadly in 'science'. What the implications are, it is hard to tell. However, if their science institution is similar to our own, we should be prepared to face problems of seniority and political manipulation (which is the reality in the human institution). This may or may not hinder scientific exchange: scholars have an enviable reputation for being oblivious to the econo-political implications of what they are doing. Unfortunately, academics—those for whom the science institution is a good career choice, and for whom the defense of the institution (and their place in it) is the crucial issue—are not. In turn this means that scientific exchanges may well be curtailed or shaped by administrative and other considerations.

8.1.4 Ethics: Individuals and Individual Behavior

Every culture is reliant on, and woven through, a social ecology—a set of sets of relations between individuals—which is mutable, and which affects and is affected by the natural and technical environments. Take, for example, the question of friendship. In the Western world this has fuzzy implications of mutual shared interest and fun, benign acquiescence to the other's foibles, occasional support, mainly emotional and physical but occasionally financial. Above all, friendship should not be based on personal advantage or calculation. In other societies, this is not the case. In Latin American societies, where the concept of friendship is closely allied to the concept of *compadrazgo* or shared rearing of a child, the relationship is *explicitly* and in the first instance, financial. In Chinese society, where friendship is one of the five major Confucian relationships, friendships is a relationship between *unequals* one of whom assumes the position of 'elder brother' to the other, with rights of support and obedience, and duties of protection and advice. Socially advantageous relationships—*guanxi*—are a matrix in which personal advantage and reciprocity ensure social position and advances. So even among humans, this seemingly simple phrase "I come in friendship" is complicated in a bi-cultural milieu. What about friendship with a non-human? What unspoken baggage (which every protagonist usually assumes is a universal) is brought to the table?

Very clearly, we need to understand the ethics of the ETI in question. Reciprocity, for instance, is a major element in all human social relationships, negatively and positively (cf. Cunningham 1998, though he argues the Golden Rule is more than mere reciprocity). On the positive side, we have concepts of balancing, friendship, and mutual exchanges. On the negative side we have ideas of vengeance and retribution. Should we consider these elements to be universal?

One of the seeming advantages of being a social scientist discussing ethics is that we need not weigh in onto the side of our favorite ethical position: we need to look at the different ethical offerings, and see how well they fit our potential social model, in this case, of relations with beings who not only may not share our ethics, but may not even share the process by which we reached those ethics. In this section, we briefly indicate some of the relevant ethical complexities within human philosophy, and then discuss the question of whether the *process* by which these ethics have been reached—biological imperative, emotional bent, or rational consideration—are likely to be universal (meaning, in this case, not humanly universal, but applying to ETI as well).

Ethics from the human genome The initial question we need to ask about ethics is whether they are genetically based. This is effectively the position of some scientists such as Edmund O. Wilson, the father of sociobiology (Wilson 1975). Some of the basic ethics that inform the various societies of which we may be members have historical bases in ancient religious or ethical dogmas which start from a non-rational premise (theological, as in Judaism, Christianity and Islam, or cosmological as in Daoism, Confucianism, and Buddhism) from which flow ethical principles.

The derivation of ethical thinking through reasoning ethical issues out has almost as long a history as religious ethics. Starting with some of the Greek thinkers such as Plato, through giants of pre-modern philosophy, notably Kant, ethical positions have been reasoned out based on apparent unemotional premises supported by reasoning.

For our purposes here, the issue is not "Which human ethics apply to ETI (or the reverse)?" but "Whether ethics are/can be universal?" That is, whether the minimal set of ethics we can agree on (if we agree) are universal, will also be accepted by ETI. There is, of course, a subsidiary question as well: whether, *if* our ethics are not those of ETI, we can or should modify our own, or cause 'theirs' to be modified? To do so, we need to have a look at the supposed universality of human ethics; not a simple task, given that ethicists have disagreed over these issues for centuries at least. The following only picks at some of the threads of the ethics discussion, and is not intended as a full-bore discussion of ethics. The only objective in this section is to point out some of the salient problems in the ethical dimension.

Some relevant ethical issues We can start with the issue of how are universal ethics *in a multi-cultural setting* arrived at? We need to keep in mind that, like religion, virtually every culture has some form of ethics. We may not like the "All strangers are meat, and consequently on the menu" ethic, but it is ethical nonetheless. Largely when discussing universal ethics, the discussion gets round to ethics that have grown from the ethical traditions in Europe (largely a product of the Enlightenment), the Sinitic world (largely a product of classical Confucian-Daoist-Buddhist thought) and the Indic world, largely a product of Hindu-Jain-Buddhist thought). What characterizes these ethics more than others is that thought is given not only to ethical conduct of the culture/nation, but that of mankind in general.[3]

Three different models for arriving at universal agreement on ethics (keep in mind: not *which* ethics, but how to arrive at agreement about which ethics) have been suggested by Fleischacker (1999): a human rights model, a necessary conditions model, and a cultural dialog model. Any universal agreement on ethics represents a paradox in- and of-itself. *If* ethics are a democratic compromise between different opinions, then by definition they are not universal. A democratic compromise means either that each party has taken from the agreement what it can, and not what it wants, or that one party has won all, and the rest have no input.

The *human rights* model starts with the assumption that there are universal ethical truths *and* that these truths are embedded in a political process which is the result of negotiation between state parties. This is an obvious non-starter, since, in practice, 'ethical' documents such as the Universal Declaration of Human Rights are neither accepted by all, are the result of horse-trading, and though they require one, have no enforcement mechanism (not so much as a critical comment, in some cases).

---

[3]Womankind, and humankind came late to the game, and only emerged as stakeholders in the ethical game in Europe and the globalized world in the late twentieth century.

The *necessary conditions* model is, to quote Fleischacker "…to pick some universal feature of the ethical life, argue that that feature is made possible only under a particular set of conditions, and conclude from there that the particular conditions must be provided for by any ethical code" (Fleischacker 1999: 112). Unfortunately, these so-called universal conditions (e.g. 'worth', 'freedom', 'person') are too vague and culturally specific.

The *cultural dialogue* model is one in which members of different cultures discuss parallel norms in their cultures and make use of those parallels when they work together. This has four elements. First it is a lengthy, time-consuming and fuzzy process. Second, it is almost always dyadic: ethical agreement between two cultures does not presuppose the same set of parallels will occur with a third culture in the mix. Third, while at allows for the growth of a universal ethic, it would do so by partly transforming either or all of the cultures involved in the dialog. Fourth, this emergence of an agreed-upon shared ethic for the two cultures occurs as individuals in the cultures concerned interact with one another and work out agreed-upon ethical standards at the micro-level which eventually transfer, possibly in modified form, to the macro level.

Whatever ethics an ETI race has, will eventually impinge upon *human ethics*, when and if we exchange information. While we can argue for the universal model (at the ethical, not raison d'etat level), it is hugely unlikely that an ETI culture will have the same ethics. Moreover, given the biological, evolutionary, technological and cultural differences, it would be very hard to find the same approaches to essential questions, at least not sufficient, in most cases, to call for a *universal ethic*. Using the cultural dialogue model, it must be acknowledged, would at least provide a method for reaching agreed-upon ethical *standards*, on the basis of which principles of behavior and law might be established. We need to keep in mind Singer's (2011) comment that one should never ask people to do anything they consider contrary to their own best interests. Only by aligning mutual or at least parallel interests, is it possible to establish a seemingly universal ethic.

Parallel interests of intelligent living beings The ethics literature is replete with terms that are considered fundamental to any universal ethical position. Before discussing these, it is worth noting that these so-called universal themes are valid for a very limited value of 'universal' largely including the intelligentsia of European, Sinitic, and Indic cultures world wide. Certainly a Muslim ethicist would add 'subject to the whim of God as expounded in the Quran as exemplified by the doings and saying of the Prophet', as would many other religious ethicists, each to his own; which would effectively limit the 'universality' of the statements.

A complete list of all the supposed major ethical values is outside the scope of this discussion, however, some of items need discussion from a xenology perspective, inasmuch as they are contentious in interspecies relations, let alone intercultural relations.

Individual freedom From the Western viewpoint freedom—to act, to believe, to think and to be—is a major foundation of ethical thinking (Brandt and Rose 2004). This has long roots in Western thought, including the Greek philosophers and the

Hebrew Bible. In almost all cases this is hedged about by numerous 'but's, depending on the society and leanings of the particular author. These hedges may be externally imposed—by other people (in Islam), training (in Confucianism), or recognition of natural law (Daoism, and in a more modern interpretation, Gaia or nature).

Freedom to choose one's path has always been problematic as a political concept, and the idea has had more lip service paid to it by advocates such as Mao Zedong, Henry Wallace, and Stalin, than actual practice. This means we must keep two things in mind. First, as MacKenzie notes, that A uses an ethically-loaded term such as 'freedom' does not mean the meaning is the same as B's, or even that A is not lying (Mackenzie 2007). Second any freedom needs to be constrained in some way not to impinge on the freedom of other individuals and the collective. Thus we come to the Golden Rule.

Golden Rule The Golden Rule appears in many forms in different cultures, ranging from European to East Asian (Cunningham 1998).[4] In practice, the Golden Rule has two principal variations, a positive and a negative version. The positive version is best articulated, and most familiarly attributed to Jesus "Do unto others as you would have them do unto you" (Matthew 7:12). The negative "Do not do to your fellow that which is hateful to you" is attributed to a contemporary of Jesus, Hillel the Elder (Talmud, Shabbat 31a). If we follow Singer's (2011) contention that ethics should be systems of rules then for our purposes, the negative formulation is more practicable as a rule: at the very least, after all, we know what we would dislike, but the likes of a non-human intelligent life form are less-well known.

Respect Brandt and Rose (2004) and Strentz (2002) suggest that respect for others on an equal and reciprocal basis is an ethical fundamental, and universal. From our perspective this represent a serious problem. For millennia, the majority of men knew that the human race was divided into two sub-types: men and women. The latter were not recognized as fully human (that is, as equivalent to *men*) until less than one hundred years ago, and in most of the human species, that recognition is little more than lip service, if it exists at all (consider most major Christian denominations, Orthodox Judaism, Islam and Buddhism). Plato and Confucius (along with Muhammad and a host of others) as MacKenzie puts it, would not accept the principle of respect and equality of all. Each of the three would have different targets for disrespect: all non-Greeks and women (Plato), those not educated to Chinese gentleman's standards (Confucius), and non-Muslims and women (Muhammad. cf. Mackenzie 2007). Mason (2005) argues that in such a case, these ethics are flawed since those who disrespect others in their own culture/society, are violating the rule of respect, and thus it is legitimate to impose the principle of

[4]I'm not aware of any similar formulation in the literature of pre-Colombian cultures, but that may be my ignorance.

'respect others' upon them: somewhat contradictory, yet very human: respect others except where they do not respect their own internal others. For the issue of ETI, this is a crucial point: can we impose our ethics on someone whose biological, social, and cultural foundations differ fundamentally from our own? And, invoking the Golden Rule, would we accept this from others? Would we accept the demand, well bolstered by rational argument, as ethical statements should be according to some (Mackenzie 2007), that goes against our fundamental assumptions: for example, from a supremely rational ETI that we stop this nonsense of religious worship?

Accountability However ethical principles are diced and sliced, a primary value in dealing with utter strangers, such as ETI, are trust (Brandt and Rose 2004) and accountability (Strentz 2002). As we shall see later when discussing trade with ETI, trust is a cardinal ethical principle that echoes down the ages, and is crucial for any kind of exchange. The more trust projected by ego (individual, culture, or species) the more freedom that ego has in dealing well with any alter. 'Trusting' in another implies that the other is accountable for actions performed by or on its behalf. Where this accountability has both extensive history and an abstract background, trust is helped along, and with it a host of relational issues.

For some humans, accountability represents a major problem. Those coming from a Christian, Muslim, Hindu and other dogmatic religious backgrounds are, by definition, unaccountable for their actions. Muslims are bound by the will of Allah, Christians by the will of the Lord. Other religions have similar 'pass the buck' systems to a lesser or greater degree. In either case, not accepting *fully* one's own actions,[5] while it makes sense *within the confines of human civilization* could be untenable for a race that does not have our theological bent. If, as Dawkins maintains, religion is a human phylogenetic aberration (Dawkins 2009), how could we justify an uncertainty in our individual accountability because of our allegiance and submission to an entity an ETI species without that particular aberration cannot make any sense of?

"Human" rights outside the Earth The value of ethics within modern global civilization often hinges on the term 'Human Rights'. The term is attractive, since it seems to encompass the needs of all of humanity. In practice, driven largely by Western cultural norms, attempting to de-privilege national government law by harnessing national governments to the cause, and codified by a body—the UN—that is a club of nations, 'Human Rights' are less a panacea than a magical potion: statements based on hopeful wishing that anyone can twist in any direction. The essence of human rights is ring-fencing certain rights that are too important to be left without guarantee or guarantor (Chandler 2001). Human rights are premised on the vulnerability of individuals who require human rights to protect them from predators and potential violators. However, in the international human rights

[5]Though Hinduism, Buddhism, and Shintô, among major religions do insist on personal responsibility, even if, in the case of some Mahayana Buddhist schools, modified by the mercy of the deity.

regime, where any and all nations are effectively violators of one or another human rights standard, one must always ask two questions. The lesser is the old *quid custodiet ipso custodes*: who will watch the watchers? The US, a serial violator of human rights domestically (viz. Guantanamo, child executions, et al.) is at the same time at the forefront of human rights action internationally: NIMBY (Not In My Back Yard) redux. In other words, a human rights system without an arbitrator is not a system. The greater question is raised by Chandler directly: will humans be the weak who need protection (likely, if other races are predatory *and* more advanced than humanity), or the strong, who condescend to offer protection (and the dependency it entails)? In the absence of a system of interstellar law (and we have no evidence pro or con) we could be either or both.

Let us reason together: Composing Ethics for relations with ETI Fleischacker's Rationalist model (1999) suggests that what I will call 'locally universal' common ethics can be attainable. That is, given sufficient time a common ground could be found. This requires that (a) both (or all) actors are prepared to agree on a modus to develop the common ethic, and (b) as in the Jewish and Muslim legalistic practice, the ethic is an evolving one, and that (c) the modes of presentation are not straightforward instructions (*pace* Singer 2011). Ethical principles are often not a clear list, but a presentation of arguments, reflections, folk tales, jokes and whatever other modes occur, or are relevant to one or another of the cultures developing the ethic. Another lesson needs to be taken from the Talmud: it takes a long time (five centuries, in the case of the Talmud), and later generations of commentators and contributors will likely bring in issues of their own time, not that of those who contributed at the start of the process. The good news, on the other hand, is that given Einsteinian reality, no one should be in any great hurry, since opinions and responsa will take years in a best-case scenario, and realistically centuries, as electromagnetic waves percolate at C from ILF culture to ILF culture.

And finally, human rights again: Against entropy? 'Human Rights' as currently conceived of by the UN and its associated actors, do not seem to play a useful part in ET morality. They put any species into one of two categories: the powerful, and hopefully benevolent, and the powerless clients who can but hope the powerful adopt a humanitarian ethic, which is in many cases, trumped by *raison d'etat*. However, I would argue, we can point to the possibility of a universal ethic, one that could motivate ETI as well. Roughly, I argue as follows. Humanity, and all ETI we are likely to come into contact with share a common habitat: the universe. The universe is about to fail. Whether we have a gravity collapse as the universe contracts, or whether there is some steady state, eventually energy and matter are likely to be distributed evenly throughout the universe: death by entropy. This will apply to all species.

What all ILFs have in common, is an inherent desire for the preservation of life (Mautner 2009), which by it's nature is counter-entropic. We thus share an ethical parallel with all other Intelligent Life Forms (Fleischacker 1999). Life itself is insufficient. As Harrison (2013) argues, cosmic, as well as local (planetary) evolution is moving in the direction of greater order. As a species advances into space, it carries that need for order along with it, since order, organization, and complexity, greater consciousness of the universe, larger and more elaborate cultures, are what developed species are about.

Hopefully, the ethic of counter-entropy will *also* include a respect for the institutions and cultures of others, for avoiding destructive warfare, and for developing mechanisms that enhance, rather than detract, from the potential contributions of the weaker. This is by no means certain, because humans are notorious for knowing required ethical conduct yet ignoring it (and the same may well be true for ETI), and because while this ethic is obvious, it may not be adopted by all (or any) ETI. Certainly this bud of potential common understanding does not imply greater democracy, kindness or altruism, as some would have it (Arbib 1979; Ball 1973; Bracewell 1974; Deardorff 1987, among others). It does however offer a parallelism and the establishment of, at the least, a common understanding of ethics.

We should therefore ask a series of questions about the *ethical* rules in a given ETI society. First, to what degree are ethical rules *comprehensive*. Second, to what degree are they *binding*. Third, to what degree are they *fixed*. Finally, and perhaps most difficult and amorphous, what are they *based* on?

1. Ethical rules are *comprehensive* to the degree they are expected to inform every aspect of an individual's or group's activities. Without mapping these out in detail (something we might well have to do when meeting a particular ETI), the more comprehensive the rules, the more will activity be circumscribed. In humans this often expresses itself in rituals but this may well not be the case among other species.
2. Ethical rules are *binding* to the degree that individuals actually perform them in the society. Most human societies, for example, have some variation of the Golden Rule. But that does not mean that it is necessarily a major component in ordering action.
3. *Fixity of rules* refers to the degree to which rules are considered to be immutable, both within a society ("Can we change our rituals today?") and for an individual ("If today is rainy, I won't greet anyone"). This relates to the question of binding, but amplifies the concept somewhat.
4. Rules may be *based* on first order consequences (genetic and biological makeup), or on derived consequences (someone said so).

If, we can ascertain the answers to these two questions, we may be on safer ground when dealing with ETI in a moral manner. Of course, we need to keep in mind that this goes both ways, and perhaps we might need to mend our ways ….

8.1.5 Religion

Religion is a pan-human phenomenon. There is no human society which does not have one. Even in societies where stringent efforts have been made to suppress religion entirely (e.g. the Soviet Union, Albania, the People's Republic of China) religion has reemerged as soon as the limits have been reduced. This points to the fact that religion is an expression of something very powerful—biological or social —in the human species.

One problem with discussing ETI religion is the ways in which we humans react to the phenomenon of religion itself. The normal human reaction is "I have religion, you have superstition". In other words, most definitions of religion have, implicit or explicit bias towards the religious experience of the definer. This is evident even in the writings with professional expertise in cultural comparisons (e.g. Spiro 1966; Hori 1968).

To avoid the ethnocentric bias inherent in these definitions, I recommend to circumscribe the phenomenon by proposing that any religion must embody in it three classes of thought and practice: cosmology, theology, and ritual. There may be other derivative or basic characteristics as well, but the three noted here—cosmology, theology, and ritual—seem to circumscribe 99 % of phenomena we would agree on as being 'religious' and exclude some borderline case: ideologies and philosophies such as Marxism and some forms of Buddhism which are at best quasi-religious.

A *cosmology* consists of a set of statements regarding the nature of the universe. Properly speaking, a cosmology can be construed as a series of statements about the nature of the physical and metaphysical universe. Religions, most of which are founded in non-scientific eras by non-scientific people, have other cosmological aspects as well. Cosmologies in the sense used here may well include moral statements.

A *theology* is a series of statements about the nature of the godhead and about the relation of the worshipers (those who maintain the religion) with that godhead. The godhead may be stated to be a single or multiple, of any sex, age, origin or other characteristic. There may be sub-hierarchies of godheads within the theology. The theology may specify which parts of the population may have relations with which god or set of gods, and how. And, as in cosmology, there may be ethical statements included in the theology.

Those activities enjoined on the members of a church '...because god wills it' or because the cosmos (however conceived of) requires them, are *rituals*. Sets of rituals may be comprehensive encompassing in detail every aspect of life (e.g. in Judaism, Hinduism, and Tirthankara). Others may be far more generalized, even vague and open to multiple interpretations.

The degree to which religion actually unifies categories of individuals has often been debated. However, at the very least, religion allows people of otherwise different backgrounds (cultural, class, caste, region) to act *as if* they are a single group. This is important in a number of ways. Travelers may cooperate with their co-religionists, there may be access to otherwise inaccessible special circumstances, and people may be provided with services on the basis of religion that they might not have access to otherwise.

Religions also impose particular social forms of organization. Sometimes, these are implicit in the nature of the religion itself. The Catholic church is organized in parishes that often serve as the basis for collective feeling or expression. Shintô shrines serve as foci for neighborhoods and their activities (Ashkenazi 1993; Bestor 1989). Muslim Sufi groups unify and exemplify collective feeling (Gilsenan 1973).

It has also been suggested by Rudd (1985) and by Dawkins (2009) that religion may be universal for *biological* reasons. In fact, if we read with a critical eye, Durkheim (1947) and Levi-Strauss (1979) are in effect suggesting that religion is biologically necessary: a step in the development and creation of intelligence as well as of groups. Modern sociobiological theory (e.g. Diamond 1992) also implies that religion is a result of the biological need for group support and unification, as well as a derivative of the dominance of males in hominid troops. Dawkins (2009) goes further, arguing that religion is merely an expression of a genetic biological structure and is, therefore, a specifically human phenomenon.

On the other hand, the formation of groups may well be a prerequisite for the organization of *any* technological society (see above). It may well be that therefore that religion is a natural, and almost automatic extension of the formation of these groups (whether we accept Durkheim's 'group spirit' idea or not). If this is the case (and, on the face of it, it seems reasonable), then we might find that ETI, whatever their technological status, will exhibit features that we associate with the phenomenon of religion.

The common conceit that rationality means irreligiosity is falsified both by empirical studies (Stark and Glock 1969) and by the many pronouncements indicating religious belief by highly 'rational' scientists. True, education correlates negatively with membership in, and participation in, organized religion, but that is a far cry from the question of support of religion in general, not to say personal religious belief.

We may say that among humans, at least, 'rationality' (measured in terms of scientific education and practice) correlates negatively with some aspects of religion such as participation in ritual or church organization and membership, but there is little evidence to suggest that it correlates negatively to the same degree with the abandonment of religious thinking or belief. The same may well be true of ETI.

Ritual and ceremony Religions embody a physical element—ritual—which is integral to all human religious practice. Rituals imply the translation of ideological elements (theology and cosmology) into material expression. This expression involves physical objects, specified actions, and so on. Rituals are extremely formal (Irvine 1979), canonical (Rappaport 1974) events. They are public, and emphasize by their solemnity and demeanor, the 'truth' of the propositions embodied in them. Ritual can be identified roughly as behavior that is non-object oriented. By adopting this definition we go back to early definitions of ritual (e.g. Benedict 1937) rather than more sophisticated ones proposed since (e.g. Geertz 1966; Segal 1983; Turner 1969). This is because, if we try to deal with more than the one species known to us well, we are forced to abandon greater specificity and narrowness in favor of greater generality. Since we are dealing with rituals of intelligent species, not only of mankind, this is not a major obstacle.

However, the religious aspects of ritual may well be peculiarities of the human species. Other intelligent species may not need to make truth statements about their beliefs in that fashion. Consider for example languages as varied as Japanese and Khoisan. In those languages, the degree of reliability of a statement is impressed, grammatically, in the form of a sentence. Thus "There is a house there: I saw it." "There is a house there: I was told so." "There is a house there: I imagine it to be so." "There is a house there: I am in it." are distinct single sentences, not compounds as in English. In Japanese, these are expressed by simply adding suffixes to a sentence. Such truth statements could, presumably, take the place of canonical representations of truth that are imbued in ritual.

The preceding also implies that when we do encounter ETI, we will have to evolve an appropriate definition for *their* variety of ritual, just as such definitions have been worked out from the specific context of *human* psychology and behavior. "Proper forks, orders of precedence, ceremonies of confirmation, all pageantry, any of the formal games we play are by any objective standard ludicrous. And since their normal application is to separate those who know and those who are from those who don't know and those who could never be, they are all the more foolish" (Panshin, *Star Well*). In other words, any ritual taken out of context represents a waste of resources, and an apparent senseless display. Knowing the full context allows for a better, more nuanced understanding.

8.1.6 The Arts

There is another intellectual activity that humans have engaged in for a very long time, in fact, probably since the species *Homo* emerged: art. Art is a peculiar intellectual activity if we think about it rationally. It has very little by way of instrumental solutions (except, perhaps via ritual), it can be resource expensive, and is very often a solitary activity, which contradicts the human drive towards functioning in groups.

For simplicity sake, we shall define 'the Arts' as a non-utilitarian activity in any of a variety of media—kinesics (body movement), aural, written, visual, lingual, nasal—in which some actor intends to manipulate the medium in order to produce a sensory, emotional and intellectual effect on the recipient. The earliest arts include visual (e.g. cave paintings and body adornments) and probably music in the form of whistles and percussion, and very likely dance (Hanna 1987).

Very clearly, there is a distinct and obvious connection between art and human physiology. Bluntly put, art affects us because we are human, and the practice of art employs our basic physiology and wiring to get us going. We need to therefore be very cautious in assuming that (a) ETI will have art, and that (b) ETI art, if it exists, will be in any way comparable to human art.

Some arts *may* be replicated under other stars. Music—orderly, almost mathematical in it's nature—may well be almost universal, even if not conducted by sound. Kaiser (2004), Roederer (2002), Vakoch (2010a, b), and Zaitsev (2008) have argued strongly for the inclusion of music as part of humankind's message to ETI. This may be so, and we ought to consider whether and how our arts will be comprehended and perhaps desired by non-humans.

Summary: The Intellect at Play We can expect that any ETI species will engage in some form of intellectual, formal endeavor. At a minimum, that would include science, because without that, the likelihood of developing a culture able to communicate across space is unlikely. However, as we have also seen, the continuum of intellectual endeavors runs the gamut from purely functional (science for the sake of technology) to the playful and non-rational.

We also need to be very careful about assigning a particular practice, including narratives and objects, to the vague realm of 'ritual' or 'art' (as happens often in archaeology) because we do not understand what the object or activity means to those who exhibit it. At the other remove, we need to keep in mind that as Thorb the Trog demonstrates (Panshin, *The Thurb Revolution*), art needs to be learned to be appreciated fully as the artist intends, and much cultural activity is essentially intellectual, a mentifact rather than an artifact.

8.1.7 Domains of Cultural Activity

The mentifact/artifact/sociofact triplet is a useful, but somewhat rough tool. It can be refined still further, provided we understand that domains, including the triplet, constantly intersect one another. Human societies can at some level be examined by looking at enduring patterns they exhibit. This is not to say these patterns of behavior and production are fixed. They are not; they mutate with time, and can be different from one situation to another. But sufficiently gross similarity exists so that social scientists speak of them as 'institutions'. Institutions are frameworks within which human activities take place to satisfy certain human social (and individual) needs. Institutions vary over time and space, and are experienced differently by different people (classes, genders, ages, statuses) within the same society. Nonetheless, they are a useful framework for parsing human activities. We can state three propositions which hold true for all human cultures:

1. All human societies exhibit a similar roster of institutions, because institutions are 'the way of doing things'; in effect, the ways certain needs are socially fulfilled in any society
2. The nature of institutions *within* a society is such, that the experience of persons of that society with an institution is roughly similar.
3. The same institution in two different societies will be expressed and experienced more differently than the experiences of persons in each of those two societies.

Admittedly, this is the weakest proposition of all: the dominant institutions probably seem very similar to those at the bottom of the heap in all societies.

To make this more real, consider a common institution such as marriage in the US and India. Both are the same institution, and yet an Indian person is far more likely to be familiar with Indian marriage customs, because they are part of a local institution, than with US customs. The obverse is also true. However, *both* societies have a set of rules about legitimate matings between people, which, roughly, can be circumscribed within the institution 'marriage/kinship/family'.

The roster of institutions described below is of course contentious It does allow for an identification of a minimal number of significant ones, on the basis that they are prominent in all human cultures. Others may be added if one thinks they are (a) significant, (b) universal among humans, and (c) sufficiently important to warrant classification on their own.

- *Marriage, kinship, and family*. This institution deals with the *social* side of biological reproduction. Among other things it ensures that economic rights of the product of a sexual union of two strangers are 'properly' (that is, by local social requirements) fulfilled, and that the individual is embedded within a primary support web (Fox 1983). We can read this human institution more generally as a system which ensures that children born within the culture are socially accepted as legitimate. That is to say, that children who are born within the rules of the institution automatically have a set of kin—parents, genitors and genetrixes, and their recognized kin—who bear some responsibility to the child, as the child does to them. Of course these relationships—who is owed what, and under what circumstances—vary, sometimes very radically between different cultures. Naturally this can break down in specific cases, but we find that in all human societies, marriage, family, and kinship are primary groups that *everyone* recognizes, however bizarre to members of one society are the practices of another (Goody 1969).
- *Economics*. Unless the society has no resources—no material goods that are restricted and controlled—there will be economic institutions which will determine what is to be produced, exchanged, and consumed how and by whom. For some societies, such as space-based ones, resources will be very different from planetary cultures. Economics deals with the realms of production, consumption, and what mediates between them: exchange. Here too, the rules differ markedly, from elaborate systems such as world capitalism (Pfaller and Lerch 2005) to the ritual exchanges of the Kula ring (Malinowski 1968).
- *Politics*. Among humans, politics can be defined most simply as the collective decision about who has the right to allocate resources. Here too it is expressed in many different ways, from national government to the rules of Calvin and Hobbes's treehouse club (Almond 1961; Lichbach and Zuckerman 1997). It should not be surprising to find political action and maneuvering among ETI as well, whether they have access to unlimited energy and materials or not.

- *Law and justice*. Which regulates how and what to do when social relations break down in some way (Hoebel 1968). Given the infinitude of ways in which people can interact with their environments, the need for dispute settlement is almost a given. While law as we understand it may or may not be universal, dispute resolution is a must. So we can predict with a great deal of certainty that any ETI is likely to have an institution (which we call law) that functions to settle disputes.
- *Religion*. Provides irrefutable statements/propositions about the nature of the cosmos, of the godhead (singular or plural), of ritual activities (justified and guided by statements of the form 'that is what the cosmos requires/the godhead demands'), and membership of churches (associations of persons following the specific propositions). Religion is a tricky issue. All human societies we are familiar with have some form of religion. That is, they have an institution which deals with ultimate causes, demands specific behaviors to fit those causes, and has lists of members, and what being a member entails (Berger 1969). All or none of these may be true of institutions in ETI cultures, but whether these are 'religious' or not requires a more clearly focused tool, as we have seen.
- *Socialization*. The means by which newcomers—infants and immigrants—are taught to become members of the society. Think schools, families, and even criminal gangs. It may or may not be an element in the kinship and family institution, it may be a completely independent institution. Nevertheless, it will exist if this culture is to maintain itself over time (Berger and Berger 1979; Rafky 1971).
- *Technology* and *knowledge*. Institutions that provide practical and nominal guides to how to create, manipulate, and think about artifacts and their underlying meaning. In modern societies this is the enormous edifice of science, that you are participating in as you read this book. In simpler societies it may be a one-on-one apprenticeship with a master canoe, temple, or house builder. If the ETI we come into contact with are a technological culture—for sake of argument, we can adopt the SETI definition "…able to construct and use large-scale radio telescopes"—then they will require some socio-cultural framework to generate, retain, and disseminate scientific and technological knowledge.

Obviously, other institutions can be proposed as well: the arts may well belong in there, as may the institutions of war and conflict. The list is not intended to restrict, but to offer a tool for analysis of societies. And, of course, for physiological, phenotypical, environmental, or other reasons, ETI may have regulated patterns to deal with issues which humans do not have. In other words, an ETI culture may well have institutions we cannot as yet predict, and when observed, not easily understand because they are a response to concerns that humans do not have.

One word should be added about transcendentals. Given that we can know little about this class of ETI, it is useful to consider a rough definition (or rather, characterization) of transcendental ETI as a race or races of ETI that are able to dispense with cultural institutions. How they do so would be very difficult to understand (but that is in the nature of transcendence). We would also have to be very careful to identify that they are doing without the need to deal with the issues that a particular institution is supposed to deal with. For example, if individuals have access to infinite energy and matter, their economic institution may consider thought, art, or some other good to be 'tradeable' and have institutions to deal with the creation, exchange, and consumption of those resources. Which institutions, if any, would not apply to transcendentals, as a 'mediocre' human myself, I cannot conceive of.

8.2 Culture and Symbols

The issue of symbols is crucial for understanding cultures. At some level, cultures are collectives of shared symbols which are so powerful for the individual that any individual in the culture acts according to them. Taken in the aggregate, that means that within certain boundaries, the collective will act as a single body.

Two features characterize human symbolizing. Symbols are able to encompass a wide variety of different, sometimes even opposed meanings, and symbols do a good job of obscuring the specificity of what they mean. Thus, to take one common example, the red flag means freedom, brotherhood, and labor to a leftists, and radical communism, upset, and criminal conspiracy to a rightist. From a purely dispassionate point of view, symbols are extremely useful in two ways. The first is that, within one's own sub-culture, one can say something like "… the red flag" and everyone else will nod their heads sagely, knowing what one is speaking about (and will assume the speaker means precisely what they do) without having to give a long and detailed explanation. The obscuration aspect is also important. If, when, saying 'red flag' everyone knew *precisely* what I meant, there would be a great chance of more than one another saying "but that is not what I mean…." And the discussion would not get any further, and probably end in acrimony. Any reader can try it, interjecting a precise explanation of a symbol—national flag, prominent monument, or DaVinci's La Gioconda—and observing where the exchange goes from there. Thus, from our dispassionate point of view, and given the Darwinian advantages symbols convey (speed of speech, smoothing over disagreements) it is highly likely that any ETI will have something like symbols.

The quality symbols have for obscuring differences is, for the integration of cultures, very important. It facilitates the ability of one individual to act with another *without* the need to examine the other's 'credentials' as it were. Without symbols, collectives would not be able to act, but would instead be constantly fissioning as members came to understand *precisely* what the other was talking about, and discovered too, that they were *not* after all, in agreement. The question that concerns us here has to do with whether this particular cultural device—an obscurational shorthand—is necessary for the existence of *any* culture, or is it perhaps a quirk of our particular species?

Assuming ETI have symbols puts the human observer into a serious morass. It is difficult enough identifying symbols in other cultures among humans (Duncan 1968; Marshack 1976), let alone non-humans (though some think otherwise cf.

Musso 2010). Many cultures have *key symbols*—critical symbols that entire realms of belief and social activity revolve around (Ortner 1973). In some cases those are very obvious, for example, the ubiquity of the cross in Catholic countries, but interpreting from observation the meaning of a man hanging from a cross is not self-evident. And any explanation of the meaning of that cross must be treated with caution until a number of explanations of meaning and observations of context have been processed (Turner 1967). To add to the problem, the value and meaning of symbols, as well as their importance and centrality can change over time, mutate, and even express something completely different.

8.3 Rationality

A sub-theme in the 'ETI are technologically more advanced than us" is the idea of ETI rationality (which is implicit in the 'benevolent ETI' thesis). This is worth examining more thoroughly. For "rationality" read the degree to which a species behavior will be determined by computing odds about the results of its actions versus the degree to which its instincts or emotional objectives predominate. At one end we would find a hypothetically supremely rational species. Their every action would be based on a calculation of the results of that action, and they would seek at all times to understand all factors related to their decision. Long and short term benefits would be weighed carefully, and some means devised to assign them relative value. At the other extreme would be a species that could not, or would not, compute the results of action in any way except the increase of satisfaction. In such a species, intellect would serve to satisfy basic appetites, either immediately, or long term without considering the costs. Both these extremes are of course ideal types, in the Weberian sense, and impossible even logically. The supreme rationalists would still have to have some basic goals which their rationality would serve, and these would be irrational. Otherwise, the supremely rational are also supremely predictable. The sensualists would require some computing faculty to satisfy their urges. But presumably species could be compared on the basis of the degree to which they approach one or the other of the two end points.

I was, in the past, involved in a multidisciplinary NASA project. One thing that impressed me throughout that work, and throughout visits to NASA and discussions with engineers in other opportunities, is the absolute belief many engineers and "hard" scientists have in the force of rationality. Put simply, most physical scientists seem to believe that if only an issue were put rationally to any human being, that human will choose the "rational" solution to the problem. What this "rationality" view tends to ignore is the very real effect of *non-rational* decision making has, not only on individual, but on collective behavior as well.

Economic Rationality? The question of the position an ETI civilization can be placed on in the 'rationality' dimension has to do with an 'economic' factor as well. I put the word in quotes because, in this particular instance, we are dealing with economics in their holistic sense, as anthropologists, rather than as economists use the term. That is, we recognize that both economic behavior and economic goals are socially defined, and rational *only* within their own contexts. Stephen Gudeman, an economic anthropologist has indicated that many of our social assumptions, including those about basics such as making a living, may well be ethnocentric and restricted to specific modes of thought and cultures. In (Gudeman 1986) he has suggested forcefully that the positivist economic model used by economists the world over, derives from a particular culture. Other societies think of their economies in different ways, and therefore *are likely to act economically* in different ways as well.

Now, it is perhaps a given that energy exchange, that is the process by which an organism procures and consumes its energy, is basic to that organism. It may be basic to *societies* as well. But if, as we have seen here, *human* motivations are not necessarily economically 'rational' even in the sphere of economics, what must be the case for ETI? This should be even more true when we come to discuss non-immediate goals, such as religion, philosophy, and other modes of thought. In other words, we cannot assume *a priori* that the motivations that inform Western technological civilization (and if you are reading this book, you are, at whatever remove, a part of this world-dominant civilization), will also drive ETI.

Why is this important? Let us consider the various SETI projects. Underlying them are what I call "the Sagan hypothesis". In this hypothesis, technological civilization, once arising, is going to be (a) interested, as we are at least peripherally, in establishing whether they are alone in the universe; and (b) be benevolently inclined, or at least willing to offer communicative data. Neither of these are necessarily true, *even* for human civilization. Weakness and small size have not stopped countries as diverse as Albania, Japan, or Tibet from closing themselves aggressively off from the world. Small size has not stopped Christian denominations, to mark just one case, from aggressively proselytizing their ideas. In other words, the "non-rational" (from the point of view of a secular humanist, a physicist, or an economist) has never been a barrier to action in human society, why should ETI be any different? Even the argument that physically visiting other societies in order to trade, make war, conquer, or proselytize (to suggest the main human interests) is too expensive does not hold water. If they (whomever they are) were interested in saving our souls as a central core of their dogmatic being, they would most likely, spare no expense to achieve their objective. Thus Gudeman's point applies very strongly to ETI as well: we must consider their behavior that has to do with economics not from the point of view of economics as most economists understand the term, but as a cultural factor, with all that implies.

Social rationality Consider, for example, the dominant trope in human reproduction. Ninety-nine percent or so of all human societies have either monogamous (one spouse per person) or polygynous (one man marries several women) marriage systems. In fact, polygyny is the base reproductive assumption in one of the worlds most dominant ideologies: Islam. The belief that monogyny or polygyny are 'natural' and sanctified by God is hugely prevalent.

Reproduction is fundamental for all human societies, and we may assume, given what has been noted above about the likelihood of reproductive strategies among ETI, that it may well be cardinal for ETI as well. Human beings are patrifocal for many reasons, but given an affluent society such as ours, there are good *rational* arguments to support polyandry or even extended polygamy (many spouses and many wives each), yet on the whole, the debate does not widely arise (Haritaworn et al. 2006). Now suppose polyandry were not only the position of an ETI, but also a point of ideological purity with them (just as patriarchy and polygyny are for an orthodox Muslim) and that they could bolster their argument with some strong and cogent arguments for the adoption of polyandry as a practice. To what degree would humans communicating with these ETI be inclined to follow the ETI's rationality? And how would relations between the two species develop, given an inherent incompatibility between ETI 'rationality' on the subject of reproduction, and human 'irrationality' on the same subject?

Our religion is supremely rational... Even the most supremely rational decisions spring from inherently irrational choices. My supremely rational colleagues at NASA were nonetheless, many of them, adherents of Mother, apple-pie, and the American Way, all of those being irrational concerns. Some were practicing Christians. Assumptions, largely unprovable, about the existence, intentions, and activities of a moot entity (I am using the value-neutral terminology developed by Guthrie 1980) AKA 'God'/'Gods', are the meat of contention among humans. Dawkins may argue that belief in god is irrational (Dawkins 2009), nonetheless, it is a powerful emotion, even if we eventually discover that it is a glandular condition. While ETI are unlikely to have the same glandular condition, the same effects could develop using other glands on other worlds.

Interstellar crusades may be more of an option than one thinks. Crusades are (ostensibly) about a battle of *ideas*, and warfare is by no means restricted to kinetic strikes. We can therefore assume that religious rationality can be identified as a point on a continuum which encompasses both dyed-in-the-wool atheists such Christopher Hitchens and Dawkins, and the most ranting religious fanatic. Given the basic propositions of religion, that is, its theology and cosmology *which are inherently non-rational*, Dawkins and Hitchens would, I know, disagree. The problem for them is that their vision, the empiricism they support, simply cannot provide sufficient empirical evidence for their claims: God, the gods, moot entities, the Great Boojum may simply operate on the basis of imperceptible, large movements that are invisible to us humans, whether because, as Goedel says, we are part of the system and therefore cannot see the whole, or because the Great Boojum is extremely subtle, as well as knowing a lot of stuff we don't. As a rational atheist, I can nevertheless see the inherent weakness in the atheist position. Many religions, *once the initial premise, whatever it is, is accepted* consider themselves, and can be considered to be, rational. Certainly much of 'normative' Judaism and Islam, based on interpretative law, and some aspects of Buddhism such as Zen, are rational follow-throughs of irrational unproven, but unfalsified propositions. Of course the propositions themselves may be wrong, certainly seem irrational, but might prove to be right (one or all, if one is truly broad minded).

8.3.1 Technology, Science, and Applied Thinking

The *degree* or intensity of technological use is a measure I am not concerned with. What is important here is whether the ETI culture is dependent upon technology for its existence (as Western culture is, for instance), or not. If *not*, this implies a great deal of cultural flexibility in the first instance. Though they might not be close to the Xenians, a model suggested by Lumsden and Wilson (1983), in which ETI may be so integrated with their technology that they are indistinguishable from it.

The term 'technology' must also include some measure for the type of technology in question. This of course depends to a great degree on whether the ETI concerned is 'normal' (that is, we assume ETI are close to our median: the assumption of mediocrity again) or exotic. Assuming they are normal, we should ask whether their technology is mechanical-electric, as ours is, primarily biological, or has some other basis. It is important to recognize that human technology is constantly evolving and that new areas to explore by science, and exploit by technology are constantly emerging. Until circa the nineteenth century, technology was a mix of mechanics, crude metallurgy, and biology. We had levers and plows, hammered out plowshares, and harnessed ourselves and our cattle to pull the plow. Since the end of the 20th century, completely new technologies, based on interactions of genes, molecules, atoms, and electrons, have become prominent. Mechanics, metallurgy and biology have not been abandoned, they are now simply one small element in a larger roster.

Kuttner (1981) has suggested that one dominant aspect of technology is the inherent tendency to *optimize* it, that is, to produce better (more accurate, less demanding of resources, more easily utilizable) technological solutions. Since there are only a finite number of such solutions to any given physical problem, and, even more probably, such solutions converge to the value of 1, technology is likely to converge in most ILF societies. If we further accept the premise (see for example Kardashev's 1964 position) that cultures are conditioned to a large extent by the technologies they employ, we are likely to find that technologically advanced ILF cultures, particularly those with lengthy technological histories, will be similar to one another.

This argument tends to ignore a number of factors. First, different environments are likely to create different needs and thus different technological solutions. A far more cogent argument is the question of the nature of ETI *knowledge*. That is, what forms of knowledge will the ETI posses, be interested in, search for. On the basis of our current cultural arrangements, it might seem to us that knowledge, its offspring technology, and the method of acquiring both are intertwined. We distinguish loosely between 'science' (both a methodology for acquiring knowledge and its results) and 'technology' (both a set of knowledge and set of techniques for applying knowledge). This distinction may not hold in an ETI society, which may not have one or the other, or may apply them in different ways.

8.4 ETI Cultures: Beyond the Human Paradigm

Is there any way to try and understand in what ways ETI cultures may function internally? One way to start is to look at the three broad approaches to culture within the human experience. As we noted above, there are three logical approaches to the study of culture among *humans*. Culture may be *entirely independent of genetics*; it may be *entirely or largely dependent on genetic imperatives*; or, finally (and admittedly, the approach supported here), *culture builds within genetic and environmental constraints*. We shall explore the implication of each of these positions. Before we do so, a word of caution is in order. All three of these positions, or none, may prove to be in existence when we do find ETI. There is reason to suppose that ETI will not be homogeneous. Nor is building on the basis of the evidence from one data point—the cultures of *Homo sapiens*—necessarily a window into absolute truth. However, this is what we have.

8.4.1 Culture Independent of Genetics

Could culture evolve with no reference to genetics? The simple answer is probably yes: certainly an AI ETI will not have genetics in its background, unless one counts it's biological precursors (and no, I don't believe World Net will always destroy it's creators). Let us consider a culture highly dependent on direct computer/organic connections. The organism's recall and processing power may well be enhanced many times by such 'wetware' connections. Artificial worlds abound wherein individuals interact through fanciful avatars, engaging in social activities ranging from economics and artistic production to sex. Since this is a virtual reality, people can experiment with a variety of genders, sexes, family arrangements: basically anything they can imagine (see for example Govers and Go 2016; Kaplan and Haenlein 2009). Some readers will of course recognize and have possibly participated in a crude virtual reality of this sort. We may consider adding a post-SETI element (that is, the culture has received ideas from another ILF), in which case, the society in question is even further removed from it's genetic roots. Finally, if the society in question has moved from it's genetic ecosystem to another—say, space, or a planet around another star—where it's genetic blueprint may actually be a hindrance, then we are likely to encounter a culture with a great deal of independence from its' genetics.

We are now at the stage in human scientific and technological development where we can conceive intellectually of such changes. The fall under the headings of virtual reality, cyborg, life in space, genetic engineering, etc. While we cannot *practice* these developments, we need to add *yet*: within the foreseeable future, some humans may be detached, to a larger or lesser degree from the imposition of human genetics. Assuming that the SETI community's common assumption—that ETI we come into contact with will inevitably be more advanced than humans scientifically—it is certainly possible, even likely, that they will be removed from their genetic beginning. This does imply bad news: a race, detached from its genetic blueprint may well be less inclined to react to others on the basis of their instincts, as humans do, and thus more welcoming of a new race they come into contact with.

8.4.2 Culture as Genetic Expression

The fundamental basis of all human cultures may be the expression of genetic dictatorship. Dawkins (2006) gives the idea an extreme twist. But even if Dawkins's overall conclusion is too extreme, all human cultural realities have a fundamental basis in genetics: we are bipedal, bifocal, bisexual descendants of apes who were probably scavengers and opportunistic feeders. We have a mixed genetic background from several different varieties of the genus *Homo*. We have developed dentition to serve a particular form of feeding, probably including use of fire, and we are relatively hairless, possibly as a means of cooling down during long hunts and/or treks. These physiological realities affect the ways we think, our objectives, and our means of expressing them, from love to aggression.

We may assume that other ILFs have similar baggage (which will obviously be different from ours), at least at some stage of their history. Understanding the genetic background of others is not at all crucial for humans, who all share this background. It is probably absolutely crucial in order to truly understand, and be able to predict the actions of any ETI race we come across in any fashion, from physical contact[6] to radiative communication.

8.5 ETI World Views

World view is "an organism's organized set of personal cognitions which constitute a model or image of reality" (Royce 1976). We shall broaden the concept in order to be able to deal with cultural aspects of world view that may be relevant to ETI.

[6]I can see physicists rolling their eyes at this statement. More of this anon: the fat lady ain't sung yet.

Humans, Royce notes, tend to view the world through epistemically limited spectacles. Though there is a multiplicity of approaches to conceptualize "reality" individuals tend to highlight one of these approaches at the expense of the others. A basic question must be, therefore, "How might highly intelligent extraterrestrial life-forms view the cosmos?" (Royce 1976: 178).

Different cultures and different intellectual traditions are bound to describe and explain their experience in terms of objects and categories of understanding substantially different from ours. The uniformity of nature does not matter, because different 'chunks' of nature will be seen. Even if we hold that given the same question asked of nature, the same answer will emerge, then a different viewpoint will likely stress different questions and will get different answers (Rescher 1985: 95). At an extreme, different physiologies may produce radically different perceptions of one's surroundings (Censor 2005).

Humans do the same within their own social limits, which makes any prediction about *human* behavior difficult enough (Berger and Luckmann 1971), let alone ETI. As such world views are developed—a mixture of reality, beliefs, and the tyranny of both genes and social expression—what appears rational and reasonable to one race, may be very different to another. In other words, ETI will probably have different takes on reality than our own. This may derive from their physiology, or possibly from their cultural background. The point here is that we *cannot* assume that what appears to one race to be a simple observation, will mean the same to another.

8.5.1 Religion, Again

Speculating on ETI religion would seem to be the most far-fetched of the ideas discussed here unless one is committed (as, presumably, many strongly religious people would be) to the belief in the universality of any particular religion. Yet it *might* be possible to consider ETI religion if we can also examine some of the bases for our own.

Religion is a collective experience, even if and when it springs from individual adherence and conceptions. In fact, as sociologists like Durkheim (1947) have pointed out, the collectiveness may well have something to do with the fact that humans aggregate themselves into groups. In other words, individual human imagination, speculation, emotion, is formatted by society into the set of behaviors called religion.

If this is the case, we can take the argument one step further, and say that a species that has collective organizational institutions *similar* to ours, may also exhibit religion. Ruse (1985) puts it more strongly: there appear to be good adaptive reasons for believing in god. If this is indeed the case, then ETI are likely to have such a belief as well. As to the form this god(s) will take, Plato's dictum holds well here too: they will undoubtedly exhibit those characteristics that are prominent for that society of that species. And whether their religion is an expression of genetic predisposition [as Dawkins (2009) argues it is for humans] or a derivative of sociation, as Durkheim (1947) says, the same sources may apply to ETI as well.

To tackle the issue of ETI religion, we need to return to the fundamentals of religious comparison. The *intellectual* side of all human religions—theology and cosmology—are, in most cases, modeled after human experience. Middle Eastern gods in modern extant religions—Judaism and Islam—are modeled on figures from a patriarchal, fiercely endoethnic past, where reliance could be placed on kin only, and where the chief of the tribe had the responsibility to lead the sheep, shear them, and occasionally slaughter a few. This has been modified through other experiences (e.g. urban life and imperium in the case of Islam, political weakness and ethnic imperatives in the case of Judaism). Cosmological ideas are modeled similarly. The idea that the world is essentially good, nurtured under the wings of a benevolent deity has been modified by historical experience, but still holds true, as does the concept of never-never land just over the horizon in Judaism and Islam. In Chinese religion, in contrast, the cosmic idea of immutable laws (the *Dao*), which, if only comprehended in essence, and followed will keep the world in balance, is prominent. The *action* side of all human religions is an involved set of activities which attempt, insofar as possible, to express, and to bring into the real, the statements that make up the intellectual side. Thus a religious person will, by praying (alone, or better, in concert with other believers) demonstrate, reaffirm, and 'make real' the presence of the deity *towards* whom one bows, prays, dances, shakes, or sings (Ashkenazi 1981). In other words, *in human societies* the deity, and the cosmology that surrounds/is part of him/her are made manifest by ritual actions.

Assuming ETI have religion, what *relevant* forms could ETI religion take? That is, to return to an early statement that informs this entire book, can we say anything useful about ETI religions? Here too, it is advantageous to understand analytically what we mean by the term 'religion' from our own experience. Taking off from Guthrie (1980) we can see that all human religions have a number of similar features. The most crucial element is that *all* religions we know of *make some form of communication to 'moot entities'*. Moot entities are entities whose existence, or powers are doubtful, because they are not apparent to every observer. Putting it in this form *can* mean that ETI to whom one deliberately addresses a communication *without knowing whether they exist or not* constitute some form of deity. This is perfectly acceptable. In fact, as someone who has studied, albeit briefly, UFO cultist, this is exactly what they are doing: communicating with moot entities from flying saucers: by definition, a form of deity. The difference between a deity and an ETI is of course that the presence of ETI's is empirically verifiable, whereas gods are somewhat less amenable to empirical verification.

Once we accept that human religions are inherently based on *communication*, we must also note that they display other common features as well: an ideology (theology and cosmology), a practice (ritual) and a social organization (church). In other words, religions are *social* institutions, amenable to tools available to social scientists. Should ETI have religion, I would predict that they would also have churches (with all the possibilities of religious difference and schism human religions display), but that may be merely a feature of my own human world view. It is certainly possible that ETI have religion, but that religion is a completely private affair, perhaps restricted to a kin lineage, perhaps chosen at preference through perusal of religious communication. What ETI religion definitely *will* possess is that it will be based on *communication* with some 'other', which, in a humanly ethnocentric manner, we can label 'deity'. Thus, at the very least, there is likely to be a component of communicating with 'others' of some sort as part of the world view.

Monotonic or multiple world views A major question we need to ask is whether the world view of any ETI enables that ETI to entertain the possibility of other world views? In practical terms we are asking whether the ETI is able to develop sufficient empathy with others to enable dealings with other ILFs. The answer has a number of consequences, such as the ETI's ability to maintain the Golden Rule and the ETI's ability to consider other ILFs 'human' whatever the perspective. First, however, it is necessary to examine the basic parameter.

Human beings, as a species, have emerged under conditions of divergent environments, which have led to divergent human cultures. This implies that humans have the *capacity* to entertain divergent world views, and indeed they do. Moreover, as anyone raised in more than one culture can testify, humans also have the ability to entertain more than one view simultaneously, though great efforts are made by everyone to achieve what psychologists call 'cognitive congruence'.

If we assume that ETI can only emerge under conditions of diversity (see Chap. 4), then it is likely that they too might entertain the *potential* for diversity in world view. I emphasize 'potential' because even humans, in many societies, are often instructed that the only world view, or sometimes the only valid world view, is the one of their own culture. The potential to entertain diverse world views may well be genetic, the limiter however, is definitely cultural. That is, given the genetic ability to entertain different world views, cultures may impose bans on doing so. It is of course possible to argue that diverse world views are a necessary concomitant to diversity which appears to be a prerequisite for the emergence of intelligence. However, we have stated earlier that there is the possibility of exotic intelligence derivatives, and thus we are forced to entertain the possibility that singular world views may well be 'hard wired' into some ETI behaviors.

The implications of the existence of one, or multiple world views may differentiate between disparate ETI cultures. Those that are unable to entertain alternative world views may also be extremely aggressive in defense of them. In an earlier section I noted that there may well be 'trigger' events which could cause an ETI to move from a position of non-aggression to one of aggression. This may well be one.

World view, as noted, involves not only a 'picture' or 'cognitive map' of the world, but also a predisposition towards action. That is, it both describes the world and demands ways to act towards it, even perhaps bring it about when it seems out of whack. Among humans, rules both restrain and compel, that is, we are able to identify patterns of thought and of action which with a high statistical certainty, limit or demand action from members of the society. This includes the realm of ethics, as well as of action. In some societies it is ethical to refrain from killing, in others killing is enjoined under defined circumstances. Acts cannot therefore be judged from an absolute or value-suspended point of view, but exhortations for action, and the degree to which they are expected to be compelling, can be.

The question of equity Sociobiology provides a suggestion as to how morality evolves, and that certain moral principles, such as kin selection, are evolutionarily guided. The argument that there will be great superficial and metaphorical differences, but a deep similarity between ILFs' moralities (Ruse 1985: 61–62) and that ETI will come to the same moral solutions as we have come to is patent nonsense. Even among the major human cultures—compare Chinese, Western, and Islamic—there are major ethical and moral differences. If an ETI society has a completely different reproductive pattern, all bets are off: what happens in the case of a species whose offspring are numerous, but whose support is given to aggregating clients? It is possible to agree with Ruse that the imperative towards greatest happiness is universal, but that tells us nothing inasmuch as we do not know what constitutes 'happiness' for an ET. For the same reason we cannot assume that ETI will necessary view the Categorical Imperative (regard one's fellow beings as ends, not means) as universal. It may well be, we may hope it is, but it is not logically necessary to develop a culture, nor for that matter to communicate with other species.

We consider homicide to be a heinous act, and it is undertaken intentionally, in most societies, only under extreme provocation. Though it needs to be added that many societies while considering homicide to be extremely serious, conduct it with a lightheartedness and playfulness that is more than a bit disturbing: consider Ifugao and Dayak headhunters, ISIS members, and UAV operators. Would this be the same, for instance, in a society in which r-selection is practiced? After all, killing, possibly even eating some of those small pests who *might* grow into sentient adults is nothing more than selection in action.

The point is, of course, that many of our fundamental assumptions about morality will be challenged by any ETI we encounter. One major fundamental view most humans tend to adhere to, to some degree, is the principle of equity. That is the idea that one must return good for good and evil for evil. In terms of world view (which is a map for action) this can have both negative and positive effects. On the negative side, many human societies practice concepts of extended retribution for slights and injuries, sometimes leading to extensive, bloody, and intractable feuds (Ahmeti 2004; Ansart 2007; Bozgo et al. 2002; Ericksen and Horton 1992; Fleising and Goldenberg 1987; Jones 1948; Newman and Lynch 1987; Richard and Cohen 1996; Tellenbach 2006). On the positive side, of course, is a world view that demands open-handed return of material and other favors. In essence, a gift must be returned (Mauss 1966), whether it is a gift or poison.

In contrast, some societies tend to be non-reciprocal. Slights and injuries are expected to be swallowed, just as gifts are to be expected without return. With the exception of the Christian 'turn the other cheek' concept, few if any societies I am aware of practice this philosophy consistently, let alone state it. This is not surprising, given that reciprocity of some sort is fundamental to human societies.

From an ETI world view perspective, however, while a reciprocity perspective (negative or positive, that is balancing injuries as well as good) is possible, it is certainly possible too, notably among advanced ETI societies, non-reciprocity might be a major intellectual structure, among other reasons because in advanced societies, resources may be freely available. This does have implications for humans dealing with them. On the one hand, such a world view would not react to quick fingered humans dipping into their intellectual, and even economic pockets: what's done is done. On the other hand, such a world view would also make little of hurting humans, their interests, or even their planet: after all, reciprocity is not expected nor anticipated.

Whatever the fundamental basis of ETI world view is (and it may well be other than the matter of reciprocity) it behooves humanity to learn, with utmost dispatch, what scratches those fundamental ETI itches. This is particularly cogent given that, at least among humans, people are not even aware what the fundamentals of their world view are, nor how they trigger action—violent, pacific, and anything in between—until a particular stimulus has appeared and been acted upon.

8.6 The Homogenous ETI Fallacy and Its Effect

ETI species, as we have noted in Chap. 5, may not be biologically invariant and homogenous. In addition, given the SETI assumption that ETI we come into contact with will be scientifically in advance of us, we also must consider the possibilities of advanced social sciences, genetic engineering, and living on other planets and in space. For example, we should assume that any ETI race has made the leap into space colonization unless it lacks a 'space imperative' which is possible but unlikely (Ashkenazi 1995). As a consequence of long-term life in space, several different societies/cultures are likely to emerge, given the enormous difference between the physical realities of life in space and on the planet of origin. Not only that, but we should also expect numerous varieties of genotypes originating from the same species, modified by science.

8.6.1 ETI Space and Planetary Cultures and Their Implications

To understand the nature of the different cultural possibilities of the major technical environments, we need to consider how these environments are likely to affect a space faring race.

Planetary Cultures Since we are all members of one or another of these, little need be said about planetary cultures. These are essentially the bedrock—genetic, social, cultural, behavioral—of any species (exotics aside). We all know what this implies: ecologies we are familiar and comfortable with, genetic variation at a 'reasonable' (whomever we are) rate, breathable air, drinkable water, edible plants and animals. Our culture is essentially based on these realities. Of course there is environmental variability but we are *intelligent*, we can figure things out and make ourselves comfortable. There is a lengthy phylogenetic history which has fitted us to our environment, and the environment to us (whether we are careless consumers, as humans are, or careful stewards, as humans should be).

Space and Other Planet Cultures The question of whether we *should* establish colonies in Solar System space is moot. It does seem as if the question is answering itself, with various agencies and even corporations working towards the fruits of space colonization: almost unlimited resources. As for any other race, for humans the colonization of its solar system space is dependent on two variables: the cost and the benefit. The latter—access to unlimited power (solar) and resources (water, metals, rare earths, hydrocarbons)—is almost certain. The costs, given the need to allocate costs while still satisfying local (that is, on-Earth) needs, are considerable; and, considering technical advances in space travel, probably unknowable at the present time.

One thing is almost certain: if an ETI race has a space culture, it will also have very large radio telescopes, which are not only easier to build in space, but also more necessary as navigation and hazard-warning devices. Other issues are less certain, and here, again, we must rely on a single point: human space colonization perspectives.

What culture(s) will emerge from space colonization has been speculated upon by numerous writers (Ashkenazi 1992; Bainbridge 1984; Heppenheimer 1978; Holbrow et al. 1985; O'Neill 1974; Turner 1985). Whatever the case, time, distance, and different ecological systems will play their part. As the Dutch population has grown in stature within a relatively short period of time through natural selection of mates (Schönbeck 2012) and the Japanese through changes in consumption (Sugiyama 2008) all within no more than 60 years, so can we expect physical, and probably sociocultural changes in a space colony months or years travel time from the mother culture. What these changes will be is impossible to project, but we should assume *ab initio* that there are more than one ETI cultures.

What has been said for space (non-planetary) cultures will be true of colonies on other planets. A race that has been long enough in space will possibly also colonize other planets in their solar system, as humans are considering the Moon and Mars (Freitas Jr 1983; Matloff et al. 2007; Seedhouse 2009). There are costs and implications for that. Consider that Mars's gravity is 38 % that of Earth. Consider changes over very few generations. Assuming you emigrated to Mars, how will your 6th generation descendant look like? The answer is not trivially the difference in gravity. Human mating selection, based on aesthetics, unconscious cultural choices and other variables, plays a major part (cf. the height of the Dutch population Schönbeck 2012): who will be a successful mate is an intricate question, and selection of successful co-mates among humans is an art, not a science. Nonetheless, given distance from planet of origin, and different ecosystems in which the colonists will live, there will be genetic changes *that are likely to be different from those occurring in the home planet*. The injection of genes and memes from the home planet is likely to be minimal, considering the likely low rate (in population terms) of travel between the two planets. This will mean a divergence in physiology, and, no less importantly, in culture. There are two other options that are highly likely to yield different cultures within the same race.

Wandering planet cultures In addition to planets that are fixed orbital features of solar systems, there is some evidence of wandering planets, that are not orbiting stars, but like stars, orbit the galactic center. Such wandering planets may serve as locales for colonization for a star-travelling species (Badescu 2011). Inasmuch as such colonies will need to control their energy and will likely lack natural day/night differentiation, and a host of other differences, we can, again, expect a different culture to emerge from different ecosystemic demands, isolation, and so on.

Slow boats Finally, of course, there is the question of STL generation or aestivation ships. Given the unlikelihood of faster-than-light travel in an Einsteinian universe, one suggestion for galactic colonization is the development of slower-than light ships large enough to carry along an active breeding population (Bainbridge 1984) or a population in some form of induced hibernation/aestivation. Indubitably, such cultures will differ from their originating mother cultures as the generations persist, and will change once again when they settle new worlds (or decide to stay in space). Once again, we are faced with multiple cultures from a single advanced race, *even if that race has developed a unitary planetary culture*.

Implications Crucially, what all of this means is that the likelihood of finding a homogeneous race of ETI, notably if they are technically advanced, should be estimated as very low. This factor also has several implications for communicating with any ETI we may encounter.

1. It seems likely that any ETI communicant that claims to be the authoritative spokesperson for its race should be regarded with a certain amount of healthy skepticism. A space going ETI race (which we will most likely encounter for reason discussed above) is going to mutate socially, and possibly physically, in many directions.
2. Similarly, any claim for complete political or other unity (world view and ethics perhaps excepted) should likely be viewed skeptically as well: different environments will breed differences, including those of interest, action, and perspective and therefore, politics.

8.7 Law and ETI

Law and government are means by which human beings formally allocate their interpersonal affairs, and by which resources are allocated. Like many other institutions we discuss here, they are found in all societies in some form or another (on law see Hoebel 1967. On politics see Smith 1966; Swartz et al. 1968). These regulatory institutions derive from a number of factors, some of which seem to be universal, others which may well not be. Universal is a need for regulating relationships between individuals. But such regulation implies the ability to infringe on others' perceived rights *and* the need to create artificial structures for the regulation of such contact. Given for example an utterly rational species, or one in which dominance is absolutely and unambiguously certain, laws will not necessarily exist, at least not in the aspect of interpersonal regulation. The same is of course true of politics, which in human societies is law's twin, each one affecting and being affected by the other.

The absence of strong *artificial* regulatory practices would have a major effect on the ability of an ETI species to deal with humans. For example, if signals of authority were unambiguous and absolutely determinate, we would find great difficulty in coming to an understanding, since among humans these are fluid and often ambiguous. This is particularly true because, as noted earlier, our own culture is based at least partly on intentional ambiguities, effectively on a series of bluffs of greater or lesser intensity and power. The reverse is also true. Conceive of a society in which ambiguity is the absolute norm. That is, members of that society must redefine for themselves and for others, their obligations and roles each time they interact. Any agreement made with such a society or its members would be worthless. "Laws" in the sense of codifications of socio-political decisions, would not exist. Instead, constant bargaining for place and for role would be the norm: something most humans would find impossible to deal with on a daily basis. Such a society, one could argue, could not possibly exist: constant flux would rend it to shreds. That is true of humans, but assume a species that has for biological reasons limited appetites? Actual social bargaining for resources would occupy relatively little of their time, the rest devoted to other pursuits *where political/legal considerations would be secondary.*

Notwithstanding the claims of legal scholars (e.g. Fasan 1986), therefore, "law" is not necessarily a universal phenomenon. It is possible, indeed likely that any society communicating across space will be forced, if only for external reasons, to develop a system of "law" (this, and the issue of trust is discussed later) but it is by no means certain that this system would be inherent, or even wholly comprehensible, to the species in question, even while they use it in their external relations.

8.7.1 Crime Across Space

What constitutes crime? Coser (1967) has suggested that crime is a social aberration. Which seems a bit unlikely. Crime is no more an aberration than any other economic activity (Anderson 1979; Burchfiel 1978; Dick 1995). Crime presupposes law, as a general principle. At this point, however, rather than getting into the discussion of law and its implications, we can define crime as activities which violate generally agreed-upon societal norms. Crime is thus inherently connected to the culture of any race, which in turn is highly dependent on their biological underpinnings and all three kinds of their ecosystems: natural, technical, and societal. The answer thus to Ruse's question "Is rape wrong on Andromeda?"(Ruse, 1985) is 'It depends'.

This raises two issues which concern humanity. First, normative behavior within a given ETI race *may* include activities that for humans, would be considered criminal. Whether or not we can expect those behaviors depends to a very great degree upon whether we (humans and ETI) have negotiated some shared rules of ethics with which both (or all, if more than one ETI race) can live with.

Second, even an ETI race may well have individuals who violate or transcend their norms and law. The ETI race in question may tolerate, enjoy, or fight those behaviors, but it is impossible to state that an advanced species will not have something akin to crime, whether it is a norm against picking pockets, stealing melons, malversation of funds, or violation of some norm we have not thought of.

What these imply is that defining the boundaries of do and don't as quickly and as detailed as possible are a major requirement in the first instances of contact. Eliciting the formal rules may be as easy as ABC, or may be extremely complicated, as it is among humans (Kuznar 2000; Michaud 1973; Xu and Chen 2005). However, that will aid in at least separating (approved) normative activities, from criminal ones.

As among humans, crime may be organized in vast networks of common criminal interest (Chang et al. 2005; Hagan 2006; Ianni and Reuss-Ianni 1972; McIllwain 1999). These networks may be local (Lacher 2012) or cultural/ethnic (Finckenauer 2007) as they are among humans. And, as in the case of humans, one must consider the degree to which these are tolerated, or perhaps woven into the social fabric for whatever reason. As Terry Pratchett notes, "The Patrician decided that if there was to be crime, it ought to be organized.": so too, acts we may consider anti-social at best—murder, robbery, assault—may be a part of normal social practices.

Crucially, from our perspective, a criminal sub-culture is nevertheless a culture, and an ETI species may well require the inclusion of such a culture into any 'multicultural' consideration as an ethical, or pragmatic practice. For humans, this may be hard to accept, and yet, we shall have to recognize that (a) ETI are not necessarily going to be a single culture, and that (b) some of the culture we encounter *and have to negotiate with* will not be perfectly palatable, even if extend our tolerance to the maximum.

8.8 Culture and the Individual

It is of course impossibly ambitious to catalog, let alone discuss, all the potential elements possible in an ETI culture. The following four topics have been selected because I believe they are important but have been overlooked by most discussions of ETI, because some of them advance the discussion in subsequent chapters, and because they are interesting activities in and of themselves. They also serve as examples of the topics of analysis we could be required to consider in interacting with ETI individuals and groups.

[Recreational] Sex, cooking, fashion, and sport are important *human* practices in that they exemplify some major aspects of being human, that is a species whose evolution is now led by *culture* rather than physiology and the natural environment. There are undoubtedly others, and the astute and opinionated reader is invited to apply some thinking towards defining what these might be. I make no apology for my own choices, determined by previous experiences and personal preference.

8.8.1 *Sex Among the Stars*

To paraphrase Diamond, sex is fun (Diamond 1997). But is it fun for ETI as well? Of course that depends on a number of factors, including the ETI's physiology (Barkow 2000). Now we can assume that in light of Occam's Razor, we should expect that satisfying a natural sex drive does produce some analog to 'pleasure' given the simple feedback behavioral paradigm. But *all* ILFs, by definition, are the product of cultural as well as natural evolution. This in turn means that sex of the recreational variety will be affected by culture. This is indeed the case among humans: consider everything from Vatsyayana's *Kamasutra* and Nefzawi's *Perfumed Garden*, to recent publications and studies of eroticism (Barker 2013; Blinne 2012; Chou 1971; Donnan and Magowan 2010; Haritaworn et al. 2006; Hoel 2015; Markowitz and Ashkenazi 1999; Martin 2013; Newmahr 2008), and the wide world of contemporary sexual practices, public and private, *all* of which are cultural expressions of non-reproductive sex.

What is crucial from our perspective is that sex is inherently tied to human physiology. ETI, with different physiology may simply not be equipped—physically or emotionally—to recreate human sexual arts. They may be able to enjoy them in the abstract—there are plenty of humans who can 'hear' a symphony by reading the notation—but not experience them. The reverse may also be true. Sex may be an important, perhaps even overwhelming component of ETI art, for example. Could we understand that aesthetic?[7] Possibly. Possibly not, but at least we might have something to trade as we exchange knowledge of experiences.

[7]Disregarding the 'blue rinse brigade': those who feel sex and enjoyment of sex are against their beliefs.

However, if we accept the concept that ETI are unlikely to be physiologically homogeneous whether they are of the same originator species or not, then sex may be a crucial item on the agenda. First, genetic and somatic variability as discussed in the previous section could mean that the gender boundaries would be, at best, fuzzy and indeterminate. This could also mean that the whole panoply of modern attempts to pin down sexuality—hetero/homosexual, bisexual, polysexual, …—would simply have no application (Even among humans this may be more a result of social determinants and cultural beliefs, than a picture of nature). Add to the possibility of protean sexual personality and behavior the idea of cyborging and computer-generated artificial reality, and there would likely be no anchoring point for provincial, sexually-defining humans.

Sexuality, as noted above, is critically important for humans. At the current stage of our cultural being, and notably under the onslaught of attempts to rigidly redefine human sexuality (whether radically, as in the West, or conservatively as elsewhere) into fixed categories, the presence of much more freewheeling sexual beings among ETI might well have a greater impact on human culture than the impact contact will have on religion. In fact, given that sexual (and other physiological) proteanism is well within humans' technological grasp, it should not be surprising that information about ETI's fluid sexual practices *could* have greater impact on the wo/man in the street, than any other discovery. The degree to which our own rather timid and restrictive sexual practices would affect ETI, is likely to be somewhat more limited.

8.8.2 *The Effects of Sexuality on ETI Culture*

Since sex is such a primary feature of all human lives we tend to ignore many of its aspects. For example, there is no human society on record in which human females have been quite equal to males. This includes both more and less primitive societies (though primitive societies have a slightly better record than otherwise). This is not strictly speaking, a matter of discrimination in role opportunities, though that is what is most commonly visible. It is rather a complex of behaviors and mental structures which runs as a thread through human society, that is, all humans. Some of the basic questions regarding sex as a social phenomenon should be examined.

(1) Human sexuality is binary.
(2) Human sexuality assigns social roles to humans, notably, we equate gender (a set of roles and their associated behaviors) with sex (a biological functional division).
(3) These roles are hierarchic and supposed to correlate with sexual difference (in other words, 'female' roles are of less prestige and reward than roughly equivalent 'male' roles in a majority of human societies).
(4) The cusp of sexual division, the sexual act itself, is of major importance in human society, with effects that reach from social behavior through artifactual constraints (consider for example the near universal clothes preferences in the

current globalized civilization: in other words, how many men among the readers of this chapter regularly wear a dress in public?).

It does not concern us in this chapter what the genesis of these differences are. Here it is important to question what the effects would be of changes in these dimensions.

The first, and most critical question is the degree to which sexuality affects role choice and role evaluation. In a case where only one sex is sentient (see Chap. 3), the role of the non-sentient sex(es) is to reproduce. Possibly they may have other functions (e.g. it is barely possible that a smaller sentient sex will live parasitically on a larger non-sentient sex. The former would provide direction, perhaps handedness, the latter would provide food, mobility, security, etc.). Possibilities such as those aside, the sentience of lack of it of a given sex is not likely to help us foresee the nature of an ETI culture. An ETI society that does not assign roles or role valuations by sex may look down on the human species with justifiable contempt, but that does not help us predict their behavior to any great degree. The other case, however, is far more informative. Where an ETI society assigns high importance to sexual role assignments and valuations, we are likely to find a pattern that extends to that species' contacts with other ETI, inasmuch as sexual reproduction appears to be so critical a model for many other behaviors.

A second question that may be no less important is the degree of importance an ETI species assigns to the fact of reproduction. The issues of nurturance and performing sex as art have been discussed earlier. Here we are concerned with the issue of the sexual act itself. Among humans it permeates thinking, language, art. It drives divisions between cultures, represents modes of thinking, and is otherwise important and significant in almost all aspects of life. There is no clear understanding of what this obsession with the act of mating has done for human society (or to it: see Donnan and Magowan 2010). What seems to be clear is the tremendous energy involved in thinking about the reproductive act (disregarding actually doing it). One can view this outpouring of energy as either a waste of effort and a factor that hinders and impedes other fields of endeavor, or as one that has been the source of great intellectual and mental ferment. Little can be said about the degree of obsession of an ETI civilization with sex. It is quite possible that humans, at least, will be completely indifferent to hyper-concern with sexuality or lack of such concern (again, disregarding frequency and energy involved in actual mating). ETI may be amused, alarmed or indifferent themselves. The question is unanswerable, it does however point to an interesting issue deriving from communication with ETI.

As I noted above, we are largely blind to the total effect of our concern with mating, on our lives. Anthropologists, psychologists, historians and others have discussed different facets, but there is no understanding of the whole. The important issue here is whether another viewpoint—that of an ETI—would not be able to see the total effect, and, possibly, would such a different viewpoint not be able to see other factors, ones that are hidden from us, because we see the trees, rather than the forest? This points towards one of the major benefits we might receive from contact with other intelligent species: another viewpoint, what we might call social parallax.

8.8.3 Cooking

Cooking is universal—say Goody and Mennel, each in their own way—cuisine is a matter of class (Goody 1982; Mennell 1985). Cuisine requires a complex society in which a class of wealthy elites attempts to demonstrate and ascertain leadership and dominance through elaborate dining rituals, expensive imported foodstuffs, and displays of consumable wealth (Cardozo and Masaaki 1998; Elias 1978). Cuisines also bring about changes in aesthetic taste, as well as the consumption of foods which are not, on first impression, edible (Dalby 2000; Rozin 1987). In other words, cuisine is part of the inter-cultural, and exploratory expression of humanity. Cooking, on the other hand, is the application of technology to available foodstuff to make food more accessible and digestible. Cuisine is generally in the domain of the wealthy and aspirational, cooking is in the domain of everyone, and influenced largely by commonly available foodstuffs and heating/preservation resources.

But is ether of these universal in the sense we use here, in other words, do ETI dine? Certainly we can assume that a technological species would use some of that technology to ease the acquisition and consumption of fuel and necessary chemicals from the body. In the human case, the application of fire makes foods more easily digestible. Even if the application of heat is not beneficial to an ETI's foodtsuffs, other technological processes might prove useful and beneficial, and thus may be indulged in as a matter of course, or as the result of technological advances.

On the other hand, cuisine. Robert Heinlein has suggested that turning the absorption of fuel into an exercise in aesthetics is a stroke of genius. After all, fuel is fuel, and maybe is so considered by ETI. Food and dining are however more important than the mere consumption of fuel (and associated required substances). For humans, dining together, along with the specific rules of who cooks, who serves, who eats, how food is served and consumed, and when, are of enormous importance (Douglas 1972). This may or may be not important for ETI for physiological or other reasons. Even the food-is-only-fuel ETI will recognize that important communicative messages are embedded in any human meal. The sheer complexity of the rules of the average meal can be staggering (Douglas and Gross 1981; Farb and Armelagos 1980): if you doubt this, try serving your standard meal in reverse order, or ask your guests to eat a Western meal with their hands. The presentation of complex and elaborate artificial standards for seemingly mundane activities characterizes cultural rather than natural activities. Non-intelligent, non-cultural species are unlikely to develop such elaborate sets of rules that stand between the supply of natural needs and their consumption. By displaying cuisines, and even humbler cooking, a species makes a claim to being governed by cultural, rather than by natural determinants. And, to a degree, as we shall see subsequently, makes a claim to being part (at least potentially) of a civilization.

8.8.4 The Peculiar Concept of Sport

Unlike the many animals (usually, but not always, juveniles) who play, humans engage in games. Unlike play, which can be any form of spontaneous individual or group activity that may bring about no immediate reward but pleasure,[8] games are a form of play in which the parties agree on artificial rules that limit and restrict the activity (Ahern 1982; Avedon 1979; Blanchard and Cheska 1985). Most games played by children are of this sort, and some games such as tipcat still played today (or were when I was a child) have origins in ancient cultures (ancient Egypt and/or India, in the case of tipcat). A subset of games are what we call 'sport': physical activities engaged under artificial rules of greater or lesser elaboration, which may provide, in addition to the physical benefits to the practitioner, additional status and potential and actual financial benefits. Some games have been co-opted into very large economic practices worth billions, and have become major economic and even diplomatic enterprises.

The love of artificial rigmarole seems to be a feature of human psychology. We see it in religious and political ritual, cuisine, fashion, and sport. There are undoubtedly many complex reasons for this preference among humans. But how likely are the same features to appeal to ETI? Given that advanced ETI are likely to conform to an artificial lifestyle demanded by culture, it is also possible that they too need artificial stimulation to engage in levels of physical activity to benefit themselves. So the idea that ETI will engage in physical exercise is not too far-fetched. The question of whether they will translate this physiological demand into an economic and ritualized set of activities is a moot one. What is very likely, is that artificial rigmarole which is a rule-determined set of aesthetic bodily behaviors would be recognized by any ILF as indication of culture. So if we are to look at any ETI and need to ask ourselves the question of whether they are intelligent, or their elaborate productions are merely the result of a super bee-like instinct, we should ask ourselves whether they engage in elaborate, expensive, and non-utilitarian activities such as sport.

8.9 Fashion and Mode

"Fashion" for our purposes is a term that has two implications. The first, the personal, has to do with the presentation of self, that, as we have seen, in likely to be a feature of all ILFs. A second implication is that fashion is a social process, perhaps even a social movement, among humans. It is to be found in fields as diverse as bodily presentation, including clothing (where the term is most often to

[8]There is no denying that all forms of play *also* have delayed benefits including the development of leadership skills, physical training, preparation for activities such as hunting and stalking, and sexual preparation. That is stipulated, and not the major point here.

be found), artistic expression, and even the sciences (Abrahamson 1996, 2009; Bort and Kieser 2011; Lask 1985; Salzrnan 1988; Sunstein 2001). In many cases, writers who have mentioned scientific 'fashion' use the term pejoratively. In some instances, it has been used as an explanatory variable supporting ideas about the diffusion of innovations in science (Bikhchandani et al. 1992) and business (Abrahamson 1991). Some theories of fashion are based extensively on human peculiarities (Kaiser et al. 1995a, b): Kaiser et al.'s interpretation relies heavily on the existence of a capitalist economy; not something we would necessarily expect to find, particularly in a theorized advanced ETI. Other interpretations such as e.g. Simmel (1957) are more universal.

The importance of the concept of fashion for xenology lies in the ways in which fashions knit together disparate communities of thought, whether in some variation of a central concept, or as a contrast to it. This is peculiarly apt in the cases of cultures which are very widespread, and thus allow (and possibly require) the emergence of sub-cultures, networks and groups that may struggle to define themselves against others, or at least to announce their membership and uniqueness. At the individual level of personal adornment, the concept is important as well. I would argue that the homogenous ETI concept so prominent in much SETI thinking is a fallacy. As a proposition, the more advanced an ETI race is (in the terms we have used before) the more likely that ETI (whatever its social forms) will be heterogeneous. This flows directly from our definition of an advanced society: a society whose members are able to fully enjoy Maslow's principle of self-realization. The concept of fashion, however, allows us to go further than that, by providing some idea of the nature of this variability. This in turn means that one of the major problems facing a widespread advanced or develop culture is the development of mechanisms to ensure unity and preempt separatism.

An early analysis of fashion (Kroeber 1919) proposed the idea that fashion has independent cycles of long (decades)-period and short (years)-period waves. This has been supported and expanded by later analysis (Lowe and Lowe 1982). These regularities indicate that fashions (not necessarily only in clothing) demonstrate regularity and respond to social factors, rather than being random. Later analysis by e.g. Bikhchandani et al. (1992), Crane (2012), Kaiser (1995a and b), Laland et al. (2011), and Schrank (1973) demonstrated the importance of the fashion concept for the spread of innovations and the development of leadership.

The social perspective on fashion is notably important in that it helps in explaining two effects of interest here. First, it helps in explaining how ideas percolate, sometimes in attenuated form, from 'fashion setters' to the general populace of followers. This, as we shall see in the next chapter, is likely to be crucially important in interstellar civilizations, as well as solar cultures. A second feature is the difference between fashion and fad. Fads are localized, short-term, limited versions of fashions. They act with a particular fashion as a referent, supporting or undermining it, challenging its perceptions and limits by exaggeration and lampooning. By doing so they create and sustain, if only for a brief time, the establishment of localized, new fashion sub-cultures, with their own elites and followers. Significantly, they do so as a challenge to the conventional fashions

(Davis 1994; Silverberg 2009; van Ginneken 2003). And, following such a challenger, either the fashions change to reflect these new outlooks/fads, or the fad disappears. This is an important model, since it accounts for both the dynamics and statics of a system of presentation of intellectual and artifactual culture.

Though some of the arguments presented in the discourse about fashion are based on particularly human particulars—e.g. a capitalist economy, human dress conventions, and so on—the general principle of waves of fashion, inasmuch as it derives from what appear to be universal issues of communication between intellectual beings, may well occur among ETI as well. We can thus expect a number of features to be present among ETI (transcendentals, perhaps, excluded). First, given the need to present oneself publicly, we can expect ETI to have forms of personal display (suited, of course, to their sensorium and culture). Second, we can expect individuals with shared intellectual or other concerns, to display mutually comprehensible symbols (however construed) showing affiliation. Third, given the necessity of dynamism in any viable social system, we should expect these displays to change in waves of greater and lesser intensity and frequency. Fourth, It is quite possible that sciences, arts, and other intellectual endeavors will be affected and influenced by internal fashions. Finally, we should expect that there will be present at any given time a mutable roster of 'fashion leaders' whose influence on these intellectual activities, for ill or good, would affect the reception of communications with other ILFs.

8.10 Body Fashion, Body Modification, and the Morphological Variety of ETI

Among humans, the biological sciences have lagged, to some extent, behind the physical sciences. In the past few decades however, notably since the discovery of DNA, things have changed. First, we know now, at least in principle, that we can make changes in the human (and other bodies) by genetic engineering. Second, there has been a convergence between the mechanical and the biological at both the macro-scale, and the nano-scale: cyborgs, medichines (medical machines), and designer bodies are now within sight (Armitage 2003; Fitzpatrick 1999; Franklin 2006; Tomas 1995; Warwick 2003, 2010; Wilson 1995). It is thus reasonable to suppose that if ETI technology is ahead of ours, then bodily modification is practicable in three dimensions: genetic modification before birth, bodily modification after birth, and cyborging (melding of machine and biological form).

8.10.1 Fashioning the Body "I"

The engineering of human embryos to remove genetic faults, and even, potentially, to improve the embryo's chance of breeding successfully (by better intellectual or physical abilities) is a scientifically viable process. Among humans, what holds it back are combinations of intellectual fears (e.g., breeding a race of superpeople, fears of Frankenstein's monster, etc.), religious fears (tampering with the work of God: unsurprisingly, much cloning work in practice is carried out in Korea, perhaps less hampered by religious fears), and the very real fear that should these techniques be available to the rich, their grip on the rest of us will be unbreakable. Post-parturition genetic modification: the injection of genetic repair kits into the bodies of people with genetic disorders, is also possible (Bedau 2002; Dudley 1990; Suzuki and Knudtson 1990). The same is true of transplanting organs from one organism to another, albeit in small numbers and with uncertain results. This does imply that radically changing the body by transplanting is at least theoretically possible (Spallone 1992). Finally, as cyborging is in ongoing development, including the implantation of medical devices which is commonplace, and of other devices (Warwick 2004, 2010). Given that ETI are expected to be scientifically more advanced, it is likely that some of them (we must allow for the possibility that religious or other sentiments may hamper such changes in one or more race, advanced as they may be) will indeed engage in radical bodily modifications. The reasons for such modification would be any mixture of reasons that apply to humans as well: the desire to improve or change progeny for better success in life, whether in new environments (e.g. life in outer space) or in given social environments, the desire to avoid illness and to extend life, better fit to the social or technological environment, proclaiming membership in a group, aesthetics, etc.

What this means is that the external forms of individuals of any ETI race we contact will have great variation. So great, that we might not be able to identify them, from their morphology, as members of the same species, and that the boundaries between sexes and possibly species, as well as the biological and mechanical will be so amorphous as to be meaningless. Though, having said that, we need to keep in mind that the concept 'species', a nineteenth century invention, is a mammalo-centric concept. Other species, including many bacteria, flora, mycellia, and even fauna reproduce across species boundaries (Roughgarden 2004), which means that ET life forms may do the same, perhaps more so than on Earth. Not only that, it is certainly possible that we might not be able to distinguish them from their machines, and, if 'true' AI has been achieved, then non-biological organisms with different mixes of the biological and artifactual will make up the people we have to deal with, confusing the observer (us) even further. If intelligence is dependent on bodily form and sensorium, and given that the motive for genetic engineering may be increased intelligence of one's offspring (or oneself?), and cyborging will likely mean better access to faster calculative abilities and memory, we should expect any race of ETI to have widely varying forms of intelligence, thinking, and even possibly ethics. It may be that a basic rule of intelligent species is that they converge towards polymorphism and heterogeneity (perhaps giving naturally polymorphic races, if any, an advantage). If that is the case, we might hazard another rule: the social sciences of such races will heavily emphasize practical arrangements to keep these divergent interests from fighting one another to destruction. In other words, we, and they, will need to ask a crucial question: what could keep this wide mixture 'together' in some form? Sheer biological necessity (reproduction) is unlikely to be the sole criteria of social association.

8.11 Beyond the Cultural: Civilizations

Human cultures tend to be relatively localized. That is, the learned patterns of behavior that we have identified as culture, are extremely sensitive to environmental considerations. Here environment is meant in the broad sense to encompass artificial and human environments such as cities, as well as natural ones, the passage of time, and even of groups of individuals. This also means that differences will occur in a culture over time (as the circumstances change) and over distance (as the environment changes). An urban centered culture such as the US can and does contain a number of sub-cultures that share many features, and yet are distinctly different. Moreover, as time passes, these subcultures change and mutate as well.

A culture also normally consists of learned behaviors, that is, the definition is skewed in the direction of the constituents of human behavior, whereas the results of such behavior are usually secondary, except where they serve as inputs for other behaviors. However, as argued before, any ETI race is unlikely to have a single culture. Inasmuch as cultures are formed by surrounding ecologies, and the ecologies of the different habitats of advanced ETI are likely to be varied, we can also assume that a number of cultures and sub-cultures will coexist. Nonetheless, we (and presumably our hypothetical ETI) need some way of ensuring that there is a spine of commonality, a basis for unity that holds these different cultures (and their different world views and interests) together in some form. Else, we will likely be faced with a situation of continuous conflict. Not unlikely, as Brin argues, but it is worthwhile exploring the alternative: some phenomenon that is centripetal, not centrifugal. To stake a claim from the start, we shall call such a phenomenon 'civilization'.

Civilizations are normally composed of a number of cultures. In the broadest useful sense, a civilization is a system of cultures in which the member cultures share some common institutions. Note that this does not mean that the member cultures all share all the same institutions. It does, however mean, that at the very least they share some mode of communication across those institutions. The institutions may be religious, economic, political or even aesthetic and philosophical. What this means is that some core symbols (Ortner 1973) are commonly accepted throughout the civilization even if their precise interpretation may differ. As we have seen, symbols, even within the same culture, have the virtue that (a) they are commonly accepted, and (b) they are open to different interpretations.

The concept of civilization is particularly important because when we come to deal with modern, very complex, and very differentiated societies, we find that they encompass a great many differences in lifestyles and learned behaviors. It is to account for those that we need the term 'civilization'. Civilizations ultimately rely on communication. The better the communication (a function partly of distance, partly of technology) the more integrated the civilization will be. Whether this also implies homogeneity seems possible, but is also unproven: homogeneity as a function of communication, as for example in the world today as result of the media, may well be a human phenomenon. It may also be an illusion to some degree: we tend to see the similarities in Coca Cola signs or golden arches, and ignore the very different contexts and languages they encapsulate (Watson 1997).

Members of a civilization might share the following: (1) similarities in language; (2) use of a given level of technology; (3) central patterns of reciprocity; (4) basic canons about decision-making and action; (5) a commonly accepted structure of consciousness (ideas about logic, world view, time, the metaphysical (adapted from Nelson 1973). Within these structures, members of a civilization may differ quite radically. They may not all have the same access to the technology, they might not agree on the details of the metaphysics, but there will be commonly agreed grounds on the issue of debate. To give an example, a member of an inner-city youth gang does not speak the same language, nor have the same access to power, to communication, or to technology, as does a stockbroker who may work and live a few blocks from him. To put it another way, when we speak of "culture" we are speaking about the general homogeneity, the shared features, of learned behaviors in a category of people. When we speak of "civilization" we are speaking of the variance within communicative limits of a category of people. Civilizations encompass variety and exhibit it; cultures encompass homogeneity and exhibit it.

8.11.1 Example: Earth's Global Civilization

An example of a civilization all the readers of this book are part of, is the current globalized one. Virtually all of the seven billion people now alive are members of this civilization. Here and there—in the jungles of New Guinea and Brazil—are small groups that have no share in this civilization. But a Congolese peasant and a Canadian millionaire are both members, albeit with differential access to resources implicit in the civilization, different perceptions and appreciations of it, different abilities to influence it, and notably, different cultures.

One way to look at the concept of 'civilization' in a way that is useful for us, is to think of it as a sort of federation of cultures. A civilization provides a broad roof under which different cultures can coexist (competing or co-operating). Civilization, unlike culture, is not mandatory: it is an ideal towards which one is trained and which requires constant conscious adherence and practice. Given the utility of the concept, we need to ask what are the essential elements of civilization, and how can we apply the term, as an analytical concept, to the study of, and communication with ETI. We deal with this issue in the next chapter.

8.12 Summary and Conclusions

Partly the question of what sorts of culture an ETI society might have rests on our definition of what culture is, something we gradually crept up on. On the social science side, the answer seems to be that culture is some version of 'anything we learn'. At the other, high-tech end of the scale, there appear to be diverse opinions. These might differ from one another only in detail,. but these details may well be significant if we want to understand 'culture' rather than 'human culture'. One such analytically based technology comes from the area of AI research: "Culture is "a society's communal data base upon which it draws to define belief systems and accepted modes of behavior patterns" (Stonier 1988). On the other hand, a society is a set of regularized relations between people. Essential to that definition in *Homo sapiens*, are ideas about sex and reproduction (which includes concepts relating to family and kin), about hierarchical relationship structures and resource allocation (politics and law), and about ultimate meanings and how to enforce their consequences (religion and ritual), because that is the sort of animal we in particular are.

In thinking about ETI, two corollary hypotheses must be considered. First, a less significant one, the closer an ETI species is to us in its biological behavior, the more it is likely to have institutions similar to humanity's. A second corollary is far more important, because less biased: unless we understand the institutions and types of institutions an ETI species has, and no less importantly, why some may be lacking, we will not be able to communicate at any great depth. Moreover, we might also founder the communication effort by not realizing that these institutions are absent.

To summarize the nature of culture, we have seen that culture has the following characteristics:

- Culture is a learned set of behaviors.
- Members of a society are socialized into a culture by a variety of means which constitute an institution: a complex of sets of activities.
- Socialization, and continued conditioning create norms which members of the culture follow unconsciously, for the most part. These norms can become codified in the form of law, but they occur in all institutions, notably production and reproduction institutions.
- Members of different cultures can, as we know, communicate with one another, even learn to become members of a different culture in the fullest sense, since they share the same overall environment, phylogeny, and needs.
- Because cultural institutions derive at their fundamental bases, from a given set of ecological and physiological requirements for species survival, culture is inherently species-specific. Other species, obviously would be able to learn these behaviors and patterns of thought, but might find it very difficult to socialize to another culture, even though the newcomer's species too would necessarily have cultural institutions, since they would express different ecological and phylogenetic bases.

This chapter attempted to organize a common understanding of the meanings and functions of culture and of some human institutions: those that seem to be critical for understanding ETI. The following chapter concentrates on trying to understand the same phenomena from the point of view of ETI civilizations.

References

Abrahamson, E. (1991). Managerial fads and fashions: The diffusion and rejection of innovations. *The Academy of Management Review, 16*(3), 586–612.

Abrahamson, E. (1996). Management fashion. *Academy of Management Review, 21*(1), 254–285.

Abrahamson, E. (2009). Necessary conditions for the study of fads and fashions in science. *Scandinavian Journal of Management, 25*, 235–239.

Ahern, E. M. (1982). Rules in oracles and games. *Man, 17*(2), 302–13.

Ahmeti, Z. (2004). *The criminal law in the "kanun of Lekë Dukagjini—the Albanian customary law*. Switzerland: Univ. St. Gallen.

Almond, G. A. (1961). Comparative political systems. In R. C. Macridid & B. E. Bernard (Eds.), *Comparative politics* (pp. 35–49). Homewood, IL: The Dorsey Press.

Anderson, A. G. (1979). *The business of organized crime*. Stanford California: Hoover Institution Press.

Ansart, O. (2007). Loyalty in seventeenth and eighteenth century samurai discourse. *Japanese Studies, 27*(2), 139–154.

Arbib, M. A. (1979). Minds and millenia: The psychology of interstellar communication. *Cosmic Search, 1*(3), 21–25.

Armitage, J. (2003). Militarized bodies: An introduction. *Body and Society, 9*(4), 1–12.

Ashkenazi, M. (1981). Analysis of modern Japanese ritual. *Transactions of the International Conference of Orientalists in Japan, 26*, 17–29.

Ashkenazi, M. (1992). Not the sons of Adam: Religious response to ETI. *Space Policy, 8*(4), 341–350.

Ashkenazi, M. (1993). *Matsuri: Festivals of a Japanese Town*. Honolulu: University of Hawaii Press.

Avedon, E. M. (1979). The structural elements of games. In E. M. Avedon & B. Sutton-Smith (Eds.), *The study of games* (pp. 419–426). Huntington, NY: Robert E. Krieger Publishing Co.

Badescu, V. (2011). Free-floating planets as potential seats for aqueous and non-aqueous life. *Icarus, 216*(2), 485–491.

Bainbridge, W. S. (1984). Computer simulation of cultural drift: Limitations of interstellar colonization. *Journal of the British Interplanetary Society, 37*, 420–429

Ball, J. A. (1973). The zoo hypothesis. *Icarus, 19*, 347–349.

Barker, M. (2013). Consent is a grey area? A comparison of understandings of consent in fifty shades of grey and on the BDSM blogosphere. *Sexualities, 16*(8), 896–914.

Barkow, J. H. (2000). Do extraterrestrials have sex (and intelligence)? *Annals of the New York Academy of Sciences, 907*(1), 164–181.

Barnard, F. M. (1969). (trans, ed). *J. G. Herder on social and political culture*. Cambridge: Cambridge University Press.

Bedau, M. (2002). The scientific and philosophical scope of artificial life. *Leonardo, 35*(4), 395–400.

Benedict, R. (1937). Ritual. *Encyclopedia of the Social Sciences*, 396–397.

Berger, P. (1969). *The sacred canopy: Elements of a sociological theory of religion*. New York: Doubledoy/Anchor Books.

Berger, P., & Berger, B. (1979). Becoming a member of society. In R. Peter Isaac (Ed.), *Socialization and the life cycle* (pp. 4–20). New York: St. Martin's Press.

Berger, P., & Luckmann, T. (1971). *The social contruction of reality*. London: Penguin University Books.
Berlin, I. (1976). *Vico and Herder: Two studies in the history of ideas*. London: Hogarth Press.
Bertka, C. M. (2009). Astrobiology in a societal context. *Exploring the Origin, Extent and Future of Life*, 1–18.
Bestor, T. C. (1989). *Neighborhood Tokyo*. Stanford: Stanford University Press.
Bikhchandani, S., Hirshleifer, D., & Welch, I. (1992). A theory of fads, fashion, custom, and cultural change as informational cascades. *The Journal of Political Economy, 100*(5), 992–1026.
Billings, L. (2006). To the moon, mars, and beyond: Culture, law, and ethics in space-faring societies. *Bulletin of Science, Technology & Society, 26*(5), 430–437.
Blanchard, K., & Cheska, A. (1985). *The anthropology of sport*. South Hadley, MA: Bergin & Garvey.
Blinne, K. C. (2012). Auto(erotic)ethnography. *Sexualities, 15*(8), 953–977.
Bort, S., & Kieser, A. (2011). Fashion in organization theory: An empirical analysis of the diffusion of theoretical concepts. *Organization Studies, 32*(5), 655–681.
Bozgo, S., Sinakoli, X., & Spahia, E. (2002). Customary law, conflict, and development in the mountains of Albania. *Mountain Research and Development, 22*(4), 341–343.
Bracewell, R. L. (1974). *The galactic club: Intelligent life in outer space*. Stanford: Stanford Alumni Association.
Brandt, D., & Rose, C. (2004). Global networking and universal ethics. *AI & Society, 18*(4), 334–343.
Burchfiel, K. J. (1978). Economic organizations of crime: a study of the development of criminal enterprise. *Criminal Law Quarterly., 20*(4), 478–512.
Cardozo, S., & Masaaki, H. (1998). *Uncommon clay the life and pottery of Rosanjin*. Tokyo: Kodansha International.
Censor, D. (2005). Relativistic electrodynamics: Various postulate and ratiocination frameworks. *Progress in Electromagnetics Research, 52*, 301–320.
Chandler, D. (2001). Universal ethics and elite politics: The limits of normative human rights theory. *The International Journal of Human Rights, 5*(4), 72–89.
Chang, J. J., Huei-Chung, L., & Chen, M. (2005). Organized crime or individual crime? Endogenous size of a criminal organization and the optimal law enforcement. *Economic Inquiry, 43*(3), 661–675.
Chou, E. (1971). *The dragon and the phoenix*. New York: Arbor House.
Coser, L. (1967). *Continuities in the study of social conflict*. New York: The Free Press.
Crane, D. (1994). *The sociology of culture: Emerging theoretical perspectives*. Oxford: Blackwell Publishers.
Cunningham, W. P. (1998). The golden rule as universal ethical norm. *Journal of Business Ethics, 17*, 105–109.
Dalby, A. (2000). *Dangerous tastes*. London: British Museum Press.
Davis, F. (1994). *Fashion, culture, and identity*. Chicago: University of Chicago Press.
Dawkins, R. (2006). *The selfish gene*. Oxford: Oxford University Press.
Dawkins, R. (2009). *The god delusion*. New York: Random House.
Deardorff, J. W. (1987). Examination of the embargo hypothesis as an explanation for the great silence. *Journal of the British Interplanetary Society, 40*, 373–379.
Diamond, J. (1992). *The third chimpanzee: The evolution and future of the human animal*. New York: Harper and Collins.
Diamond, J. (1997). *Why is sex fun? The evolution of human sexuality*. New York: Basic Books.
Dick, A. R. (1995). When does organized crime pay? A transaction cost analysis. *International Review of Law and Economics, 15*(1), 25–45.
Donnan, H., & Magowan, F. (2010). *The anthropology of sex*. New York: Berg Publishing.
Douglas, M. (1972). Deciphering a meal. *Daedalus*, 61–81.
Douglas, M., & Gross, J. (1981). Food and culture: Measuring the intricacy of rule systems. *Social Science Information, 20*(1), 1–35.

Dudley, W. (Ed.). (1990). *Genetic engineering: Opposing viewpoints*. San Diego, CA: Greenhaven Press.

Duncan, H. D. (1968). *Symbols in society*. New York: Oxford University Press.

Durkheim, E. (1947). *The elementary forms of the religious life*. London: George Allen & Unwin.

Elias, N. (1978). *The history of manners*. New York: Urizen Books.

Ericksen, K. P., & Heather, H. (1992). Blood Feuds": Cross-cultural variations in kin group vengeance. *Cross-Cultural Research, 26*(1–4), 57–85.

Farb, P., & Armelagos, G. (1980). *Consuming passions: The anthropology of eating*. New York: Washington Square Press.

Finckenauer, J. O., & Chin, K. (2007). *Asian transnational organized crime and its impact on the United States*. Washington DC: National Institute of Justice.

Fitzpatrick, T. (1999). Social policy for cyborgs. *Body and Society, 5*(1), 93–116.

Fleischacker, S. (1999). From cultural diversity to universal ethics: Three models. *Cultural Dynamics, 11*(1), 105–128.

Fleising, U., & Goldenberg, S. (1987). Ecology, social structure, and blood Feud. *Cross-Cultural Research, 21*(1-4), 160–181.

Fox, R. (1983). *Kinship and marriage: An anthropological perspective*. Cambridge: Cambridge University Press.

Franklin, S. (2006). The cyborg embryo: Our path to transbiology. *Theory, Culture and Society, 23* (7–8), 167–187.

Gardner, J. (2003). *Biocosm: The new scientific theory of evolution: Intelligent life is the architect of the universe*. Makawao Maui HI: Inner Ocean.

Gilsenan, M. (1973). *Saint and sufi in modern Egypt: An essay in comparative religion*. Oxford: Clarendon Press.

Goody, J. (1969). *Comparative studies in kinship*. London: Routledge & Kegan Paul.

Goody, J. (1982). *Cooking, cuisine and class*. Cambridge: Cambridge University Press.

Govers, R., & Go, F. (2016). *Place branding: Glocal, virtual and physical identities, constructed, imagined and experienced*. Berlin: Springer.

Gudeman, S. (1986). *Economics as culture: Models and metaphors of livelihood*. London: Routledge and Kegan Paul.

Guthrie, S. (1980). A cognitive theory of religion. *Current Anthropology, 21*(2), 181–204.

Hagan, F. E. (2006). Organized crime and 'organized crime': Indeterminate problems of definition. *Trends in Organized Crime, 9*(4), 127–137.

Hanna, J. L. (1987). *To dance is human: A theory of nonverbal communication*. Chicago: University of Chicago Press.

Haritaworn, J., Lin, C.-J., & Klesse, C. (2006). Poly/logue: A critical introduction to polyamory. *Sexualities, 9*(5), 515–529.

Harris, M. (1979). *Cultural materialism: The struggle for a science of culture*. New York: Random House.

Harrison, A. A. (2013). Cosmic evolution, reciprocity, and interstellar tit for tat. In D. A. Vakoch (Eds.), *Extraterrestrial altruism: Evolution and ethics in the cosmos* (pp. 3–25). Heidelberg: Springer.

Heppenheimer, T. A. (1978). *Colonies in space*. New York: Warner Books.

Hoebel, E. A. (1967). *The law of primitive man*. Cambridge, Mass: Harvard University Press.

Hoebel, E. A. (1968). Law and anthropology. In M. H. Fried (Ed.), *Readings in anthropology* (pp. 447–461). New York: Thomas Crowell.

Hoel, N. (2015). Engaging Islamic sexual ethics: Intimacy, pleasure and sacrality. *Sexualities, 18* (1-2), 80–99.

Holbrow, C., Russell, A., & Sutton, G. (Eds.). (1985). *Space colonization: Technology and the liberal arts* (Vol. 148). Geneva, NY: American Institute of Physics.

Hori, I. (1968). *Folk religion in Japan*. Chicago: University of Chicago Press.

Ianni, F., & Reuss-Ianni, E. (1972). *Family business: Kinship and social control in organized crime*. New York: Russel-Sage.

Irvine, J. T. (1979). Formality and informality in communicative events. *American Anthropologist, 81*(4), 773–791.
Jones, V. C. (1948). *The Hatfields and the McCoys*. Chapel Hill: University of North Carolina Press.
Kaiser, A. (2004). Sound as intercultural communication: A meta-analysis of music with implications for SETI. *Leonardo, 37*(1), 36–37.
Kaiser, S. B., Nagasawa, R. H., & Hutton, S. S. (1995a). Construction of an symbolic interaction theory of fashion: Part 1. Ambivalence and change. *Clothing and Textile Research, 13*(3), 172–183.
Kaiser, S. B., Nagasawa, R. H., & Hutton, S. S. (1995b). Construction of an symbolic interaction theory of fashion: Part 3. Context of explanation. *Clothing and Textile Research, 14*(1), 54–62.
Kaplan, A. M., & Michael, H. (2009). The fairyland of second life: Virtual social worlds and how to use them. *Business horizons, 5*(6), 563–572.
Kardashev, N. S. (1964). Transmission of information by extraterrestrial civilizations. *Soviet Astronomy, 8*(2), 217–221.
Kroeber, A. L., & Kluckhorn, C. (1963). *Culture: A critical review of concepts and definitions*. New York: Vintage Books.
Kroeber, A. L., & Parsons, T. (1970). The concepts of culture and of social system. In E. A. Hammel & W. S. Simmons (Eds.), *Man makes sense* (pp. 84–87). Boston: Little, Brown, and Co.
Kuttner, R. E. (1981). The nature of extraterrestrial communication. *Cosmic Search Spring, 3*, 20–21.
Kuznar, L. A. (2000). Application of general utility theory for estimating value in non-western societies. *Field Methods, 12*(4), 334–345.
Lacher, W. (2012). *Organized crime and conflict in the Sahel-Sahara region*. Washington DC: Carnegie Endowment for International Peace.
Laland, K. N., Atton, N., & Webster, M. M. (2011). From fish to fashion: Experimental and theoretical insights into the evolution of culture. *Philosophical Transactions of the Royal Society of London B: Biological Sciences, 366*(1567), 958–968.
Lask, B. (1985). Editorial-fads and fashions. *Journal of Family Therapy, 7*, 313–314.
Levi-Strauss, C. (1979). *Myth and meaning*. London: Routledge.
Lichbach, M. I., & Zuckerman, A. S. (1997). *Comparative politics: Rationality, culture, and structure*. Cambridge: Cambridge University Press.
Lovelock, J. E., & Lynn, M. (1974). Atmospheric homeostasis by and for the biosphere: The Gaia hypothesis. *Tellus, 26*(1–2), 2–10.
Lumsden, C., & Edward, O. W. (1983). The dawn of intelligence. *The Sciences, 23*(2), 23–31.
Mackenzie, J. (2007). A reply on behalf of the relativist to Mark Mason's justification of universal ethical principles. *Educational Philosophy and Theory, 39*(6), 657–675.
Malinowski, B. (1968). In Tewara and Sanaroa—mythology of the Kula. In R. A. Georges (Ed.), *Studies on mythology* (pp. 72–101). Homewood, IL: The Dorsey Press.
Margulis, L., & Lovelock, J. E. (1974). Biological modulation of the Earth's atmosphere. *Icarus, 21*, 471–489.
Markowitz, F., & Ashkenazi, M. (Eds.). (1999). *Sex, sexuality, and the anthropologist*. Urbana: University of Illinois Press.
Marshack, A. (1976). Some implications of the paleolithic symbol evidence for the origins of language. *Current Anthropology, 13*, 455–477.
Martin, A. (2013). Fifty shades of sex shop: Sexual fantasy for sale. *Sexualities, 16*(8), 980–984.
Mason, M. (2005). A justification, after the postmodern turn, of universal ethical principles and educational ideals. *Educational Philosophy and Theory, 37*(6), 799–815.
Mauss, M. (1966). *The gift*. Ian Cunnison (Trans). London: Cohen & West.
Mautner, M. N. (2009). Life-centered ethics, and the human future in space. *Bioethics, 23*(8), 433–440.
McIllwain, J. S. (1999). Organized crime: A social network approach. *Crime, Law and Social Change, 32*(4), 301–323.

Mennell, S. (1985). *All manners of food*. Oxford: Basil Blackwell.
Moore, S. F., & Barbara, M. (1977). Introduction: Secular ritual: Forms and meanings. In S. F. Moore & B. Myerhoff (Eds.), *Secular ritual*. Amsterdam: Van Gorcum.
Musso, P. (2010). A language based on analogy to communicate cultural concepts in SETI. *Acta Astronautica, 68*(3-4), 489–499.
Nelson, B. (1973). Civilizational complexes and intercivilizational encounters. *Sociological Analysis, 34*(2), 79–105.
Newmahr, S. (2008). Becoming a sadomasochist: Integrating self and other in ethnographic analysis. *Journal of Contemporary Ethnography Journal of Contemporary Ethnography, 37* (5), 619–643.
Newman, G. R., & Lynch, M. J. (1987). From feuding to terrorism: The ideology of vengeance. *Contemporary Crises, 11*(3), 223–242.
O'Neill, G. K. (1974). The colonization of space. *Physics Today, 74*, 32–40.
Ortner, S. (1973). On key symbols. *American Anthropologist, 75*, 1338–1346.
Pfaller, A., & Lerch, M. (2005). *Challenges of globalization*. Rochester, NY: Transaction Publishers.
Rafky, D. M. (1971). Phenomenology and socialization: Some comments on the assumptions underlying socialization theory. *Sociological Analysis, 32*(1), 7–20.
Rappaport, R. (1974). The obvious aspects of ritual. *Cambridge Anthropology, 2*(1), 3–69.
Rescher, N. (1985). Extraterrestrial science. In E. Regis (Ed.), *Extraterrestrials: Science and alien intelligence* (pp. 83–116). Cambridge: Cambridge University Press.
Richard, N., & Cohen, D. (1996). *Culture of honor: The psychology of violence in the South*. Boulder, CO: Westview Press.
Roederer, J. G. (2002). Biological conditions for the emergence of musical arts in a civilization of intelligent beings. *International Atomic Energy Agency (IAEA) Bulletin, 34*(1).
Roughgarden, J. (2004). *Evolution's rainbow: Diversity, gender, and sexuality in nature and people*. Berkeley: University of California Press.
Royce, J. (1976). Consciousness and the cosmos. In J. Christian (Ed.), *Extra terrestrial intelligence: The first encounter* (pp. 177–195). Buffalo, NY: Prometheus Books.
Rozin, P. (1987). Psychobiological perspectives on food preferences and avoidances. In M. Harris & E. Ross (Eds.), *Food and evolution: Toward a theory of human food habits* (pp. 181–205). Philadelphia: Temple University Press.
Ruse, M. (1985). Is rape wrong on Andromeda? An introduction to extraterrestrial evolution, science, and morality. In E. Regis (Ed.), *Extraterrestrials: Science and alien intelligence* (pp. 43–78). Cambridge: Cambridge University Press.
Salzman, P. C. (1988). Fads and fashions in anthropology. *Anthropology Newsletter, 29*(5), 31–33.
Schrank, H. L. (1973). Correlates of fashion leadership: Implications for fashion process theory. *The Sociological Quarterly, 14*(4), 534–543.
Silverberg, M. (2009). *Erotic grotesque nonsense: The mass culture of Japanese modern times*. Berkeley CA: University of California Press.
Simmel, G. (1957). Fashion. *American Journal of Sociology*, 541–58.
Singer, P. (2011). *The expanding circle: Ethics, evolution, and moral progress*. Princeton: Princeton University Press.
Smith, M. G. (1966). A structural approach to comparative politics. In D. Easton (Ed.), *Varieties of Political Theory*. Englewood Cliffs: Prentice-Hall
Spallone, P. (1992). *Generation games: Genetic engineering and the future for our lives*. London: The Women's Press.
Spiro, M. E. (1966). Religion: Problems of definition and explanation. In M. Banton (Ed.), *Anthropological approaches to the study of religion* (pp. 84–126). London: Tavistock Publications Ltd.
Stark, R., & Glock, G. Y. (1969). Dimensions of religious commitment. In R. Robertson (Ed.), *Sociology of religion: Selected readings* (pp. 253–261). Baltimore: Penguin.
Steward, J. H. (1973). The concept and method of cultural ecology. *High Points in Anthropology*, 322–332.

Stonier, T. (1988). Machine intelligence and the long-term future of the human species. *AI and Society, 2*, 133–139.
Strentz, H. (2002). Universal ethical standards. *Journal of Mass Media Ethics, 17*(4), 263–276.
Sugiyama, S. (2008). Meat diet and use of milk in the history of Japan. *Yakushigaku Zasshi [Journal of Japanese History of Pharmacy], 43*(1), 91–94.
Sunstein, C. R. (2001). On academic fads and fashions. *Michigan Law Review, 99*(6), 1251–1264.
Suzuki, D. T., & Knudtson, P. (1990). *Genethics: The ethics of engineering life*. Cambridge, MA: Harvard University Press.
Swartz, M. J, Turner, V. W., & Tuden, A. (1968). Political anthropology. In M. H. Fried (Ed.), *Readings in anthropology* (pp.409–446). New York: Thomas Crowell.
Tellenbach, S. (2006). Blood feuds and how to bring them to an end: Experiences from Turkey. *Conflicts and conflict resolution in Middle Eastern societies* (Schriftenreihe des Max-Planck-Instituts für ausländisches und internationales Strafrecht : Reihe I, Interdisziplinäre Forschungen aus Strafrecht und Kriminologie ; Bd. I 13), pp. 637–643.
Turner, V. (1967). *The forest of symbols*. Ithaca: Cornell University Press.
Turner, V. (1969). *The ritual process: Structure and anti-structure*. Chicago: Aldine.
Turner, B. (1985). Asking sociological questions about Lagrange Five. In C. Holbrow, A. Russell & G. Sutton (Eds.), *Space colonization: Technology and the liberal arts* (pp. 56–72). Geneva, NY: American Institute of Physics.
Vakoch, D. A. (2010a). An iconic approach to communicating musical concepts in interstellar messages. *Acta Astronautica, 67*(11–12), 1406–1409.
Vakoch, D. A. (2010b). What does it mean to be human?: Reflections on the portrayal of pain in interstellar messages. *Acta Astronautica, 68*(3–4), 445–450.
Van Ginneken, J. (203). Fads and crazes. *Etnofoor, 16*(2), 132–137.
Warwick, K. (2003). Cyborg morals, cyborg values, cyborg ethics. *Ethics and Information Technology, 5*, 131–137.
Warwick, K. I. (2004). *Cyborg*. Urbana: University of Illinois Press.
Warwick, K. (2010). Implications and consequences of robots with biological brains. *Ethics and Information Technology, 12*(3), 223–234.
Watson, J. L. (Ed.). (1997). *Golden arches east: Mcdonalds' in East Asia*. Stanford: Stanford University Press.
White, L. A. (1959). *The evolution of culture*. New York: McGraw-Hill.
Wilson, E. O. (1975). *Sociobiology: The new synthesis*. Cambridge Mass: Harvard University Press.
Wilson, R. R. (1995). Cyber(body)parts: Prosthetic consciousness. *Body and Society, 1*(3-4), 239–259.
Wittfogel, K. A. (1956). *The hydraulic civilizations*. Chicago: The University of Chicago Press.
Xu, J. J., & Chen, H. (2005). Crimenet explorer: A framework for criminal network knowledge discovery. *ACM Transactions on Information Systems (TOIS), 23*(2), 201–226.
Zaitsev, A. L. (2008). The first musical interstellar radio message. *Journal of Communications Technology and Electronics, 53*(9), 1107–1113.
Zubrin, R. (2000). *Entering space: Creating a spacefaring civilization*. New York: JP Tarcher.

Chapter 9
ET Civilizations

"Culture" as we have seen, is a useful term to circumscribe a set of learned behaviors of a group of actors, which can conveniently (and conventionally) be divided into social institutions that provide behavioral responses to certain needs of individuals within their social, technical, and natural ecologies. Cultures are effectively local—whether a few hundred or many millions participate—and tend to fission into sub-cultures. As a rule of thumb, the greater the communicative distance between the actors, the more the actors will need to adjust to different ecosystems (again, the natural, the technical, and the social), and more likely it is that different cultures will eventually emerge. What do we need the term 'civilization' for? After all, culture neatly circumscribes what we mean: patterned learned behavior. Nevertheless, notably when discussing a far flung culture, we must acknowledge that the term 'culture' is somewhat restricting (or useless!): it presupposes a homogeneity that for good reasons, we cannot assume exists in a far flung, possibly interstellar, society.

Intelligent extraterrestrials (certainly those able to build large radio telescopes) *may* be planet bound for any of many reasons. For example, physiological inability to travel or live in space might be one, lack of resources for space exploration another, the type of solar system a third. However we can assume (and indeed, this is the assumption of many SETI writers, see e.g. Kardashev 1964; Stephenson 1981) that intelligent, highly technological races will seek to colonize their solar systems (if only for access to the resources in space), and possibly the galaxy (Ostriker and Turner 1985; Turner 1984; Walters et al. 1980).

In this chapter, we shall examine a concept often used, and worse, misused, in the discussion of ETI: civilization. Just as with 'culture', the intention is to find a useful definition, or circumscription of the term, which will help us in understanding the nature of ETI.

The issue of civilization is a ticklish one. Like the term 'culture' it is used in two very different senses. The popular meaning has to do with refinement, 'elite' behaviors, propriety. In a sociological sense it has been used extremely broadly, including what we have called here "a culture" (e.g. Warner 1958) or very narrowly

M. Ashkenazi, *What We Know About Extraterrestrial Intelligence*,
Space and Society, DOI 10.1007/978-3-319-44456-7_9

to refer strictly to technically complex social systems (Merton 1936) or to any complex of behaviors that appear to be comprehensive and distinguishable from others (Umesao 1986).

9.1 A Critique of the Term 'Civilization'

Over the past decades, the term civilization has generally fallen out of favor for social scientists. There are two reasons for this, which need to be mentioned, before proceeding to revive the concept as a useful heuristic device for our purposes here. "Civilized" has meant (depending on culture, and usually, class) being able to eat with knife and fork, not eating other people, knowing how to read and write classical Chinese, knowing how and when to get drunk, wearing a tunic but no trousers, or a number of other markers. Second, like the term 'culture' the term 'civilization' is a highly charged word. Used as an adverb or adjective, it served as a boundary separating the good—the civilized—from the bad—the barbaric.

The use of the term 'civilization' virtually vanished from social science discourse during and after the seventies of the previous century (though undergoing something of a comeback in the 2000s). Partly this was due to embarrassment: the use the term had been put to was largely pejorative, either providing a supposed hierarchy of worth among different cultures (*we*, you understand, were civilized, while *you* were not. More often then not, 'we' were white, Christian, Europeans. 'You' were everyone else). There were some exceptions to the rule: Warner (1958. Original publication date 1937) described the Yolngu, an Australian aboriginal people as a 'civilization', notwithstanding their paucity of technical sophistication or economic complexity. This, however, does not help us much: the Yolngu were a small, relatively homogeneous group that today we would describe as a culture.

Therefore, before discussing ETI civilizations—a term that has been thrown about with great abandon by the SETI community—it is necessary to see whether the term serves any useful function for xenology or not. If we do not do so, we stand to be in a position of prejudicing our discussion from the outset. For analytical reasons, we need to find a set of practical socio-intellectual phenomena which (a) *allow* for the adherence of members of different cultures, (b) permit or even encourage forms of change, (c) allow for great flexibility, (d) function over long ranges of distance and time, and (e) help in establishing and maintaining a sense of commonality, even union, among participants in these phenomena.

9.2 The Nature of Civilization

In order to fulfill the requirements for a socio-intellectual phenomenon that will transcend the centrifugal nature of cultures, we need to make some inductive propositions about features that will help in ensuring centripetality in very large populations. These features cannot be based on force (economic or military), and perhaps not even on morality in the conventional sense, for reasons discussed before. In a developed or advanced species, we should expect that the burden would be taken up by a mix of self-correcting technological and intellectual features.

I argue that a civilization is a (socio-cultural) structure composed of three structuring branches: a *complex set of standards* set forth in a fuzzy set of agreed-upon *texts and narratives*. It is a *learned style*, of behaviors for individuals and groups. Finally, a civilization might have a *physical component* to it, which, I argue, needs to be addressed very carefully in the case of inter-planetary/stellar civilizations.

9.2.1 Complex Set of Standards

The Oxford dictionary defines a *standard* as "a level of quality or attainment; … something used as a measure, norm, or model in comparative evaluations". Both of these qualities are apt here. Importantly, standards are to be attained: they are cynosures, not givens, and the higher the standards, the more investment is needed to attain it. Among other things, standards (movable and varied to be sure) are what fashion (or rather, fashionable people) tries to achieve.

Intellectual Standards A civilization can be characterized, most directly, by conceiving of it as an idea, in effect, an idealized map of social, cultural, and material reality (extrapolated from Nelson 1973). It consists of three sets of standards: *aesthetic* (that is, a set of measures by which reality can be judged as pleasing for its own sake); a set of *ethics*, that is standards of behavior to which all those who profess to be members of the civilization are expected to pay at least lip service, since they are laid down by traditionally significant authorities in the far, and perhaps mythic past; and a set of commonly accepted *knowledge*, which members of the civilization are expected to be privy to.

Praxis Ideas, noble or base as they may be, are hardly sufficient: anyone can profess to have ideas, or to believe in standards. Ideas must be maintained, preserved, and practiced, which brings us to the second element of civilization: a *praxis*, or behavioral dimension. Members of a civilization, basing themselves on intellectual standards, have a common set of agreed upon *behaviors* that exhibit adherence to the standards of aesthetics and ethics. This set of behaviors can cover the entire range of social and material relationships from completely agonistic, through neutral, to completely amoristic (in other words, the civilization has standards for both love and war, and everything in between, provided it is done *with style*). The behaviors defined might include those towards *all socially significant others*, from friends to enemies, equals and unequals, as well as towards particular *sets of objects*, based on how they fit into the accepted aesthetic and ethic. I would argue that the more comprehensive the defined behaviors are, and the fewer items/persons excluded, the stronger the civilizational impulse is likely to be.

Canon Finally, the ideas and behaviors that make up the civilization are described and discussed in a corpus of texts and narratives, which are available to, commented upon, and emulated insofar as possible, by members of the civilization. Among developed and advanced ILFs, this canon is likely to be stored and available in widespread form, through electronic or other communicative means. We *do* know that in the human proto-civilizations (Rome, England, US) and civilizations we know of (China and Globalized world), an essential component has always been access to written/printed forms of knowledge of the ideology and praxis required of a member of the civilization.

In practice, the ideas, praxis, and canons of a civilization are spread differentially throughout a civilization's population. Virtually everyone *knows* about them, if only in general. A great number of members follow *some* of the behavioral and other expectations. A much smaller number live their lives according to these ideas, practice the expected behaviors, and have knowledge of the texts. A much smaller percentage are fully knowledgeable about all three areas of civilization, adhere to them and perform them as full socio-cultural displays, and are erudite expounders and discussants of the canon or large parts of it.

Where such a socio-cultural structure—a civilization—exists among technically advanced ETI species, we can expect to see two other phenomena cropping up as well. First, we are likely to see that the small percentage that are truly knowledgeable about, and practice the civilization, form an elite of what the Chinese called *junzi* (usually translated as 'superior person' or 'gentleman' though it is used here without any gender or other connotation). To become a *junzi*, whatever one's social status at birth, and as long as one is born with the inherent desire and ability to *become* a *junzi* is simply a matter of learning how to think, behave, and know, as other *junzi* do. The initial standards and value having been set by unquestioned authorities (usually in the mythical past), one proceeds to study, and, by doing so, becomes a part of the elite that holds the civilization together. As a type case, membership in a civilization is not dependent on personal physical or social membership qualities, but simply on one's adherence to the principles of the civilization. Not unnaturally, not many individuals want to spend time and energy becoming full members of the club: sometimes a little knowledge goes a long way, and it is sufficient to be able to put on the external fashions of the civilization.

On the social side, the '*junzi*', 'gentlemen', 'cognoscenti' or whatever they are called become a social network with very widespread nodes, all of whom share a viewpoint, and all of whom because they are invested heavily in the civilization, are also major supporters. Since they are a network, it is not surprising that they are likely to support one another in other activities. In many cases they become the elite: those who can interpret the civilization to those less knowledgeable. They also become a cynosure: those who are not as knowledgeable as the experts, look upon them as models and examples.

On the intellectual side, since the gentlemen *know* the things that *are important* (and they repeatedly demonstrate that by their display of knowledge of the canon), they shape, and become, intellectual leaders, and possibly political and economic elites, or a part of the elite. Significantly, while the gentlemen can translate their moral and ethical qualities into political and other leadership, this is not a given: their status rests on their support of the civilization, not the specific cultures or cultural institutions within it. And because civilization is imbued in a mix of individuals, visible practices, and long-term recordings which are easily accessible, it has both an inherent self-correcting function (the *junzi* as a whole are likely to examine any innovation to civilizational practices very carefully, thoroughly, and broadly), while allowing for changes and evolution of concepts and practices (even if that sometimes means only nominal adherence to some of the canon). The concept of junzi is to my mind crucial to understanding any world, solar, or inter-stellar civilization, and we shall return to it subsequently in this chapter.

9.3 Learned Style

It is important to understand that becoming a *junzi*/gentleman/cognoscento is a *learned* process. One cannot be born a *junzi*: a born *junzi* is a contradiction in terms, and use of descent as a membership criterion leads very quickly to a destruction of the civilization and the emergence of an aristocratic elite (as happened in most human civilizations). Of course those with more economic resources at their disposal are more likely to have the time and leisure to devote to learning. This might be a feature in developed ETI cultures, but, by definition, not in advanced ones, where the issue of resource availability is a non-issue. Nonetheless, as we know from human history, where opportunities are offered, determined people with few resources can and will take advantage of them.

Given that what defines a *junzi* (we'll use *junzi* since it is non-gender biased) is learning and practice, it might be useful to characterize what this means socially. The *junzi* in any given civilization are a subset of a broader range of *practitioners* of the civilization they are embedded in. They *may* also fulfill political, economic, military, artistic, sexual, or other roles as central or peripheral figures: the degree to which any individual has civilizational attainments is neither dependent upon, nor strongly correlated with other social roles.

Practitioners span a spectrum from *junzi* (individuals who by their deportment, participation in maintaining standards, and knowledge of the praxis are recognized as civilizational experts by other *junzi* and by the pubic as a whole) at one pole, to civilizational ignoramuses at the other.[1] By their activities, deportment, and active teaching, *junzi* 'infect' others with civilization, which can transcend their daily lives.

In sociological practice, we can roughly divided the persons carrying civilization, its practitioners, into three categories. *Experts* are *junzi*, who know, discuss, expound and may teach, civilization, its practices, standards, and canon. *Dabblers* are those individuals who while not experts themselves, practice and strive to attain civilizational standards, perhaps only in one or two media: artistic, communicative, performative, or some other criterion. The vast mass of the population are most likely to be *accepters*: they know about the standards of civilization, they may try to some degree to follow these standards, they may be aware of some of the canon or all of it, but overall, they have little investment or involvement beyond the passive. Of course it is possible, notably among advanced ETI, that most, even all of the population are experts, though even there we should expect to find those among them who are looked up to as authorities in all or some of the civilizational practices.

Since civilization is a learned *style*, rather than an absolute, we can expect that the principles which affect fashion and style will apply here as well, to a lesser or greater degree. Modes will change, within the framework of the encompassing canon. These changes will likely be cyclical, as old or worn practices are cycled out to the margins, and new ones are cycled in. In an interstellar civilization, we can expect these waves of style to take many centuries and millennia since the fastest propagation will be at light speed across space. And, as in style, those *junzi* who can justify and sustain an innovation of praxis, will likely emerge as civilizational opinion-makers (not to be confused with political or economic leaders, though they may be that as well).

9.4 Possible Material Implications

The nature of civilization as a blend of intellectual ideas and their expression within the framework of agreed-upon standards, may well have several material implications worth exploring, because they are important for human futures.

First, the presence of a civilization as defined can be logically translated into other features of ETI societies. We can hypothesize, that the more advanced an ETI society is, the higher the percentage of individuals able to devote time to self actualization (see the earlier discussion of Maslow) and the higher the number of civilization experts. Parallel to that, we would expect that the facilities for interpersonal and public communication would cover the entire populace. Moreover, and looking at the other end, ETI we come into contact with that have a high penetration of

[1] I feel uncomfortable using the word 'ignoramus' which, by its history and definition is pejorative, but cannot find a more appropriate English term. Closer to what I mean is a combination of two states defined in the Passover Haggadah: the child who does not know—tam—and the child who "does-not-know-how-to-ask".

standards, practice and access to a single (possibly complex) canon and who express the need to attain certain universal standards, are likely to have an excess of energy available to them, whether they are Kardashev 'civilizations' or not.

Second, the members of a civilization *may* have a political commons of some sort: a world, solar-system, or interstellar government. Given the length of time edicts from the government could percolate, or political agreement could be reached, this seems highly unlikely. It seems far more likely that the standards expressed in the civilizational canons will provide frameworks, and the standards of civilization will provide bases, for political activity within each culture that makes up the civilization. After that, each culture or political demesne would have to work out its own arrangements. That is to say, that we would likely find many different, perhaps cross-cutting polities, operating on the basis of common standards, with little or no enforcement across different cultures (of course, this is more true of interstellar civilizations than of planetary ones).

Third, we would likely find that whatever the shape or form of the members of the civilization, they will share an appreciation of their common intellectual heritage, imbued in the canonical texts from which standards are derived for all. The chain from canon, through standards, to praxis which attempts to reach and maintain those common standards, would be the common basis holding the civilization together.

9.5 Keeping It Together: No Politics Today, Thank You

As we have seen, what keeps a civilization together is a circulation of ideas and standards, and the (potential for) exchanges between members of the civilization. Exchanges would likely largely be intellectual and through messaging, though physical movements of individuals would certainly be conceivable, depending on the realities of physical travel, which in turn depends on energy availability and biological features. However, the exchange of ideas is perpetuated easily enough by electromagnetic radiation, even if members of the civilization are perpetually 'behind' temporally: by the time they have adopted a new element, the place of origin may have moved on.

Whether such a civilization is capable of mobilizing (force, energy or food supplies) is debatable. And the potential descent into a Galactic empire (or any other wide-spanning political unit) is *possible* (depending on issues such as the psychological and physical makeup of the race in question, and the interstellar distances under consideration) but highly unlikely in an Einsteinian universe.

Distance again: Empires In the absence of any evidence that the Einsteinian realities of the universe can be violated, we need to consider the assumptions that accompany a multi-star system association of ETI. To complicate things even more, notably in order to deal with arguments put forth by (e.g. Ball 1980; Bracewell 1960; Ćirković 2006; Newman and Sagan 1981) which imply some sort of political system, we need to discuss politics as well. At the onset, it is important to accept that any kind of interstellar political system (AKA 'empire') seems unlikely. *Messages* from HQ would percolate at the speed of light C; but empires require two other kinds of transmission: enforcement (power projection), and redistribution. Any expeditionary force intended to enforce the might of the empire would be at a great disadvantage, acting years from home with planetary resources available for defense by 'the rebels', and plenty of time to identify approaching punitive forces and prepare for them. Force projection by an expanding empire would work *once* on every system captured: incorporation into the empire would teach those occupied the technologies and organization needed to resist, and a second expeditionary force would be at a major disadvantage. Stationing a large garrison force would be self-defeating, since such a force would gobble up the resources the empire demands in the first place. Of course, there is always the option of covert or overt genocide—a US Cavalry against the Indians scenario—but that does not change the numbers, it merely substitutes members of the conquering race for the autochthones, and the same question persists: how does one ensure imperial control and continuity?

An empire is a *political* construct, composed of a number of cultures held together by political fiat, that is, mixtures of threats of force, political negotiation, political alliances among elites, bribery, commercial benefits, marriage alliances (notably among the elites of the different cultures) and occasional violence. More often than not the empire is based on a particular culture which fairly early on in the game got the military, economic, and political jump on the others within the empire. *Most* of us live in a political empire. Britain admits the idea freely—an empire established by the English, and consisting of English, Scots, Welsh, Cornish and a number of other cultures (AKA 'ethnicities') who accept, some more, some less willingly, the cultural domination of the English. The US is too an empire, even if it does not admit it: the traditional reference to the Founding Fathers and the thirteen founding states notwithstanding, it is characterized by numerous cultures, ranging from Yankees in the northeast, through southerners, Cajuns, Texans, Hoosiers, Californians and many others, let alone Native Americans, some of whom have a record of agitating, and even fighting, against the dominance of the central culture. Crucially, *empires are political constructs*, and, as we follow a Weberian understanding of the term, are based, ultimately, on the potential use of force. And, as we have seen, the potential for projecting force in an interstellar medium is low to unlikely, whatever the fictional literature on galactic empires may say.

Overall, therefore, I would argue that in the (almost certain) *absence* of FTL communication and/or travel, multi-star political unitary structures (AKA empires) are hugely improbable. However, political unity is not a *sine qua non* of interstellar relations. And, indeed, most SETI writers who have discussed advanced *civilizations*, imply, but do not claim political unity, which is the form we humans recognize as necessary for unitary action, notably organized relations with upstart humans. The question, however, remains: without political domination of some sort, what could hold a large association of many star systems and/or intelligent races together?

Tradition A potential answer comes from the study of culture and tradition. 'Tradition', which in common usage means something that has been sanctified by usage from time out of mind (keep civilizational canons in mind), is, in most cases, a form of reiterating certain practices and statements in an attempt (usually by the elites, local or national) to ensure their way of life is not contested by other interests (Hobsbawm and Ranger 1992). Tradition, moreover, comes in two major forms, what Redfield (1955) and Shanklin (1981) call "Great" and "little" traditions. Roughly, for our purposes, Great Traditions are maintained by the elite (more often then not, the literate elite) of a social, political, or economic association. Great Traditions are inherently conservative, and the Great Tradition's purpose (if a mentifact may be said to have a purpose) is to ensure the continuity of the unit, and of course of its major social producers: the political, social, cultural, and economic elites. 'Little traditions' on the other hand, are those traditions that emerge, and are maintained by the smallest social/cultural/economic/socio-political units (AKA 'communities' or 'villages' in Redfield's writings), partly because they reflect local, rather than imperial realities, and partly as a bulwark against the Great Tradition's pressure to conform. In practice, there is a great deal of interpenetration, with some elements of the Great Tradition finding local expression in the little traditions, and sometimes elements of the little traditions becoming an element in the Great Tradition (cf. The development of a national cuisine in Japan, Ashkenazi and Jacob 2000 and Cwiertka 2006).

What is significant for the discussion here is that a Great Tradition can serve as a cultural glue that holds, an interstellar civilization together, however tenuously. Alexei Panshin (*The Thurb Revolution*: Chap. 5), also noting that interstellar empires cannot function as a political unit, is spot on:

> [An] empire is a gallimaufry of cultures of which the so-called High Culture is only one. Cultures are media of expression, like languages. … The High Culture is largely artificial, the native lifestyle of a bare few, understood and practiced by only a small percentage more …. On the other hand, it is a lingua franca. It is the medium of expression of men of birth and education, and those who travel. It is the chief of the invisible forces that hold [an] Empire together. … there is no universal culture. … [R]eal unity is impossible—the universe is too large. The common experience of the opening of space end[s] forever the possibility of one common culture.

Substitute our *junzi* for Panshin's "bare few" and one can see how Redfield's concept fits neatly into the concept of civilization. In an interstellar ETI civilization, physical reality ensures that communication is slow and error-prone, and that if at all, only a few select members can travel. This in turns implies that the emergence of some form of 'Great Tradition/High Culture' is almost inevitable, if the civilization is to endure as a unit. Socially and politically, we would expect differences, perhaps even major ones, between the units of the civilization: the various cultures that make it up, whether they are physically present on planets or other elements of solar systems, or on interstellar travelers: spaceships, space stations or even whole worlds. However, even a slight population movement of individuals/groups versed in a Great Tradition, and carriers of it, will be a major factor in ensuring that a civilization survives as a coherent whole, if at all. For that matter, as (Wilkinson 1995) argues convincingly, the interpenetration of cultures and learned behaviors that comes about as the result of conflict, also helps create a single civilization in opposed cultures (in direct opposition to Huntingdon 1997).

Great Tradition in the Galaxy and relations with Homo Sapiens The presence of a Great Tradition in an interstellar civilization also provides a partial solution to the issue of communicating with such a civilization. Great Tradition is not inherent. It is *learned*, and thus, however difficult, is *learnable* that is, it has a canon, standards, and praxis. We have examples in human civilizations which are worth contemplating.

For the past 2000 odd years, the political culture of the Chinese empire was based on a premise articulated by scholars of the Spring and Autumn and the Warring States periods (roughly, 741-403 BCE). The idea had two parts. First, in order to maintain order and prosperity, humans needed to be ruled by superior individuals—gentlemen—who had an inherent propensity for propriety, thought and action. Second, such individuals could, and should be *trained* to become gentlemen. We have to gloss over the fact that Chinese thought was inherently misogynistic, though in principle the same applies to women. Gentlemen—the elite—could not come about without training (Chang 1955; Cooper 1986; de Bary 1957; Friese 1963; Krack 1963; Menzel 1963; Nivison 1963). The entire management of the empire was supposedly in the hands of such scholar-administrators. Moreover, while the national Great Tradition was what anyone who wanted to get on in the world needed to know, and at least pay lip service to, local traditions were allowed to flourish freely (so long as they accepted the Great Tradition's dominance). So great was the power of the system, that even when the empire was conquered by barbarians, as happened occasionally, the civilizational system still functioned and flourished, incorporating the barbarians into the system. In fact, individual barbarians, so long as they were able to pass the examinations, could be incorporated as acceptable gentlemen, as happened to two Italians—merchant Marco Polo and Jesuit priest Matteo Ricci—at different times under different dynasties.

What this implies is that *should* humans come into contact with an interstellar civilization, it is very likely that there will be 'little traditions' as well: local in an astronomical sense, as the traditions of a solar system and of its constituent cultures. These may be based on a specific race within the civilization, or on local living conditions (planetary or otherwise). These 'little traditions' may *not* be accommodating to strange, possibly barbaric/edible/physically repulsive/... humans. However, there is also likely to exist a Great Tradition—which can be seen as a semi-conscious, constructed tool to keep the civilization together—which is likely to accept humans, *provided* humans, individually or collectively—learn the language of that tradition and demonstrate adherence to it. This does not provide absolute protection from either exploitation or acts of dominance, including mass violence as members of the Jewish community in Europe who adopted the Great Tradition of the Enlightenment (the European intellectual movement of the late 17th and 18th centuries, which emphasized reason, individualism, and humanism) learned to their cost. Some of them became members in full standing, in some cases, of the European elite, which did not stop demonization and persecution of the Jews as a whole. On the other hand, the Chinese Great Tradition accepted and absorbed the Jewish elite in their midst, who mastered and incorporated the Great Tradition to the point that the community as a whole virtually disappeared with the morphing of it's elite into *junzi* (Plaks 1998).

9.6 Holding a Civilization Together: Conscious Work and the Work of Conscience

Unlike culture, civilization needs to be maintained by actors who have a vested interest, and a shared set of trained reflexes, consciousness and identity. A culture essentially holds itself together (or, at least is held by individuals) because it is the net result of the different ecological systems (that is, social, natural, and technological) working together, and there is constant pressure to conform from the three ecologies working together (or the median result of different ecological pressures). A civilization, on the other hand, needs to be practiced consciously. In some cases the civilization is 'practiced' by a small number of individuals who serve as a model for all the rest (the Chinese *junzi* never numbered more than 10 % of the population) or it may be carried through the efforts of large percentages of the population, via technical means and intentional dissemination.

9.7 Models of Civilizations

Most writings dealing with ETI civilizations have concentrated on what their culture implies for, or how it can affect, our own. This automatically means that the types of civilizations that have been proposed in the SETI literature are those with some pretension to, or ability at, interstellar contact. That is, by and large, stable civilizations possessing a high technological profile.

It is impossible to discuss the internal *nature* of possible ILF civilizations or cultures. The possibilities, once the initial hypotheses are made, are simply too numerous, the variations too dependent on factors we cannot imagine, let alone enumerate. I focus therefore not on the civilizations per se, but rather on the possibility of *types* of civilizations, and their likelihood.

The discussion of the forms of ETI cultures/civilizations in the literature has been extensive, not to say lively. Obviously, the greater one's imagination, the more can be said on the subject, inasmuch as we are continuously faced with Fact One. Nonetheless, some reasonable rules of argument can be established, and within the various opinions expressed, one can find essentially three approaches to the attempt to define the form and activities of ETI civilizations: predicting on the basis of technological development, universal physical laws, or extrapolations from our own civilization.

9.8 Extrapolating from Biological Bases

We know too little about exobiology to base any discussion of ETI culture on it. However, as has been shown in previous chapters, some things can be assumed, others deduced. Obviously, many of the factors discussed in the chapter on ETI biology will come into play in the formation of the ETI civilization as well. The possibilities are too numerous to allow much valid speculation. However one intriguing, overarching possibility must be considered, something that has been discussed in this chapter already: the degree to which an ETI civilization will be determined by *genetic* or by *cultural* factors. That question has not been settled even for humanity, but we can examine the logical extremes as postulates, and see what the implications are. One extreme type of culture or apparent culture can emerge when the basis of technology appears as a genetic blueprint. How can a civilization be simultaneously genetically fixed and transmitted by culture? Lumsden and Wilson (1983) suggest that an ETI species would teach and learn all that they know to their offspring, the transmission would be limited to fixed sets of knowledge. In other words, such a species would entirely lack what we term "imagination" or a speculative ability, as well as curiosity. In the discussion of intelligence it was noted that these appear to be important factors in human intelligence, but the same might not be true for the evolution of other life forms, however unlikely this seems to modern humans. At another extreme a species could evolve with no, or very few genetic predispositions. Such a species could theoretically absorb any language, any song, any code of ethics, with equal ease. Their genes would direct the construction of their body and brain but not their behavior. Their minds would be entirely a product of the accidents of history, the place in which they live, the foods they encounter, and the stray inventions of words and gestures (Lumsden and Wilson 1983).

In the first case, we would have to assume a lengthy period of relative biological stability, or else an absolute, invariable genetic cull of some sort. Individuals in that environment would live to breed *only* if they reacted immediately, and thoughtlessly to specific and unvarying stimuli. And, of course, we would need to account for the question of where the knowledge came from in the first case? At the other pole, a species under constant and unremitting adaptive pressures, say on a young planet with high crustal activity, would have to be extremely flexible to survive. Adaptability would be all, genetic fixing a liability. Nevertheless, we would need to understand how any nurturance could take place, and, significantly, what its contents could be? After all, if every individual is *sui generis*, the likelihood of a culture emerging, let alone a world or solar civilization, would not appear to be good.

It is hard to see how intelligence, let alone technology could emerge under either of those two extremes. In the case of the genetically-fixed species, we may well find a non-intelligent technological species, perhaps even more technologically advanced than humanity. Given sufficient time, such would not be unreasonable. In the other case, we might find a non-technological species having great emphasis on immediate problem solving, and almost no faculty for planning, predicting, or dealing with the long-term. In either case, we may well have difficulty deciding which, if any, is "intelligent".

What the above *does* teach us, is that the biological basis of an ETI civilization would have to be a species that, however bound by its physiology or surroundings, is nevertheless capable of change and innovation and the transmission of a set of commonly accepted (for the species) realities.

9.9 Extrapolating from Social Bases

We may be able to make some assumptions about an ETI civilization if we look at a combination of physical and social laws and rules (keeping in mind the difference between the two terms, in the physical and social sciences respectively).

The C limit is one such reality. As noted before, it will likely force interstellar civilizations to duplicate many functions to reduce waiting for information exchange. Habitats separated by interstellar distances will become culturally independent, and cultural evolution will thus be promoted. Bainbridge (1984: 661) raises the interesting possibility that internal processes—tensions, social or religious movements—could cause a large civilization to embark on some civilizational mega-project. This seems a likely possibility: examples such as the Grand Canal in China constructed by emperor Qin-shi Huang-ti abound in human history. Indeed, ignoring the physics for a moment, a solar Dyson sphere (or, physically more likely, a ringworld (Niven 1970)) will validate Kardashev's argument.

One answer to the question of the relationship *between* ETI civilizations has been provided by Kuttner (1981) who, in much the same argument as management convergence theory (Dore 1973), argues that the inherent tendency to optimize technology would lead to convergent technological evolution, and thus to equivalent capabilities, which would mean, in effect, that peer civilizations would have little to say to one another, in as much as no new phenomena will be uncovered by the exchange. By implication, communication between them would be at a minimum. Ignoring for a moment the lengthy tradition of tourism among humans which may well have an ETI analog, we should need to keep in mind that possibility that while *overall* capacities might be similar, variations in technological solutions are likely to exist. Civilizations will include a large degree of technical redundancy, so that learning a new technique to do the same thing *does* have benefits, which in turn implies that interstellar civilizations will engage in at least some form of interaction. Being human, as we are, and with our history as a guide, we need to ask an obvious (to most humans) whether such communication will be peaceful or aggressive?

After all, civilizations in our experience have ignored one another only when the distance (as e.g. between Chinese, pre-Columbian American, and Mediterranean civilizations) was too great to allow for aggression or even much contact.

9.10 Extrapolating from Technology: Convergence or Divergence?

One approach—technological descriptions, of which the Kardashev analysis is one example—attempts to define ETI civilizations on the basis of their assumed technology. Two general rules seem to have been applied. First, descriptions of ETI technology have been based on extrapolations of human technology in the present. Second, minimal social extrapolations have attempted to understand a civilization as a function of its technology. This is by no means easy, as anthropologists know. Any attempt to create generalizable rules on the basis of a technology must have many 'ifs' and 'buts' (see for example Steward 1963).

Technology is an obviously attractive way to portray ETI civilization because our own is so highly technologized. It is easy to assume, that the pervasiveness of technological gadgetry, which has increased geometrically in the past century, will extend to the foreseeable and unforeseeable future. There are possibly limits to the development of technology, but these seem far in the future for humans, and, the argument goes, if a technology is conceivable by humans, it is executable by any ILF as a matter of time. The nature of a technological society is conditioned, by definition, from the nature of its technology. What that implies is that an understanding of the directions in which a technology *is likely* to go, may well serve as a clue to the nature of an ETI culture.

A good, if rather simplistic way to start is to list a roster of traits that an advanced civilization may have at its disposal. The problem with such predictions is that they tend to reflect the wish list of the human writer. The most extreme such case may well be Tough's (1986) list of the characteristics of ETI civilizations. It includes: (1) Virtually unlimited energy; (2) increases in brain power and ILF-machine linkages; (3) FTL transportation; (4) infinitesimally fast communication over infinite (or close to infinite) distances; (5) organic connections to other members of the race; (6) extreme weapons technology. Two problems derive from this. First, the extrapolative basis is very weak. There is, for example, little scientific basis for assuming either (3) or (4). The second problem has to do with the human linkage: all of these characteristics (and others that will be discussed subsequently) appear to be a human wish list, based on human preferences. In fact Tough virtually says so himself: The second item on his list is "technology indistinguishable from Clarke's Law." For those who are unaware of Arthur C. Clarke's law, it goes "A sufficiently advanced technology is indistinguishable from magic." In other words, we can and should define a sufficiently advanced ETI civilization by its ability to supersede what appear to be the physics of the universe.

Kardashev's classification falls into this category as well. As noted previously, from our perspective this is an extreme example of the equation of culture or civilization with a technology, but it tells us virtually nothing about the civilization beyond its technological prowess. Certainly this sort of approach is attractive. It effectively absolves us from the need to examine ETI civilizations in any way but the technological. After all, if anything such a civilization does is 'magical', surely social laws need not apply?

A far more sophisticated approach to ETI technology has been proposed by Bainbridge (1984). He suggests that the effects of technology can be subsumed as three possible states of change: technological *failure*, that is, less use of technology to the point of its utter non-use; technological *stagnation*, where no technological change occurs; and technological *progress*, where technology changes, perhaps even evolves, either continuously or quasi-continuously (Ball 1980: 658). The technological failure possibility may end in destruction, in which case it amounts, in effect, to a termination boundary for L. The possibility of technological stagnation is not discussed often in the literature: the assumption being, presumably, that technology continues to change and advance. We have, for instance, no clear examples of technological stagnation from human history. The oft cited Chinese and Egyptian examples demonstrate *slow* technological development, not stagnation (see for instance Needham 1969). Alternatively, technology is expected to lead to chaos and destruction through over-technologization or war.

Bainbridge points out that technology can be seen as a series of processes, that occupy a civilization. True, this effectively emphasizes an evolving technology rather than the other two possibilities. Nonetheless, this is the sort of technology we, the human species in the early twenty first century, are most familiar with. Indeed, we tend to equate technological *change* with technology itself and with modernity. The list of processes Bainbridge is suggesting is, and that is not surprising, a derivative of what occupies our own society. Five domains of activity characterize the civilizing process, he argues. These are expected to be characteristic of any ETI civilization encountered. The five are: (1) acquiring and processing information; (2) planned improvements in intelligence; (3) planned engineering of biological and social entities; (4) integration of the domains of machine, plant, animal and people (5) increased mechanical and biological efficiency (Bainbridge 1984). One *technological* consequence of this would be the ability to exploit heretofore unused planetary and spatial habitats. One social consequence also seems likely: great heterogeneity in social arrangements, *if*, as seems likely, there are multiple technical solutions to these broad issues, and *if* we can assume, as a broad universal law, that different technologies influence or create different social arrangements to carry them out.

Two things need be said about Bainbridge's approach in contrast to Tough's. First, all his suggestions are logically circumscribed continuations of processes already inherent in what human science does today. In that sense they are extremely conservative, unlike Tough whose proposals are far more radical. Second, unlike Tough's list, Bainbridge's proposal allows us to extrapolate to some degree what the social and cultural aspects of such a civilization might be, as we have seen from the earlier discussion of the concept 'civilization'.

Tough and Bainbridge have largely concerned themselves with societies that are technologically superior to humanity. Implicitly, at least, these societies are actually or potentially not bound to one planet. This is not surprising since the perspective is largely a SETI one. The limitations inherent in SETI (that an ETI society will be technologically more advanced) does not concern me here. Campbell's (1981) categorization of technologies, with all its potential flaws is far more apposite. He suggests (a) that not all ETI need be technological, and that (b) technology, or what he calls "the technological life style" may be divided into a number of cases, which *may* [emphasis mine] be developmental stages. Prototechnology, the non-systematic use and modification of tools, which humans share with some animals is the initial stage. This is followed by tool manufacturing technology, then pyrotechnology, which exploits fire, and thus fire generated tools such as metals and ceramics. Facility and energy control technology follow. The highest two stages are information technology and investigation technology whose sole purpose is the examination of ultimate causes (Campbell 1981). An ETI civilization may be found at any one of the latter stages. The details of this scheme are arguable, particularly since it derives, not accidentally, from the specific case of *Homo sapiens*. Technologies need not be developmental, and stagnation (to follow Ball's classification) may well be a norm in more than one society.

The historical technological-developmental sequence of humanity is not a requirement. Metallurgical sciences are an example. We tend to think of metallurgy as starting from copper/bronze, proceeding to iron, then to steel, and finally to other metals. This is the normal sequence in the three major centers of modern civilization: the Fertile Crescent area, China, and South Asia. It is hard to see how the *sequence* is connected to the eventual dominance of these areas. In other areas, the sequence was quite different. African metallurgy skipped the copper/bronze period entirely, and there is evidence for native invention of iron-work in prehistoric times (Childs and Herbert 2005; Schmidt 1997). In the Americas, the Inca were using sophisticated techniques to work platinum, and all pre-Columbian states were well-acquainted with gold and silver metallurgy without ever making use of basic metal tools.

A distinct technological approach is suggested by Rescher (1985). In trying to modify the Drake equation he suggests a number of refinements to the variable p_i (the probability that a technological species will arise). Effectively, Rescher argues that alien science may be truly alien, because the machinery of formulation (the descriptive means used: the mathematics or geometry), the orientation of science (the things that interest them, and what they concentrate on) and the conceptualization of science (the points of view of the description) could be widely variant from the ones accepted by human civilization. Because of that, it is likely, Rescher argues, that the 'science' of an alien civilization is oriented around concepts that are truly alien to us (Rescher 1985: 88). In support of Rescher, Egyptians as well as Greeks, Maya, and Chinese had organized systems for doing all those things, though they differed in the subjects of their interest and the form of data storage.

The factual information they collected and systematized were far different. The Egyptians concentrating on the afterlife and geometry, the Greeks on political sociology, the Maya on astronomy and metaphysics, the Chinese on history and human management. All of those societies however (Rescher only mentions the Egyptians, but the others would seem to fit his criteria as well) also had excellent factual information about agriculture, metallurgy and other practical issues.

Rescher also argues, along with Ball, that an ETI civilization may be technologically sophisticated and yet not dynamic. This has two aspects. First, at what rate must change occur in order for the society to be considered dynamic? Obviously, no change in astronomical time frames (that is, billions of years) would be considered more than a bit ossified. Species (on Earth) rarely survive beyond ten million years, so absence of change over biological time scales would seem to be a good measure. But civilizations appear to endure even less. Our longest lasting civilization (though I am stretching the word somewhat) was the Acheulian culture: thirty thousand years perhaps. The shortest is the current nascent global civilization, less than a hundred years. To be reasonable, let us arbitrarily allot some time frame of two thousand years. Within that time frame, *if there are no significant technological advances* we could assume that the society is technologically non-dynamic (though it may be dynamic in other ways).

We tend also to equate technology with intelligence, but could technology occur *without* intelligence? Two possible scenarios should be considered. First, a species so intelligent that it has, or is capable of, "working everything out." In other words, as Larry Niven says, a sufficiently intelligent and rational organism would have no choices: its actions would *always* follow the best possible solution to any problem. Second, a species could evolve a sophisticated technology without any intelligence whatsoever (Raup 1985). We have such species on Earth. Bees, termites, ants all function technologically (they create and manipulate sophisticated artifacts) without, apparently, displaying any intelligence. If we extrapolate from that example, rather than from the technological one of humanity, we must acknowledge the possibility of technological-hives or organisms able to do much of what we can, including building spaceships, radio, and other devices, purely out of instinct. The generation of radio waves is an insufficient criterion for associated intelligence (Raup 1985).

Our technology is heavily dependent on a body of organized knowledge and method we call science. As can be seen in many cases, technology can emerge without science. For our purposes here, "science" is a systematic, organized body of methods concerned with the regularities of nature. Specifically, as we know it today, it is a human mentifact supported heavily by tools, and derives in large part from a specific history. Rescher (1985: 100–101) points out that this particular set of circumstances will not repeat itself, and therefore any ETI science will have different emphases and will be a specifically ETI mentifact, and consequent set of artifacts. Moreover, we would find it extremely difficult to compare sciences. If 'they' are doing something substantially different, then the concepts of "more" or "less" advanced have no referent, since they require comparison. ETI science, whether they are an older or younger civilization, may not be comparable to our future science, on which many of the analyses described here are based. Whatever we predict about science relies on our particular version of it (Rescher 1985: 104).

Rescher's argument is not new even in the human context. For many years an argument raged about the uniqueness, or otherwise, of Japanese management methods. Some theorists, including notably (Dore 1973) have argued that there is, in effect only one "good management" method, and that the successes of the Japanese were merely the better application of general principles. These "convergent theorists" have claimed that eventually successful managers will converge on a similar pattern of doing things. In contrast, divergent theorists exemplified by such as Marsh and Mannari (1976) have argued that there are diverse methods of good management, and that different methods can exist as potential alternatives to one another. Substitute science for management, and the same argument can be seen to exist in discussions of ETI. With the exception of some, such as Rescher, most *physical scientists* at least, seem to be arguing, on the basis of what they know of natural law, for *convergence* Explicitly, such arguments have been advanced by Tipler (1985) and by Kuttner (1981) and implicitly, by others. Most social scientists, seemingly either ignorant of the physics (which is unlikely) or, (what is more likely) looking at the problem from the point of view of their own disciplines, see divergence as a very real possibility. I would suggest that ETI science, when we encounter it, will very possibly be *unknown* to us, but not *unknowable*. This unknown is not in terms of the *result*—they are describing and analyzing the same world around us/them—but in terms of the methods and emphases they bring to the enterprise. We know far too little about the universe, including its basic physical rules, to be able to circumscribe all knowledge. At the same time, it seems that natural rules, perhaps even social ones, are comparable, perhaps universal. Whatever understanding of these rules one ILF species can tease out, can be teased out, using a different methodology, by another.

The conclusion, from the point of xenological methodology seems clear: it would be very hard, perhaps foolhardy, to equate a civilization with a particular set of technologies. Perhaps there is some ultimate conversion point, but that, insofar as our knowledge of the social-intellectual-physical matrix goes at present hardly justifies defining ETI civilizations on the basis of their presumed scientific level or technological abilities.

9.11 Physically-Based Models

A second approach, similar to the technological characterization of civilizations and often barely distinguishable from it, is descriptions of ETI *interstellar* civilizations on the basis of physical facts. The most prominent fact is of course the C constant. The question asked here is, what forms could a civilization take given the physical background (the interstellar medium) and its limitations, notably distance. and the absolute limiter C. In other words, can we determine the 'shape' of an ETI civilization based on the physical limitations imposed by the universe?

A number of authors have attempted to do so. The significance of these schemes, from our point of view, is not their physical or technological validity, so much as the social and cultural prerequisites and consequences of universal physical realities. A well developed scheme of that sort has been proposed by Stephenson (1981):

1. All intelligent species evolved on or near the surface of a ± Earth type planet.
2. In order for the evolution to take place, there must be shifting ecological pressures.
3. Species are resistant to pressures and able to expand.
4. A species will expand until it reaches some boundary that exceeds the expansionary pressure of the species.
5. The technical ability of a species proceeds by a series of steps marked by harnessing new types of energy sources. The division into steps is not exclusive and boundary conditions may obtain.

The technical progress of a civilization is best considered, Stephenson argues, as a series of steps, rather than a linear continuous process, which most evolutionary biologists would agree to. If this is the case, the emergence of a civilization can be examined as a series of conditions that must be overcome if a culture is to be able to conduct SETI. At any stage a culture may either fail or succeed to pass the boundary. This also helps in understanding what a dynamic civilization is: one that passes from boundary to boundary in a time that is short compared to astronomical time-spans. While this may be true for SETI (and, to be fair to Stephenson, SETI variables are the core of his argument) there is no universal reason to suppose that all his assumptions hold water. Species may *not* expand, for example (assumption 4) but fluctuate intentionally or not between two Stephensonian boundaries. They may have a perfectly satisfactory source of energy *for their purposes* whatever those may be.

More complicated is the issue of C. We can accept C as an inviolable boundary (at least until proven otherwise) for communicative velocity. From the perspective of an interstellar civilization, the velocity of an object with mass is far less, in practice. This would seem to mean that mass exchanges, let alone living mass exchanges within an interstellar civilization would be prohibitive.

On the colonization and settlement front (that is, the physical expansion of a species throughout the galaxy) a great deal of thought has been given to potential settlement strategies (for example Bjoerk 2007; Bracewell 1974; Cotta and Morales 2009; Oliver 1976; Sagan 1963; Turner 1984; Vallentine 1985). Not all of these strategies involve biological organisms. Of those that do involve biological expansion, it is recognized that biological differences also imply differences in acceptance of time scales. That is, a life form with extended life-span, aestivation techniques, or other biological realities, might not consider lengthy physical travel to be anything more than a temporary inconvenience: "The range of biological alternatives makes it extremely difficult to place any lower physical limit on the speed of interstellar settlement which is relevant on the timescale of galactic evolution. … differences in lifespan of ETI, or their size, would reduce spaceship size and other considerations. Another alternative might be possible if radical biological modifications are possible, is to cover the surface of an immature ETI with a protective film, attach such a unicellular organism to some solid material such as a 100 gm meteorite, and fire it into space. At a given time, it matures and reproduces if it finds a suitable habitat" (Singer 1982: 57–58).

We thus reach the point of recognizing that not all space-colonizing species will necessarily develop a single civilization, and that interstellar distances may not hinder the development of one (or more). The point of all this is simple: given the huge potential variations in species biology, as well as in the potentials they have for developing different cultures, technologies, sciences, and understanding of their surroundings, it is very presumptuous to suggest that cosmic distances would be the sole framework we could adduce for the nature of an ETI civilization. Cosmic distances simply do not completely trump biological and cultural imperatives: they merely provide a structure, to which intelligent species are expected to adapt any way they can.

9.12 Extrapolation Models

The third approach to interstellar civilizations is one which extrapolates from human society and attempts to predict what the social parameters of an ETI civilization might be like. All three of the extrapolation approaches tend to be idealistic, almost utopian: in more than one case, an ETI civilization is an idealization of what *the writer* would like to see (apparently) in human society.

Nevertheless, if there are general social laws applicable at some level to *all* social organisms (an assumption that is part of the fabric of this book), we could predict negative limits ("this could not take place") for ETI cultures. Unconsciously, perhaps, certainly with no great reflection on the analytical consequences, general positive universalities been assumed in many attempts to characterize ETI civilizations. Beyond the almost science fiction-like extrapolations of desired human moral community such as e.g. Angeles (1976), Ball (1973), Deardorff (1986), Kuiper (1977).

Some authors even explicitly provide characterizations of galactic civilizations that reflect human future ideals. "Interstellar expansion, though physically possible has never occurred, because all sober reasons for expansion are terminated, and emotional reasons for anything are excluded, in any surviving civilizations, long before the technical ability for expansion has been reached" (Hoerner 1975: 708). Others, such as e.g. Tough, speculate on the nature of ETI civilizations by producing a wish list, based essentially on human liberal yearning, including "Loving cooperation, altruism, mental health, ability to relate harmoniously with members of their own society. … Excellent skills at interaction with other species. … Control over biological reproduction and evolution." (Tough 1986: 491–493). Not only that, but "Ultimately, every well-behaved star of the galaxy is bound to become populated by a stable stellar civilization that has achieved a high level of ethical completion" (Tough 1986: 498).

Tough's idealism must be counterpoised by another view (parenthetically, both of these views emerge from the debate over the Fermi paradox). Brin suggest, in counter-distinction to Tough, that it is competition for livable space, over the span of mega-years, that would shape galactic civilizations. Brin's view is uncompromisingly Darwinian: competition throughout the Galactic disk brings about military-driven secrecy, concealment, and obscuration. The violent competition between spacefaring species develops insular civilizations that compete with one another without mercy. *ETI, ETI lupus* (Brin 1983). Brin's view may be criticized as being too overtly Darwinian, as well as for the assumption that great space battles will obscure the presence of ETI rather than displaying them like fireworks in the night. However, its importance for the discussion here is the suggestion that numerous mechanisms may be at work and that an attempt to circumscribe ETI civilizations need to keep in mind that there is a danger to all human definitions of what ETI civilizations might be like.

To a degree, the definition of civilization we use here *could* in theory be classed within this approach, except for the fact that it depends not on extrapolation so much as on identifying relevant, and apparently universal realities. On the intellectual side *something* is needed to keep a civilization together, and allow it to last. Occam's razor suggests that the simplest feature, and, I would add, the most cost effective, is what could keep ETI civilizations together—a sharing of mentifacts and ideas, expressed practically in shared forms of expressive behavior. In addition, since all organisms require some form of energy to function, the demand for easily usable energy is likely to be *the* major preoccupation of every organism as it fights entropy and develops from simplicity to complexity. For the same reasons, this will be true of complex non-biological organisms such as social systems as well. We should not be surprised if ETI collectives can be located on a Maslowian dimension characterized by the sharing and use of energy: a scrabble for the magic number of consumable kilocalories at one end (not a civilization) to providing a surplus at demand to fulfill the 'self-realization' wishes of *all* members of the civilization at the other (an advanced civilization).

9.13 Nomadism Through Space and Time: Change as Civilization

Humans, perhaps conditioned by the adjectival use of the term 'civilization' tend to think of civilization as a settled entity. However, even in Earth's limited area, there have been widespread *nomadic* civilizations that have persisted for centuries. Inasmuch as the area, or rather space, available to sophisticated, technological species is much greater, we need to consider the potential of nomadic, as well as settled, civilizations.

'Nomadism' in this context must be read very broadly. Not only could we have, potentially, entire civilizations founded upon a space-based (not even solar system based) way of life, but we ought to consider one of the corollaries of our view of 'civilization'. Remember that a civilization needs to be *maintained*. That is, someone(s) need to consciously use and propagate [whatever they think are] the cardinal principles of the civilization. Inasmuch as a civilization is a *conscious* construct, it is potentially possible that 'management' (whatever that means in the context of an advanced civilization: the canon, a conscious body, or internal social rearrangements and preferences) may decide that continuing, cyclical and/or locational change is the best way to go.

The idea that *change* rather than stasis is the way to go has numerous advantages (and disadvantages) the principle among them is that nomadism presents a constant difficulty: an internal frontier offering constant challenges. If an entire civilization relocates (moving worlds, or world ships) to space, it will be faced with constant challenges, both to individual cultures within it (for sake of argument, a culture may equate to the population of a given world ship), and to the civilization itself as a largely conscious socio-cultural construct.

Nomadism may not necessarily be a spatial phenomenon. Since civilization is dependent on texts, deliberate destruction (or, since I love books, the sequestration) of the canonical texts of a civilization, will force the creation of a new canon, and thus, contribute to the rejuvenation and change of the civilization, if this is deemed, for reasons beyond current human reasoning, to be desirable and possible.

9.14 To Conclude This Section

Since the advent of SETI, there have been numerous attempts to characterize and understand possible ETI civilizations. Characterizing civilizations is important, because, as we have seen, 'culture' as a sociofact is insufficient to deal with the complexity raised by social associations spread widely over time and space. Technology-based characterizations of ETI civilizations have the disadvantage that they tell us virtually nothing about the social and cultural makeup of such civilizations. Which, in turn, means that we would find it extremely difficult to understand them. Using cosmic-physical parameters, specifically the universal C, as a measure of civilizations, suffers from the same weakness as technology-based characterizations, and, in addition, is weakened by the initial foundations of this book: time and space are *subjective* in their *importance* for different species. Finally, a great number of writers have extrapolated from human experience or, more often, from their own wishful thinking, in characterizing ETI civilizations on the basis of an assumed natural morality and universal benevolent ethic, up to and including lists of supposed (actually, desired) characteristics.

I argue in contrast that:

(a) 'Civilization' is an important term, inasmuch as some parsimonious explanation must be provided for maintaining social, intellectual, and possibly material associations *notwithstanding* the very real problems of communication and exchange created by distance and the universal velocity limiter 'C'.
(b) The most parsimonious potential social 'glue' is that provided by the presence of individuals and a system who have mutual interests in maintaining a Great Tradition as a repository of texts, ideas, intellectual standards, and behaviors that are self-consciously 'special' and that can, and must be learned and practiced in prescribed ways.

As the discussion on nomadic civilization asserts, civilizations will mutate and change through time and space, so that we cannot assume that they will be the same throughout their lifespan L, or that they will necessarily be spatially fixed. At a certain level of technology, for all we know (whether because of the predatory nature of the universe, as Brin hypothesizes, or for some other reason) civilizations may be intensively nomadic, not tied to any particular set of star systems or area of space.

9.15 AI Civilizations

One aspect of assuming that ETI and human technology will be convergent (which we take from either the first or the last of the extrapolations about ETI civilizations: technological descriptions or extrapolations from human society) is the issue of AI civilizations. Many writers have considered the possibility of true AI: artificial intelligences with full personal consciousness, and the difficulties the concept imposes philosophically, as well as practically (Adams 1986; Barbrook 2007; Bedau 2002; Bray 2009; Hawkins and Blakeslee 2007; Lehman 1981; Marien 2011; McNally and Inayatullah 1988). Given the possibility that AI might be developed, it is reasonable to suggest that AI might act together in some social form. The form might, of course, be nothing more complex than swarm behavior, but might also lead to a complex set of organizational practices, and the consequent emergence of AI civilizations would be a real possibility. The nature of such a civilization is hard to predict, because it borders on what we have called here the 'exotic.' We can, for instance, assume that such civilizations will be bound by physical demands for energy. Insofar as the collaboration between individual and collective units and the whole follows the same rules we have discussed for civilizations, we could call this an "AI civilization". However, we cannot predict to any degree how such a civilization would function: More than biological beings, AI would likely be a software complex on a physical substrate. AI would be, to quote a popular film's title, souls in a machine. Presumably the machine itself would be plastic and mutable, which implies great differences between the AI perspective on the physical world and biological perspectives. We cannot assume that AI motivations, in even the remotest way, will be understood by us. Of course, it is possible, as some have noted, for humans to merge into an AI structure, with humans and AI existing in either biological or mechanical form as convenient (Dror and Harnad 2008; Lupton 1995; Schank and Birnbaum 1994; Wilson 1995). It also seems likely to some (Michaud 1982; McGowan 1966) that the existence of what appear to be infinitely improvable artificial organisms, will spell the end of biological reproduction and culture, and bring us into the realm of transcendence.

9.16 The Problem of L

How long could an ETI civilization last? L, the length of time a technological civilization can broadcast radio signals receivable by a radio telescope (the classical SETI definition), is not, strictly speaking, an issue for our purposes. Nonetheless, it is an intriguing question, because it reflects back on the question of *types* of civilization. To change the formulation a bit, let L = the length of time a technological civilization can sustain itself. Any civilization which manages to sustain itself over truly long periods of time, must be able to overcome certain limitations. Having done so, it may truly have something to communicate of value. But. Its very survival may depend on some innate quality, not on generalizable circumstances that could be applied to another species.

Accepting the definition of civilization used in this book, changes the problem a bit, and brings us to the philosophical conundrum of my great-grandparent's axe: "This is my great-grandparent's axe received by me from my own parent, as my parent received it too. My grandparent changed the blade, which was damaged. My parent replaced the handle. Nonetheless, this is my great-grandparent's axe, and will go as such to my child, though I have replaced the binding."

Whether one considers an item (social, material, and even perhaps mental) as 'the same' after it has undergone changes to fit it for changing circumstances is an ethical and philosophical point more than a concrete one. So yes, the discussion of L *in SETI* has validity: if the change involves stopping interstellar radio broadcasts, then the social entity emitting the broadcasts originally, is no longer a civilization. In 'real' (whatever that may mean) terms, it continues as before, and if it previously had the characteristics of a civilization, it may still retains those.

In our characterization, virtually all of the characteristics are mutable over time. The personnel—the experts or gentlepersons—that carry and exemplify civilization will die and (possibly) be replaced by others with slightly different approaches to the standards and how to uphold them. Partly as a consequence, styles and standards will change (and, as we have seen, possibly in predictable ways), and possibly the canon itself will be modified as new and updated commentaries, and perhaps new texts will emerge and become central to the civilization itself.

Inasmuch as interstellar communication will likely be at C, messages between one race (us?) and this civilization may take millennia. By the time we have replied, and they have replied to our reply, the same individuals will not be involved. Possibly even the corporate entity—the International Astronomical Association, or the SETI Institute, for sake of argument—will no longer exist. Standards and styles will have changed, possibly radically. So, like the eponymous river, we shall (and they shall) never encounter 'the same' civilization again. Yet, at some remove, they will nonetheless be the same civilization we have exchanged messages with before (assuming they survive physically).

What has been said here implies that the question of L has validity *largely* for the SETI program much less for the broader questions of xenology. *Of course* Brin may be correct, and a state of open predation may exist, with cultures and intelligent races hiding or being consumed constantly, lowering L. Of course it is possible that races reach some transcendent horizon or mutate into non-biological machine intelligences. And, of course, they may blow themselves up, be hit by a meteor, or any of a number of catastrophes. But L constitutes a bit of a snark: it has many solutions, perhaps even some improbable ones we have not thought of, and to my mind, until further information comes in, is not really worth worrying about too much.

9.17 "Fact One": Expansion as an Example of 'Rationality'

The Fermi paradox has been referred to before. To reiterate, if we assume that some (at least) ETI civilizations will be technologically ahead of us, why has Earth not been contacted by them? To put the matter with greater extremity, assuming a Slower Than Light (STL) propulsion system, and a will to colonize space, it has been calculated (Newman and Sagan 1985) that the entire Galaxy could be colonized in ten million years. Given the assumption of mediocrity in addition to the others, it is therefore eminently possible that Earth should have been colonized before the human species could emerge.

The argument above, which has been cited by Tipler (1981) and by Hart and Zuckerman (1982) as proof that ETI do not exist, bears further examination. First, this is not unnaturally a physical-technological assumption. Whether the straight-line colonization scheme proposed by Tipler, or the more sophisticated "drunk walk" analysis used by Sagan and Newman is used, little attention is paid to any other variables than demographic growth, interstellar distance, and propulsion speed and range.

If we are to explain why Earth has *not* been colonized, given the possibility of an earlier technological species in the Galaxy, we must provide some reasons other than the absence of ETI. Brin's (1983) argument is very compelling: given even *one* expanding civilization, the entire Galaxy is likely to be colonized eventually (in ten to one hundred million years), *even* if the expansion impulse is occasional, random, and non-directive. To be fair, Brin suggests several cases in which Earth would not have been colonized. One is a fallowing program, in which technologically advanced species purposely leave some sectors of the Galaxy fallow for new intelligent species to arise: a sophisticated "Zoo hypothesis." A second alternative is a decline of interest in colonization. A third may be that Earth is in the inner core of an expanding colonization shell which has somehow missed this particular planet.

Let us, however examine some of the social aspects of continued colonization, rather than the technical ones. The first question we need to ask is "Why expand?" Species that are subject to natural evolution will expand into available ecological niches as a derivative of their evolutionary drive to reproduce. Where there are evolutionary pressures (better nutrition from unfamiliar sources, predators, different climatic pressures) the species will eventually speciate: two species instead of one. Is this true of intelligent beings, who, almost by definition, are subject to cultural evolution? There are, of course, ecological niches for any developed race. Plenty of them, in fact: planets, space itself, ice comets. They all present niches for a culturally opportunistic race. So there are pressures to migrate to other star systems. However, this represents a demographic conundrum. *If* this race is subject to *biological* pressures (over reproduction leading to over population), *migration is unlikely to solve the problem*: the capacity of a starship, however large, is unlikely to reduce demographic pressures. Moreover, constructing a sufficiently large starship will inevitably be at the expense of the populace, which may be reluctant to allow that. And if the race is sufficiently foresighted to understand that it is in a Malthusian trap, it should be sufficiently foresighted to curb its population. If there are more than one faction, and the race is combative, violent conflict would seem to be inevitable, with dire consequences, considering we are discussing a space-going race.

The previous mini-scenario is one of many possibilities we can think of. Indeed, many have been considered by various authors. The critical point is not that there are many reasons a race may *not* occupy the entire galaxy. It is the question of whether technological development (which would allow, perhaps encourage unlimited spatial expansion) is paralleled by social and ethical development. If yes, the race is likely to find other toys to play with than expanding exponentially: seen the first thousand star systems, seen them all. If they do *not* develop socially and ethically, they are far more likely to war among themselves (which, among other things, would bring us the news at C).

It must be said that this balancing of technological and social development is only one potential solution to the Fermi paradox (for those interested, the most comprehensive survey of solutions to the Fermi paradox is to be found in Webb 2010). It does show my bias for social rather than merely technological or physical solutions. The principle point is one that has been a foundational axiom of science for centuries: absence of proof does not constitute proof of absence. I would accept any number of solutions to the Fermi paradox, none of which would prejudice the SETI effort. Once again, our data set is too small to be significant.

References

Adams, S. T. (1986). Artificial intelligence, culture, and individual responsibility. *Technology in Society, 8*(4), 251–257.

Angeles, P. (1976). ETIs and the problem of intelligence. In J. Christian (Ed.), *Extra terrestrial intelligence: The first encounter* (pp. 161–175). Buffalo, NY: Prometheus Books.

Ashkenazi, M., & Jacob, J. (2000). *The essence of Japanese cuisine: An essay on food and culture*. Richmond: Curzon Press.

Bainbridge, W. S. (1984). Computer simulation of cultural drift: Limitations of interstellar colonization. *Journal of the British Interplanetary Society, 37*, 420–429.

Ball, J. A. (1973). The zoo hypothesis. *Icarus, 19*, 347–349.

Ball, J. A. (1980). Extraterrestrial intelligence: Where is everyone? *American Scientist, 68*, 56–63.

Barbrook, R. (2007). New York prophecies: The imaginary future of artificial intelligence. *Science as Culture, 16*(2), 151–167.

Bedau, M. (2002). The scientific and philosophical scope of artificial life. *Leonardo, 35*(4), 395–400.

Bjoerk, R. (2007). Exploring the galaxy using space probes. *International Journal of Astrobiology, 6*(02), 89–93.

Bracewell, R. L. (1960). Communications from superior galactic communities. *Nature, 186*(4726), 670–671.

Bracewell, R. L. (1974). *The galactic club: Intelligent life in outer space*. Stanford: Stanford Alumni Association.

Bray, D. (2009). *Wetware: A computer in every living cell*. New Haven: Yale University Press.

Brin, G. D. (1983). The 'Great Silence': The controversy concerning extraterrestrial life. *Quarterly Journal of the Royal Astronomical Society, 24*, 283–309.

Campbell, B. (1981). Evolution of technological species. In J. Billingham (Ed.), *Life in the universe* (pp. 277–296). Boston: MIT Press.

Childs, S. T., & Herbert, E. W. (2005). Metallurgy and its consequences. In A. Brower Stahl (Ed.), *African archaeology* (pp. 276–300). Oxford: Basil Blackwell.

Chung-Li, C. (1955). *The Chinese gentry: Studies on their role in 19th C. Chinese Society*. Seattle: University of Washington Press.

Ćirković, M. M. (2006). Macroengineering in the galactic context: A new agenda for astrobiology. arXiv:astro-ph/0606102

Cooper, E. (1986). Chinese table manners: You are how you eat. *Human Organization, 45*(2), 179–184.

Cotta, C., & Morales, Á. (2009). A computational analysis of galactic exploration with space probes: Implications for the Fermi paradox. arXiv:0907.0345

Cwiertka, K. (2006). *Modern Japanese cuisine: Food power and national identity*. London: Reaktion Books.

de Bary, W. T. (1957). Chinese despotism and the Confucian ideal: A seventeenth-century view. In J. K. Fairbank (Ed.), *Chinese thought and institutions* (pp. 163–203). Chicago: The University of Chicago Press.

Deardorff, J. W. (1986). Possible extraterrestrial strategy for earth. *Quarterly Journal of the Royal Astronomical Society, 27*, 94–101.

Dore, Ronald. (1973). *British Factory–Japanese Factory*. Berkeley: University of California Press.

Dror, I., & Harnad, S. (2008). Offloading cognition onto cognitive technology. arXiv:0808.3569

Friese, Heinz. (1963). Qualification for office: Expertise or character. In J. M. Menzel (Ed.), *The Chinese civil service* (pp. 77–83). Lexington, Mass: D.C. Heath and Co.

Hart, M. H., & Zuckerman, Ben (Eds.). (1982). *Extraterrestrials: Where are they?*. New York, NY: Pergamon.

Hawkins, J. & Blakeslee, S. (2007). *On Intelligence*. New York: Macmillan.

Hobsbawm, E. J., & Ranger, T. O. (Eds.). (1992). *The invention of tradition*. Cambridge: Cambridge University Press.

Huntingdon, S. P. (1997). *The clash of civilization and the remaking world order*. New York: The Free Press.

Joseph, N. (1969). *Science and civilisation in China* (Vol. 1–7). Cambridge: Cambridge University Press.

Kardashev, N. S. (1964). Transmission of information by extraterrestrial civilizations. *Soviet Astronomy, 8*(2), 217–221.

Krack, E. A., Jr. (1963). Sponsorship and the selection of talent. *The Chinese Civil Service*, 84–91.

Kuiper, T. B. H., & Morris, M. (1977). Searching for extraterrestrial civilizations. *Science, 196*, 616.

Kuttner, R. E. (1981). The nature of extraterrestrial communication. *Cosmic Search Spring*, 20–21.

Lehman, W. (1981) Frankenstein unbound: Towards a legal definition of artificial intelligence. *Futures, 13*(6), 442–57.

Lumsden, C., & Wilson, E. O. (1983). The dawn of intelligence. *The Sciences*, 23–31.

Lupton, D. (1995). The embodied computer/user. *Body & Society, 1*(3–4), 97–112.

Marien, M. (2011) Humans and machines: A new world order? *World Future Review, 3*(1), 126–30.

Marsh, R. M., & Mannari, H. (1976). *Modernization and the Japanese Factory*. Princeton: Princeton University Press.

McGowan, R. A., & Ordway, F. I. (1966). *Intelligence in the universe*. New York, NY: Prentice-Hall.

McNally, P., & Inayatullah, S. (1988). The rights of robots: Technology, culture and law in the 21st century. *Futures, 20*(2), 119–36.

Menzel, J. (Ed.). (1963). *The Chinese civil service*. Lexington: D. C. Heath & Co.

Merton, R. K. (1936). The unanticipated consequences of purposive social action. *American Sociological Review, 1*(6), 894–904.

Michaud, M. A. G. (1982). The last question. *Cosmic Search 1982*, 6–7 (First Half).

Nelson, B. (1973). Civilizational complexes and inter civilizational encounters. *Sociological Analysis*, 79–105.

Newman, W. J., & Sagan, C. (1981). Galactic civilizations: Population dynamics and interstellar diffusion. *Icarus, 46*, 293–327.

Newman, W. J., & Sagan, C. (1985). Nonlinear diffusion and population dynamics. In B. R. Finney & E. M. Jones (Eds.), *Interstellar migration and the human experience* (pp. 302–309). Berkeley: University of California Press.

Niven, L. (1970). *Ringworld*. New York: Ballantine.

Nivison, D. S. (1963). The criteria of excellence. *The Chinese Civil Service*, 92–106.

Oliver, B. M. (1976). Proximity of galactic civilizations. *Icarus, 25*, 360–367.

Ostriker, J. P., & Turner, E. L. (1985). The inclusion of interaction terms into population dynamics equations of interstellar colonization. *International Astronomical Association*, 85–475.

Plaks, A. H. (1998). The confucianization of the Kaifeng Jews: Interpretations of the Kaifeng Stelae inscriptions. In J. Goldstein (Ed.), *The Jews of China* (pp. 36–49). New York: ME Sharpe.

Raup, D. M. (1985). ETI without intelligence. In E. Regis (Ed.), *Extraterrestrials: Science and alien intelligence* (pp. 36–43). Cambridge: Cambridge University Press.

Redfield, R. (1955). The social organization of tradition. *Far Eastern Quarterly, 15*, 13–21.

Rescher, N. (1985). Extraterrestrial science. In E. Regis (Ed.), *Extraterrestrials: Science and alien intelligence* (pp. 83–116). Cambridge: Cambridge University Press.

Sagan, Carl. (1963). Direct contact among galactic civilizations by relativistic interstellar spaceflight. *Planetary and Space Science, 11*, 485.

Schank, R., & Birnbaum, L. (1994). Enhancing intelligence. In J. Khalfa (Ed.), *What is intelligence* (pp. 72–106). Cambridge: Cambridge University Press.

Schmidt, P. R. (1997). *Iron technology in east Africa: Symbolism, science, and archaeology*. Bloomington: University of Indiana Press.

Shanklin, E. (1981). Two meanings and uses of tradition. *Journal of Anthropological Research, 37* (1), 71–89.
Singer, C. (1982). Settlements in space and interstellar travel. In M. H. Hart & B. Zuckerman (Eds.), *Extraterrestrials where are they?* (pp. 46–61). New York, NY: Pergamon.
Stephenson, D. G. (1981). A classification of extraterrestrial cultures. *Journal of the British Interplanetary Society, 34*, 486–490.
Steward, J. H. (1963). *Theory of culture change*. Urbana: University of Illinois Press.
Tipler, F. J. (1981). Extraterrestrial intelligent beings do not exist. *Physics Today, 34*, 9–81.
Tipler, F. J. (1985) Extraterrestrial intelligent beings do not exist. In E. Regis (Ed.), *Extraterrestrials: science and alien intelligence* (pp. 133–50) Cambridge: Cambridge University Press.
Tough, A. (1986). What role will extraterrestrials play in humanity's future? *Journal of the British Interplanetary Society, 39*, 491–498.
Turner, Edwin. (1984) Galactic colonization and competition in a young galactic disk. In IAU symposium 112: The search for extraterrestrial life: Recent developments.
Umesao, T. (1986). Keynote address: The methodology of the comparative study of civilization. *Senri Ethnological Studies, 19*, 1–8.
Vallentine, J. W. (1985). The origins of evolutionary novelty and galactic colonization. In B. R. Finney & E. M. Jones (Eds.), *Interstellar migration and the human experience* (pp. 266–276). Berkeley: University of California.
von Hoerner, S. (1975). Population explosion and interstellar expansion. *Journal of the British Interplanetary Society, 28*, 691–712.
Walters, C., Hoover, R., & Kotra, R. K. (1980). Interstellar colonization: A new parameter for the Drake equation? *Icarus, 41*, 193–197.
Warner, W. L. (1958). *A black civilization: A social study of an Australian tribe*. New York: Harper.
Webb, S. (2010). *If the universe is teeming with aliens—Where is everybody? Fifty solutions to the Fermi paradox and the problem of extraterrestrial life*. Berlin: Springer.
Wilkinson, D. (1995). Central civilization. *Civilizations and world systems: Studying world historical change* (pp. 46–74). Walnut Creek, California, USA: Altamira.
Wilson, R. R. (1995). Cyber(body)parts: Prosthetic consciousness. *Body & Society, 1*(3–4), 239–259.

Chapter 10
Parsing ET Civilizations and Cultures

I have argued in the previous chapters that we do know something about the *general* nature of all ILFs (at least until proven otherwise), including the reproductive, social, cultural, and civilizational aspects. In essence, we have a kit of tools which should allow us, off the bat, to examine and start understanding ETI we may contact. We may be able to take this further, and in this, and the following chapters, we shall explore how we can refine this toolkit in order to see what else we know about ETI. From here on, however, as Burl Ives sang, the valleys get deeper, and the hills higher: much more speculation than fact, though hopefully based, as always, on human scientific knowledge about social phenomena, and our old friend, Occam's razor.

Having spent a great deal of ink and space in distinguishing culture from civilization, and pointing out the importance of this distinction, in this chapter we shall, pro-tem, blur that distinction. Now that we know the relationship between these two associative forms as an abstract reality, we need to recognize that if and when we contact ETI, we will not be contacting an abstract, but a concrete, more often than not in the form of individuals or a group of individuals. Analyzing their cultures and civilization will be important, but we need a quick and dirty way of telling "…those with feathers that scratch from those with teeth that bite…" as Humpty Dumpty said. So in this chapter we shall attempt to estimate what forms ETI cultures/civilizations might take from the human standpoint.

The basis for our analysis is the parallels that can be made between the race *Homo sapiens* and ETI races. Obviously, such a parallel—less than an analogy, more than a correlation of features—has its dangers and limitations. Nonetheless, anthropology, which is as close a discipline as we have to xenology, suggests some pathways to understanding what parameters may be common to all ILFs. We turn therefore to examining possible parameters of ETI civilizations.

There are three *general* assumptions and their corollaries that all good cultural analysis must take into consideration. First is the organismic, or ecological

M. Ashkenazi, *What We Know About Extraterrestrial Intelligence*,
Space and Society, DOI 10.1007/978-3-319-44456-7_10

assumption: species are the product of their environment. Second, intelligence, once developed as a biological solution to a specific set of problems, becomes the source of solutions to other problems. Third, once intelligence becomes a primary adaptive tool, cultural, rather than natural ecology become more and more important, to the point (potentially) of becoming the main factor in evolution.

Having reached this far, it becomes possible to examine the consequences of these assumptions on ETI civilizations. To start, we need to make a scaling assumption that will help with the method. As Bar-Yam (2002) notes, differences in scale of analysis are crucial for proper understanding of complex phenomena. In our case, we can make a (largely artificial, and definitely pro-tem) distinction between 'culture' and 'civilization'. On the whole, certainly at interstellar scales, we have seen that civilizations encompass or encapsulate cultures. We have also seen that what ties together disparate cultures within a civilization, is the adherence of an elite (and through the elite, much of the population) to principles of the civilization. In other words, to follow Bar-Yam's perspective, cultures and civilizations are the same phenomenon viewed at different scales. In this chapter we will adopt the idea that cultures are a finer-grained scale view of civilizations. In order to provide useful data for contact with ETI, we shall need to ask a series of questions that would be able to provide a framework for understanding and communication. These questions are parametric: we want to understand the other, perhaps in comparison with ourselves, perhaps in comparison to some other 'other'.

A quick review and summation of the crucial points of civilizations and cultures is in order.

Culture is a set of learned behaviors that are essentially 'imposed' on individuals by a process of socialization that we can expect to be present in any culture, human or otherwise. For analytical convenience, we look at any culture as if it contains a set of *institutions*. Institutions are patterns of behavior that tie together members of a culture, and that include fuzzy sets of rules that can be conceived of as answers to problems of living together. In human societies these institutions include patterns that deal with reproduction, production, and consumption (*kinship and marriage, economics*), resource allocation (*politics*), conflict and law (*war* and *law/justice*), intellectual activity (*religion, science and knowledge*) and even leisure (*entertainment, music*). These institutions interpenetrate one another, have no fixed, objective boundaries, and are largely an analytical convenience.

Civilization is a set of standards based on an agreed-upon (though fuzzy) set of texts and narratives, otherwise a *canon*. It is a *learned set of behaviors* for individuals and groups who uphold, promote, and advance these standards and their basic canon. A civilization *may* have a material component which is based on the ability and desire of the civilization to supply energy needs to [all] members to a degree that is well above the minimal energy requirements of each member, and perhaps [to provide sufficient energy to supply all the wants of each member]. The terms in brackets provide a measure of the *quality* of the civilization as an organizing system of complexity.

10.1 Civilization: Personal Parameters

When was the last time you saw a family? Think hard. Was it this morning, at breakfast? Yesterday as you jogged in the park? Last night, on TV? Surprising as it may sound, neither you, nor I, nor anyone else has ever seen a family. What we *have* seen, is a group of individuals, of roughly the right sizes, modes of dress, and expected variations in behaviors and modes of expression that correlate with those sizes, modes of dress and expression we expect of a 'family'. If the smallest one speaks in what we assume is a childish voice, and the largest one has a mustache, and … we assume they are a family. Even if they are bunch of non-kin actors behaving in a certain way that makes us *think* they are a family.

There are two lessons to be learned from this. First, any ETI individual we meet/communicate with is at one and the same time an *individual*, a member of a *group*, of a *culture*, and probably a *civilization.* So who are we communicating with, really? Suppose the individual we are communicating with is a rogue member of the society? A pirate? A member of an extreme sect? A solipsistic anarchistic hippie artist? Whom does that person represent? Second, whomever and whatever the role of that individual in its society, it is nevertheless a member of its society, and it does represent (struggle against it as it might try) its social origins. We cannot assume, as we do with a TV family, or even a group of unrelated people we see walking down the street, that we see a 'family' simply because they fit our preconceptions of what a family should 'look like'. Thus too, we cannot assume that a broadcast we receive from an ETI source is from a government, a scientist, or the ETI equivalent of SETI.

Contact of any sort with a member of an ETI race must start from a form of the assumption of mediocrity: that individual is the median for its race, culture, civilization. This of course is problematic if the individual has a fringe social role in its society (and those parameters are not human, but alien). As a consequence, we need to look at any facts known as a point on a spectrum, whose edges and mode must be identified. Human characteristics are almost always measurable on a bell curve. Any quality—music-playing ability, literacy, mechanical ability—can be plotted for any human population on this curve. A small number of people have almost none of whatever quality X is being measured. More people have some of X. Fewer have a greater degree of X, and a very small number are X experts.

The same is likely to be true of any quality we identify in any ETI individual. The question always breaks down into two sub-issues: what is the quality we have identified, and where on the bell curve does that individual stand? To take an example. Is the individual we have contacted savvy about its culture's technology? If so, is he normal (in the statistical sense) or an outlier? Whatever information we get from any given individual (and, for the sake of reality, any transmission originator is an individual) must be assessed on the basis of those two questions. 'Culture' helps in assessing the answer to both those questions.

10.2 Civilization: Cultural Parameters

The crucial parameters *for a human observer* derive from our understanding of institutions. It rests on the assumption that *whatever the biological or other bases of an intelligent species, certain issues need to be addressed systematically, and these systematic solutions to problems are the equivalent of the human 'institution'*. This by no means that ETI solutions, institutions, or regularities will be like human ones. We have, at present, no knowledge of ETI biology, phylogeny, or environment to allow us to make judgments about the *content* of their institutions, We merely know they will have observable patterns we can label 'institution'.

From a human point of view, what the content of these institutions is—the mandated ways of behavior and social exchange—are givens, to which humans should not, and cannot deny or make ethical criticism. Institutions are the product of lengthy historical development at any given time point, and they derive, as we have seen, from a series of continual developmental and selection pressures, including the culture concerned and the ecosystems it is based on and affected by.

I would suggest that a number of parametric answers help us in understanding what and who we are dealing with. These are based on our understanding of institutions. One word of warning is necessary, however. While it is likely that institutions are universal, since they are reifications of certain requirements of any organism, the *list* of human crucial institutions may not be a complete one for another species. The list presented here is a parsimonious one *for humans*. ETI may have, in addition, other institutions just as important, supporting needs we do not have and cannot conceive of. And what we know or are able to know of those institutions, will necessarily be fuzzy, that is, possibly a reply with a wide degree of internal variation and flexible boundaries.

1. *Reproductive* parameter. Are social behaviors mapped onto biological reproductive processes? Not at all <——————> Completely
2. *Economic* parameter. What are the culture's critical resources and what is the demand for them? Very Low <——————> Very high
3. *Political* parameter. What is the complexity of political rules, positions, and access to power? Very simple <——————> Very complex
4. *Law and justice*. In cases of dispute, is access to mediation/arbitration universal? Everyone has equal access <——————> Access is differential
5. *Religion* parameter. Are there dissimilar statements of cosmology, theology, and ritual? Single set of statements <——————> Multiple sets of statements
6. *Socialization* parameter. Sources of socialization available to individuals? Multiple sources <——————> Single source
7. *Knowledge* parameter. Are sources of knowledge Unitary <——————> Multiple

This quick roster (and, as yet, we have not tackled the issue of *how* we extract this knowledge) provides the human observer with a 'chart' as it were of whom we are dealing with. This chart is deliberately fuzzy, not clearly defined, and lacks specific questions. What we need to do is to utilize what we *do* know about ETI in order to gain comparative, systematic insights into what we *should* know. We can elaborate a bit.

The answer to (1) gives us some insight into the degree to which the race we are communicating with has managed to transcend (or not) it's biological origins. In other words, to what degree is this race's behaviors framed by its genes. This *may* also indicate the degree of flexibility they have in other areas, as well as their ability to incorporate non-members of the race into the compass of 'honorary' members. The economic parameter gives some insight into the kinds of drive imposed upon this race as a whole. More importantly, it also provides a set of red lines about how to deal with them: obviously, certain lines are not to be crossed, or if crossed, will trigger potentially hostile responses as they defend what they see as critical resources. As a general rule, authoritarian regimes have very few intercessionary groups between the ruler/top management/boss and the ruled. Political forms which allow and encourage intercessionary groups—governmental, quasi-government, civil society, and so on—typically are less authoritarian, more democratic. This is neither a judgement on authoritarianism, nor on democracy. Nor can 'authoritarian' be equated with 'primitive' since a developed culture may develop artificial means of control and feedback. The point however is that human references are a pull between authoritarianism and populist forms of politics (see for example the wonderful description of rhythmic shift between aristocratic and egalitarian systems in Highland Burma, Leach 1954). And we find it unpleasant to deal with either extreme. An ETI's viewpoint, on the other hand, might view our obvious ambivalence and ambiguities as distasteful, or even worse. We all want to have our cases heard in case of dispute, and one of the ideological components of world civilization is equal access to justice.[1] An ETI race, with internal biological differentiation may see things differently, or perhaps a qualified individual value system has evolved, as, e.g. better access to justice for individuals qualified in a particular field of activity. Be it as it may, we learn where humans (including human legal personae such as corporations) could fit into the scheme of things, which in turn implies the kinds of exchanges we can expect, and the possibilities of receiving what *we* humans, consider fair treatment. Parameter (5) may well be a human affliction, put upon us as a genetic joke by dice-playing cosmos (Dawkins 2009), which ETI are not afflicted by. If so, the location of our hypothetical ETI race is on the far left of the dimension: there is only one set of statements, and they are a null set. On the other hand, other races may have religion(s) as well, in which case, if only one set of religious statements is possible/allowed/available, we humans could be on the short end of the stick, and we would need to pry (gently) to see if other statements are *acceptable* at all. Parameter (6) returns us, in a sense, to the political question: does the culture concerned allow for multiple, contrasting, and possibly colliding viewpoints? If not, we need to be very careful with this ETI culture, since

[1]Not really available in the vast majority of cultures that make up our current nascent world civilization, and not in the 'developed societies' either, but it is ideologically on the menu.

our rather freewheeling human ways (notwithstanding demands for absolute conformity by Islamists, Americanists, and doctrinal Communists) lead us in other directions. *We* are socialized by in-groups, family, schools, media, each with its own (sometimes clashing) message. How would this play with a culture in which one single socialization method is practiced? The same is true of the final parameter. If all knowledge is derived from a single source, it is *possible* that we are dealing with a transcendent society, one that has integrated some form of IT deeply into the fabric of the culture. Or it is possible we are dealing with a society in which one center has control of all knowledge, which could have some unpleasant implications, notably since we humans could be an additional source of knowledge which could upset the apple cart.

10.3 Civilization Parameters

A civilization, as we said, is a complex, largely intellectual construct. A civilization may not, in practice, be able to mobilize personnel, resources, or force to accomplish a specific purpose as a normal state of affairs. But it may be perfectly capable of doing so should the need arise, and if the civilization elite are able to fit the need into their set of notions. Thus, as we have said, if the ethical demands of the civilization include the demand to supply energy at some level to a constituent unit (cultures, solar systems, or individuals) and these resources are available, it can be done. So we do need to look at some parameters that might help us understand better how to characterize the peculiarities of any individual ETI civilization (and, if they are sentients building a radio telescope, we could bet on them also having at least a planetary civilization).

10.3.1 The Intellectual Standards of Civilizations

A civilization, by definition, has a set of ethical/moral, behavioral, aesthetic, artistic and other standards. These are codified in the form of a fuzzy set of 'sacred' propositions. Sacred in this case means that they are codified and rely upon the sacrality of tradition (Hobsbawm and Ranger 1992) and of supposed 'venerable persons' (Guthrie 1980) who 'originated' the standard. In practice, the question of who originated the standard 'really', is unimportant. What is important is the degree to which the actors (in this case, the civilization elite) accept these standards as unquestionable in general (though, of course, dispute about specific issues and practices often exists).

For an outside observer (e.g. a member of human civilization) wishing to communicate with a member of a civilization, certain questions are crucial for understanding and even potentially survival and communication over time. Several sets of questions corresponding to the major aspects of civilization need to be asked.

In the first instance, some understanding of the *standards* the civilization in question holds itself to, must be understood. These standards flag those individuals and corporate actors who are civilized (and thus can be, at the very least, 'marginal' members of the civilization), from those who, by virtue of an absence of traditional knowledge, and an inability or unwillingness to learn them, would be considered inferior and outside the pale.

Crucially, these standards need to be elicited by asking variations of the questions "If X does Y, would you [the civilization in question] consider X to be acting properly/appropriately or not?". This set of questions is problematic because it is a *confirmatory* question: it effectively only confirms/disproves the observer's existing knowledge/assumptions about cultural behaviors she is aware of. An additional set of exploratory questions is needed, of the form "What are the appropriate behaviors expected of X in dealing with you (singular and collective)?". The aggregate of those responses allows analysis of the standards one is held to as a member of the Great Tradition concerned, as well as the permitted range of variation—the fuzzy index—of those standards. Unbiased questioning of such sort between members of different cultures is a well-known, and quite accurate technique (e.g. Frake 1964).

10.3.2 Praxis: Doing Civilization

Civilization, as said, is not only a set of standards, but also a set of practices and behaviors. These might be linguistic, performative, behavioral, decorative, or what have you. Clearly in a non-human civilization, notably one that is mono-species, there will be physical variations that humans might not be able to sense, let alone copy. Body scent, gestures with organs humans do not possess might be impossible to replicate. Nevertheless, one would expect that over time an associative roster or glossary of the association of certain circumstances (social, technical, or natural) with certain practices will become apparent. To take an example from human civilization, in all cultures the presence of a social other is acknowledged in some form. This might be a handshake (worldwide), a bow (Sinitic cultures), or even a thump on the other's chest (South Sudan). This is more often than not accompanied by stock phrases: "Hello", "Have you eaten", "Peace" and so on. From the perspective here, the specific content, important though it is in other forms of analysis, is less crucial than the fact that almost *every* human culture has this practice. So too, we can expect other civilizations to have identifiable stock behavioral repertoires—meta statements—which will tell us what appropriate behavior for a member of the civilization consists of.

By and large, the best way to identify behaviors is to observe and catalog them, and, again, to create glossaries of behaviors and their contexts. In practice, of course, given the universal velocity limitation C, this is probably not practicable in the flesh. Nonetheless, even electromagnetic radiation messages can convey images and sounds, which, for all their weaknesses (e.g. deliberate or culturally-biased falsification: how often does one see a human actually defecating on broadcast TV?) may still be the best we have.

10.3.3 Canon: Fixed Texts in an Age of Change?

Finally, we need to obtain the canonical texts which structure the behaviors and standards of the civilization. Without these, an observer is in an invidious position: the observer is unable to participate in a dialog "among civilized people" because of limited ability (if any) to either understand, or originate cultural signals based on the canon.

It is important to note that while the term 'texts' is used here, texts do not need to be written: they can be in any recorded form. While in most human civilizations written texts are the chief canonical form (for example, the Bible), pictures (Van Gogh's "Sunflowers"), music (Beethoven's Symphony No. 9) and performance (the *Ramayana*) are all canonical texts, the knowledge of which flags a person as a member and practitioner of world civilization. Less 'textable' items occur as well: pizza, Coca Cola, and sushi are a type of 'text', in the sense used here.

Synesthetic texts—fixed and immutable, that are to be repeated under the appropriate circumstances—have been a teaching feature of Asian arts for centuries. Complex repeated motions called *waza* in Japanese, are combined into canonical forms called *kata*, which have three purposes: as canonical forms, they are texts of how the art concerned must be performed (though different interpretations are possible and encouraged); as fixed bodily movements, they are an important element of learning the praxis; and they embody the standards of the particular art, from flower arranging, through the preparation and drinking of tea, to fighting (DeCocker 1998; Singleton 1998; Ikuta 2008). Similar kinesics exist in various cultures, perhaps less consciously, in knowing for example, the intricacies of English afternoon tea, French wine tasting, or the proper behavior at a baseball game. Know them, and you demonstrate that you are 'civilized' for a given value of civilization/culture. Know them not, and you are either novice to be taught, or a barbarian to be pitied, sniggered at, or ignored.

Modern human civilization—the globalized world—is a civilization based to some degree on change. As has been said earlier, this is a civilization which is in the process of coming into being, and we are living through the birth pangs. It is certainly conceivable that due to a catastrophe—from global warming and flooding, through war or pandemic—the current globalized civilization that is coming into being will fail and disappear. This of course refers to the question of L, about which more again later. As it stands in 2016, this is a civilization which values change very highly. Driven obviously by technological and scientific advances, most of us have experienced more change in our lifetimes than our grandfathers did in theirs. That *may* be a feature of all world civilizations. It may also simply be a phase, after which certain texts will make the cut into the canon, and others will remains as obscure texts, or those treasured by a particular culture within the civilization, but not be a part of the canon of human civilization.

10.4 Practitioners

Who are the practitioners of civilization, the actors who hold the civilization together? In human historical civilizations this would have been easy to answer. In the paradigmatic Chinese civilization, the literati (Friese 1963; Cheng 1973; Guimaraes 1918; Kracke 1957)—men who had studied and memorized the canon, were able to write a good hand, and were expected to conduct themselves in definite specific ways—were the repository and guardians of the civilization. While they were an elite, they were not necessarily political leaders. *Many*, due to the Chinese political system, did indeed become the administrative spine of the Chinese state. Many others—those who failed the official examinations, those who took them but did not assume office, or retired from it—were not. In Japan, the political leadership—the *bushi* (warrior) class, adhered nominally to one set of standards (*bushido*, the way of the warrior) but other standards—knowing how to serve and drink tea (*chanoyu*), knowing something of the Confucian classics, and even farm—were acceptable canons of behavior as well. In modern Japan, the standards of cultural behavior are paradoxically maintained to a very great degree by two sets of professional performers: sumo wrestlers and geisha entertainers (Ashkenazi 2003; Crihfield 1978; Dalby 1983; Fischer 1966; Havens 1982). We have no such definite category of people in the nascent world civilization, but presumably that may come about.

Practitioners of civilization may fulfill different roles in a civilization. The role may be formalized, as with Chinese literati, or expressive, as sumo wrestlers and geisha. It may be something we are familiar with, such as the bards of old Norse culture, or modern music masters. Crucially, these experts are expected to maintain and observe (at least in public) a rectitude on the one hand, as examples, and the right to point out the necessary standards on the other. Moreover, once they are experts in the standards of civilization, they become *de facto* members of a floating elite, welcome by their peers everywhere within the civilization, lauded as experts in standards the civilization holds dear, and courted to teach, train, and 'emit' civilization for those who wish to be a part of that greater thing. In historical Sinitic civilization, Confucian and Buddhist experts circulated between China (the fountainhead), Japan, Korea, and to a lesser extent Okinawa and Vietnam, reinforcing the lessons their local peers taught by example and precept. In Norse culture, bards did the same.

In modern world civilization, still a work in the making, artists, scholars and academics, architects, and religious figures circulate geographically, developing followings of greater or lesser value and influence.[2] Civilization is maintained by an urbanized lifestyle. It is in urban settings that institutions such as universities, theaters and so on can flourish, because the practitioners of civilization—scholars, artists, writers, literati, sumo wrestlers—can interact with one another creating standards and a praxis, which are later dispersed throughout the civilization. I would agree with the essential point: the standards, texts, and praxis are developed within a relatively dense community of interacting practitioners. In a historical perspective, these communities could only exist in urban settings (where population density eases intra-communal communication and the ferment of ideas). In a modern setting, however, where communication relies more and more on technology, to the point that, if desired, inter-personal contact can be virtualized within an IT framework (Curtis and Nichols 1993; Grady2003; Hiltz and Turoff 1978; Hogan 2010; Lipnack and Stamps 2000; Lusted 1992) personal physical proximity is likely to be substituted by virtual reality and other forms of IT, those currently known to humans and those not. Given the universal velocity limitation, personal visits would be possible (more possible with races that have the necessary biological qualifications) but rare. At the solar system level, the practitioners could meet more often in the flesh, but most interpersonal contact, as well as group contact, would likely be computer mediated. Certainly when it comes to contact between humans and an ETI civilization, the civilization standards, praxis, and canon would most likely be transmitted, queried, and discussed in some form of computer mediated communication.

10.5 Those that Have Feathers and Scratch, and Those that Have Fur and Bite

Here I want to broaden the discussion, and briefly touch on some dimensions derived from human societies, which would be worthy of exploring for understanding any ETI civilization. These dimensions have the advantage, methodologically, of being finer grained than the analysis in the previous section. They are intended to do two things: sharpen our understanding of the minutiae of civilizations, and provide a template for looking, in ETI civilization, for issues that concern us humans.

Rather than primarily an exhaustive directory of the (assumed) ETI cultural features, it is more useful to suggest continua of action which might inform *any* culture—and then to examine these suppositions for their social prerequisites and implications. Applied to human cultures this method has been used by Steward (1986) to look at social factors such as marriage patterns, by Harris (1988) to test hypotheses regarding food choices, and by Smith (1974) to consider political

[2] As a minor case in point, scholars, ranging from Levi-Straus through Stephen J. Gould and Albert Einstein, to cite only a few, are household words throughout the ‘civilized world’.

systems. In the case of ETI societies there is no broad range of empirical data to support my conclusions. This matters little, inasmuch as I am intent here on constructing the first steps in a *methodology* for analyzing ETI, rather than describing ETI: something that is currently impossible in any case. Thus the details of supposed ETI culture are far less important than the potential processes by which these details could validly and reliably be reached, and the complex that each feature might imply. Empirical data, when we have access to it, will further modify and refine, perhaps replace, many of the parameters discussed here.

What we cannot do is assign probabilities to any particular feature or cluster of features appearing. What we *can* do is suggest broad ranges of possibility, that is, parameters of features which appear to be significant. Some of these fundamental realities are derived from a purely human perspective. Others are truly species-independent, even, perhaps, independent of the normal-exotic distinction that has been made throughout this book, and truly universal.

10.5.1 Standards

The standards of a civilization are the norms to which everyone who is a member of the civilization is expected to adhere, publicly as well, perhaps, as privately. Essentially these are known by consensus, and series of internal codes that those socialized into the civilization know, even if not fully articulated. For example, British imperial civilization of the nineteenth and twentieth century had a standard of "gentleman" by which individual behavior was measured. Those who displayed the behaviors (particular language use and pronunciation, dress codes, proxemics, indifference to danger, particular expressions of humor) were judged to be 'gentlemen' and were assumed to originate from a particular class, have gone to a public school, be of pure English stock (Scots, Welsh, and English-derived Irish were allowed to be gentlemen too, but not Indians, Jews, Americans, or southern Europeans. Cf. much of the work of Kipling, all of Jane Austen, and virtually all of Saki), and so on. So standards of behavior make widespread assumptions about the individual, the individual's background, and how they fit into the social structure.

Ethical standards One crucial issue that we must concern ourselves about are ethical/moral standards. We need to ask how, within the standards adhered to by the civilization, are inter-actor relationships governed. At one end of the spectrum, ethical standards are non-existent. Pure amorality is assumed, and actions, whatever their outward expression, are conceived of as expressing and driven by the wishes of the individual. At the other pole stand strongly governed ethics. That is, every civilized behavior is supposedly motivated by ethical prescriptions, which can be very detailed. In practice one can assume that any civilization will be located somewhat closer to the center than the absolute poles.

The implications for communication by humans are vast. Closer to the amoral pole means that assumptions of benevolence, assumed by some writers about ETI

(Deardorff 1986), will get no traction. Closer to the prescriptive pole would mean that humans would need to ensure that they learn the standards very quickly if we are to be accepted in the realm of civilized persons, and if we are to participate in the interstellar civilization concerned.[3]

Obviously, the content of the ethical/moral standards are important as well. There is a vast difference between "Love thy neighbor" and "Love thy neighbor with salt and pepper". Given the absence of data, we cannot even speculate on the content of the ethical principles, merely on their presence. We can argue for a general useful continuum of dominant ethical principles. At one pole are societies that adhere, to as great a degree as possible to the 'Do not do unto they neighbors that which is hateful to thyself', considering the difficulties pure adherence creates (see above, Chap. 8). At the other pole is an authoritative standard (exhibited among humans by early Christian churches, and modern Islamic Salafists) of 'Do what you are commanded to do, for that is universally best'. Neither position has any 'pure' value, as each represents logical and empirical problems. Nonetheless, some information about where the ethical standards of the given civilization stand, would be critically important.

Aesthetic standards For human beings, aesthetic standards are so important, that we see them in every society we know of. All societies have standards of personal beauty (though sometimes confined to one sex or gender, sometimes different standards for different genders). This leads to aboriginal customs of personal decoration, ranging from the use of cosmetics (for both/either men and women, Casal 1966), personal adornment (Gosden and Marshall 1999; Buss 1998), and bodily modifications (Caplan 2000). In addition, humans seem to have an innate drive to decorate their surroundings: housing, gardens etc. To the point of developing an entire discipline—architecture—for aestheticization of the machinery of shelter. Aesthetic standards clearly depend heavily on the species sensorium. A non-visual species is as unlikely to develop a visual aesthetic as humans are unlikely to develop ultrasound music. That having been said, however, there are two issues that would help us understand a non-human civilization's aesthetic standards.

First, how *fixed* is the aesthetic? Is it a general standard "We have standards in general, but we recognize that change occurs" or is it an *absolute* standard? If it is the first, there is room for bidirectionality: humans will be expect to contribute to the aesthetic standards, indeed, may produce fashions that change the aesthetic standards. This would have a number of implications, not the least the fact that the artistic institution in any society is deeply tied to the economic one (in other words, opening for human cognoscenti to set, and teach, aesthetic standards). This does not mean, by the way, that the aesthetic is open to all winds of change, but that it

[3]One danger of course is that due to strict physiological requirements, we may not be able to. In such a case, a *strict* prescriptive set of ethics will relegate us to second class wannabes. A civilization more experienced in inter-civilization communications may allocate room for the physically incapacitated (e.g. Humans).

accepts the possibility of standards, fashions, and modes: a strong core of standards that can be interpreted in many ways.

Second, which artifacts and sociofacts are aesthetic-dependent? Among humans, as said, virtually any and all aspects of life have become aestheticized somewhere to some degree (consider your toilet room, body, military uniforms). At one end, zero aesthetic perception. At the other, a life led in obedience to aesthetic rules, with every artifact (and possibly, sociofacts and mentifacts as well) required to be designed/displayed/presented according to some aesthetic standards.

Without entering the debate of what aesthetics are, we can see that for the human communicator, knowledge of the (undoubtedly fuzzy) boundaries between what must be done/presented aesthetically and what not, are going to be difficult. This means that a great deal of attention must be given to understanding the underlying principles, even if we do not experience them with our own sensorium.

Performative standards The issue of performative standards was discussed earlier, during the discussion of role presentations. Among humans, proper presentation of self, that is, minimum performative standards are an absolute requirement for being able to function at least minimally, within any culture. Yes, in many societies, notably those that have a great deal of contact with the world civilization, allowance is made for novices, children, newcomers, and the incompetent. But to be accepted as a full member in society, one must know when and to whom to defer with voice tone and body position in European societies, and to whom to bow, how deeply, and using which expressions of respect in East Asian ones.

So once again we need to ask what are the rules—proxemics, voice, decorations, scents, or anything else—that are the standards of proper civilized behavior in our ETI civilization. Not only that: the entire repertoire of statuses and their performative interrelations needs to be known. Given the importance of proper presentation, as we discussed above, it is unlikely (though not impossible) that an ETI civilization has dispensed with performative standards. What is far more likely is that the cognoscenti will *emphasize* these performances to display a sign of their membership in the group of experts.

It is also important to know *which* of an individual's role sets require performative standards. Do they apply in the home? During expressive interactions? At work? And in relation to what situations are they triggered, or not?

Rationality Under multiple pressures—Greek didacticism, Jewish polemics, early Christian argumentation, and the rise of European Enlightenment—the Western-dominated world civilization has a premium on rational discussion. This may or may not be the best way to go about inter-actor exchanges, it certainly is not the only one. In modern Japanese society, a great deal of emphasis is given to unspoken, emotional communication between civilized individuals, while avoiding any show of conflict. In other societies, instead of rationality in discourse, beauty and shape have a premium. We thus need to ask what standards of rationality are to be employed by the civilized among and ETI culture. At one pole, we have purely rational standards of discourse, and a set of standards that ensures that e.g. emotional. aesthetic, and other non-rational standards creep in.

For human observers, the purely rational would seem to be easier to identify. This is true provided the basic assumptions and starting points are known as well (as they might, from the canon). Non-rational standards are by their nature more difficult to parse and understand. Crucially, the observer must be able to make some sort of distinction between those arguments that are based on some logical set of rational procedures, and those in which a purely non-rational influence enters.

Expansion Analogies are terrible analytical tools. Yet even so, their appeal is so great that they are used, even here, with great abandon. Many SETI writers are American, and whether consciously or unconsciously, the myth of expansion into uncharted territories and lands seems to be accepted for ETI as well without a murmur. We need to question this.

Stephenson (1981) argues that expansion throughout available ecological niches is one of the barriers that an intelligent species must overcome. Expansion across the planet, into the solar system, and finally settlement across star systems. In fact, many discussions accept that ETI will, driven by some force of nature, expand and create multi-star system associations of some form (Finney and Jones 1985; Turner 1984; Walters et al. 1980).

Ehricke (1979) has assumed that all societies will likely have an extraterrestrial imperative: an inbuilt, evolutionary drive to explore space. Exploration may well be a derivative of *Homo sapiens*'s primate heritage. Indeed, it has been argued (Kashdan and Silvia 2009) that curiosity has a positive survival value. But confusing a genetic predisposition with the actual investment in space exploration seems unnecessary. We may, however, assume, that some fraction of ETIs will indeed be potential or actual spacefarers. Some proportion of those, perhaps even close to 100 % may also be space *colonizers* with an interest not only in the exploitation and study of space, but in the establishment of long-living colonies on other planets or in space (Ashkenazi 1995a, b).

Migration by space-faring species has been modeled both from a simple diffusion approach by Newman and Sagan (1981) and via Monte Carlo iteration by Jones (1981, 1982). These arguments assume that colonized areas will expand under the population pressure to fill available living sites (presumably including not only habitable and 'terraformed' worlds, but also space habitats converted from planetoids orbiting near sources of solar power and volatile gases).

Overpopulation is probably the most basic problem the human species faces at the beginning of the 21st century. From it flow numerous other ills such as heat gain (the "greenhouse effect" and "global warming"), food scarcity, economic over-concentration and scarcity, and of course political disagreements. It *may* be assumed that an ETI society will have solved such problems, but this is not necessary. The invention of a method for interplanetary transportation, or for that matter interstellar, would not solve population problems to any appreciable degree *unless* it was accompanied by some other motivating factor. Colonization for the sake of relieving population pressure would appear to be prohibitive.

Nonetheless, the need for providing newer pastures, or at least an outlet for population pressure that *seems* to the ETI population concerned to relieve some of the pressure, is conceivable. Of course, if such a policy were adopted, whether using FTL or STL drives, the home planet would soon find it impossible to send out additional colonies, as it runs out of resources. But from our perspective, such a policy when adopted initially, might be continued until contact with humanity were achieved. But population pressure is a poor motivator for expansion particularly since population pressure and expansion are part of a zero-sum game: as population increases, there are less resources for expansion, and vice versa. There are, we must consider, other motivators.

Consider ants. Limitations and control of fertility is no problem. The creation of a new nest is a function of a species need exercised through a queen. It is the queens who are likely to fight, and therefore a new queen must leave. Could not the same mechanism evolve among ETIs? Moreover, it is also conceivable that such a motivation may not be biological at all, but cultural. Aristocratic systems tend to evolve what Morris (1967) has called *demesnes*. Simply put, aristocrats love running things their own way, and hate having someone around telling them what to do. In the case of our ETIs, therefore, colonization might be motivated by internal cultural assumptions and political necessity. not by reasons of biological or ecological demand: the people that rule have made an assumption and everyone else is following suit.

We can assume that there might be excellent reasons, whether biological, scientific, or economic, for interstellar expansion. There are also, however, some good reasons for limiting expansion. First and foremost, of course, is the immense cost associated with the distances we are concerned with. And of course there may be natural limits to expansion: mature cultures may be interested in other things than expansion (cf. Papagiannis 1983; Hoerner 1975).

The foregoing discussion leads to the formulation of our first parameter. Any ETI civilization, viewed either at a particular point in time or over a historical period is either *expanding* or *stable*.[4] Expanding civilizations are spending an appreciable proportion of their resources in exploration, and, what is more significant, colonization of other environments. These may be on their planet (the human race, it must be recalled, barely uses the oceans, and certainly does not inhabit them), near-planetary space, near-solar space, or interstellar space. The extreme (a very unlikely one) is a species that expends one hundred percent of its resources on expansion. A stable civilization is one that, for whatever reason, does not expand beyond the limits it occupies. A (hypothetical) purely stable civilization would not engage in any expansion whatsoever, not even into available ecological niches on its home planet. Both extreme cases would seem to be unlikely, but intermediate cases would allow comparisons too.

[4]From this perspective, shrinking is equivalent to stable, except the costs are higher (transportation back to the home worlds, or killing off the population of the colonies).

There are multiple social, economic, political, and military reasons for colonizing one's home system (Ashkenazi 1992a, b, 1995a, b; Heppenheimer 1978; Launius and McCurdy 2007). Colonizing other solar systems is more iffy, considering the long time delay and huge efforts involved. We must, however, accept the fact that other civilizations may not have the same perspective, whether for material reasons, or ideological ones. As such, we need to accept that ETI civilizations may be located somewhere on an expansionist-stay at home dimension. Some may have, for one reason or another, a standard that says 'we must increase and expand'. This may include physical expansion, or intellectual, civilization standards to be expanded throughout the universe.[5]

To summarize the issue of standards Standards come in a variety of forms and shapes. What they exhibit in common is some fuzzy measurable set of criteria by which an observer can measure a behavior or practice to see whether it measures up. In any civilization, there may be many or few standards, and some standards will nest within themselves more detailed standards. As a general rule, at the civilization/Great Tradition level, a few general or specific standards with overriding canonicity will have been created and maintained. At the culture/little tradition level, there may be a plethora of standards maintained on the basis of local tradition, but making reference to, and subordinating themselves (perhaps only in lip service or convoluted interoperation) to the grand standard they profess to be a part of. Social phenomena are not neat!

10.5.2 Praxis

When we, as humans, look at the praxis of an ETI (or any other) civilization, we need to ask several question. In reality, these questions are the same we would ask of most social situations we are strangers to.

Membership criteria 'Who' is a crucial question. Who can become, and who is, considered a member of the civilization and can be, or is, an acknowledged cognoscento: someone who knows and tries to live the rules? Participation in the civilization may be extremely restrictive and inherent, or broad and achieved. That is, does one have to be born into a specific biological/kinship group to become a member of the civilization in all senses, or can this be achieved solely by hard work and study? On the one hand we have European cultures, which rarely if ever accepted those who were not bred of their own: Africans, Asians, Jews, Indians, need not apply. Women could be accepted, often only in specific roles (as e.g. Sappho of Lesbos, Empress Theodora, and Carolyne Herschel). In contrast, Chinese

[5]In the modern world civilization, 'televangelists' and 'techno evangelists' each for their own reasons proliferate as part of the CMC revolution. The issue of *why* certain philosophies (Christianity and Islam are prominent religious ones) feel that they must expand is complex and interesting, but beyond the scope of this book.

society was open to scholars of 'all' origins, *provided* they carried themselves (that is, behaved publicly and privately) as a literatus should. Women, until the rise of e.g. geisha in modern Japanese culture, were completely out of luck as expert practitioners of civilization standards.

A related question is the practice of becoming a cognoscento. Is it a formal process (as in the case of most Chinese *literati)*? Or can anyone do it by force of will and application alone (as Van Gogh, Louis Armstrong, and Ramakrishnan)? The ability to learn new patterns, which is inherent in humans, is limited where formal requirements are part of the equation, as in e.g. the British requirement that a 'gentleman' go to the right schools and have the right ancestry (in practice, white western European). If these are not available to an individual, or s/he is not equipped with this from the start, they are relegated to second class existence, even if they excel in the praxis of the civilization.

So we need to ask of an ETI civilization, "Who is able to become a practicing cognoscento of civilization?". The standards are likely to be well know, as are the canon, but if there are additional restrictions, and particularly if these restrictions are tied to physiological or species uniqueness, then humanity is going to be a second class culture within such an interstellar civilization.

Differentiation: What are the options A crucial element in all advanced and developed societies that distinguishes them from less advanced and less developed societies has to do with the number of statuses available. As a general rule *among humans* the freedom to choose one particular status over others is a critical measurement of what we call 'freedom'. I was raised in a society in which my ability to select a status was dependent largely (not wholly: parental wealth and status, certain types of membership also play a role) on my choice. I *chose* to become a scholar. Others in the same circumstances did not. In, for example, North Korea, I would not have had that luxury. Nor would I, had I been born a woman, in a misogynistic and culturally limited society such as Saudi Arabia. So one standard of praxis that we need to inquire about is the number of available statuses in a given civilization, and even among the cognoscenti who embody the civilization: what kinds of choices does an individual have?

On the one hand, there might be only a few civilization statuses available, as in the classic Chinese system: independent scholar, student for examinations, mandarin (government official), retiree. In other civilizations there may be vast reams of choice available—from author to zoologist—who constitute members of the 'civilized' elite.[6]

Civilizations whose cognoscenti allow only minimum status/role differentiation would possibly demand greater external displays of praxis: as we know, symbols (which forms of dedicated praxis certainly are) keep groups together by sweeping differences under the rug. High differentiation allows for greater permeability; civilizations which allow for greater variety, would of course make it easier for

[6]No one in a civilization which has both Kim Kardashian and Stephen Hawking as civilization icons has any right to doubt this.

humans and other newcomers to fit in, as they search for an appropriate niche of praxes.

Authoritarianism An important measure of civilized practices is the index of the relation between horizontal and vertical social ties. In authoritarian societies, vertical ties to the leader are supposed to be highly authoritarian in that ideas, requests, and demands from a superior are treated with respect bordering on, and sometimes exceeding, reverence. This happens in the case of individuals carrying on the standards of civilization as well. In a well known study, Francis Hsu showed how the schools of the traditional arts in Japan—a form of civilizational cognoscenti—were structured onto vertical structures with a single individual, the *iemoto* or 'root of the house' attaining a status of unquestioned and unquestionable civilized dominance, a *font et origo* of all cultural activities. This *iemoto* system is still prevalent in almost all Sinitic civilization practices, and echoes of it are to be found in traditional arts to this day, in China, Korea, Japan and abroad, in activities as varied as tea ceremony, doll-making, and martial arts (Hsu 1975).

The schools, adhering as they do to a standard set of practices formed supposedly in hoary antiquity,[7] many of which are intentionally spectacular, and canons, written in some cases (Okakura 1964), expressive in others (Redmond 2008), are perfect examples of the creation of civilization leaders, whose actions exemplify, and disseminate cultural standards. Non-*cognoscenti* may not practice the standards of civilization, but they accept them as a cynosure. However, such schools *among humans* tend to create hidden struggles over power expressed in struggles over the legitimacy of changes in the standards and the praxis.

Here too, the human observer of an ETI civilization needs to ask how authority is distributed among the cognoscenti, and to what degree is it flexible or rigid: can anyone play, once they have reached a recognized level of expertise, or are positions reserved to certain defined class of persons? What is the nature of authority invested in masters of the civilizational standards and praxis?

Fight and aggression Human World Civilization constantly proclaims, as a standard, an adherence to peace and nonaggression. It is repeated by the standard bearers of civilization, political leaders, and leaders of the arts. It is, of course, nonsense. Violence may be declining (Gat 2013; Pinker 2011) but it is still very much an aspect of humanity (Aijmer and Abbink 2000). Whether violence and fighting are biologically or socially determined is not the issue. We can say, as a middle-way, that fighting and violence are one of a series of responses available to most humans in social situations. And, indeed, the *proper* application of violence has always been one of the aspects of the gentleman. Confucius and Meng-tzu saw violence as counterproductive and undesirable, yet expected gentlemen, if driven to it, to be successful in dealing it out (Lewis 1990).

[7]See, for example, the repeated claims of many Chinese boxing schools that their practices derive from the commands of Boddhidarma, a (possibly mythical) Indian missionary in the Shaolin monastery (Shahar 2008).

The question we need to ask is whether violence is sanctioned in an ETI civilization, and if so, what behavioral standards are to be applied? In other words, is war a gentleman's sport, as it was in Europe during the eighteenth to early twentieth centuries? Or an out-and-out no-holds-barred contest as some military strategists see it (cf. Liang and Wang 1999). If so, what are the standards, and who can play? The implications for humans are obvious. Notwithstanding the *general* unlikelihood of interstellar military action, that does not mean it is impossible. Nor does it mean that standards for practices such as a *code duello* (Kingston and Wright 2010), or a coup system (McGinnis 1990) are nonexistent. What it does mean is that violence, and the agents of violence are regulated by a civilizational canon, and, in order to be acceptable, must follow certain behavioral principles.

So at one end we would have the acceptance of all forms of violence, from war (and possibly genocide) through small group warfare, individual acts of violence, to no violence at all, as standards of behavior. This does not mean that violence will not exist, but that its practice is opprobrious, or that it is practiced only under specific behavioural restrictions (as was the case in eighteenth-twentieth century Europe). Two points must be emphasized. If humans are an analogy, then the practice of violence against non-civilization members is not restricted by codes of conduct, and, regulating violence, and *forbidding* violence are two distinct kettles of fish. In other words, if only for our own safety, we need to learn the rules about violence as quickly as possible, and (something humans actually do quite well) develop specialized civilized practitioners of violence.

Contact/communication forms Humans have a range of communicative forms, the meta-rules of communication, that are important for establishing the tone and structure of communicative exchanges (see the earlier discussion of roles). These meta-rules are essentially a form of praxis that are likely to apply in any ETI civilization as well. In this case, it is of course possible that one end of the dimension is "We have no particular meta rules. Go ahead. Say what you want." That is of course possible, but I would suggest, that in a civilization composed of many cultures, each with is own notions of status, and the external displays of status, this would lead to confusing and crossed messages, as one party to an exchange reads something into a message of the other that was not implied by the sender. In other words, it is highly likely that communicative meta-codes—politeness, expressions of supposed amity and respect, and other forms that are common in some human cultures and less common in others—will emerge. Given one of civilization's consequences—smoothing over interpersonal communication between people from different social, economic, political, and group standing—it is highly likely that one major behavioral requirement of an ETI civilization as well, would be codifying the 'presentation of self' as Goffman put it.

Here too, we will find a continuum. At one end, a very restricted (or possibly, though unlikely, an absence) of meta-rules for the presentation of self, deference, and demeanor. It is conceivable that these will be restricted to certain media as well: consider the differences, among human, of the communicative meta-rules between written letters and e-messages (Baron 1998; Haugh 2010; Murphy and Levy 2006;

Waldvogel 2007). At the other end, we would find very detailed practices concerning the meta-rules of communication for every and any set of communications, whether material messages (letters, gifts), electronic ones, or even interpersonal behavior. For a human from a low context culture, such as e.g. Israel, this implies a rather large burden. For one from a European background it could be slightly burdensome, and for someone from a high context culture e.g. Javanese or Japanese, no real burden at all.

10.5.3 Heterogeneity and Homogeneity Within ETI Civilization

Human societies are characterized by the degree to which the society is differentiated into distinct groups. This has a number of major ramifications at the political, technological, and social levels. We tend to assume, generally, that all humans in a modern society are more or less homogeneous (I exclude for a moment racist and elitist views). Any given ETI may be far less homogeneous Polymorphic species (natural or artificially developed), might have members that are not intelligent, and therefore cannot be entered into the variable. An ETI society may be so rigid that no mobility is possible. Nonetheless, this set of measures provides us with an important index: an index of similarity between humans and the ETI species. To the degree that human society and the ETI society are similar, they will find it easier to communicate. To the degree that they differ, they will find it difficult.

Societies with an excess of energy tend to become role-differentiated as surpluses are accumulated. To look at the issue from the other end, a high degree of role differentiation in a human society implies a surplus of resources. To simplify calculations, all resources can be measured in units of energy. That is, the energy that is necessary to acquire them, and the energy that they can be translated into, whether mechanical, human, caloric, or whatever. As a rule of thumb, the more statuses available, the more the excess that can be distributed and must be distributed among these statuses.

One more factor must be considered. Authoritarian societies, no matter what their coloration, seek to minimize the abilities of their members to aggregate in varied ways. That is, the more authoritarian a society, the less it is interested in having its members organize themselves into groups that might claim the allegiance of the members of the society. The major distinction between democratic and authoritarian human societies boils down to this one essential difference: the direct or indirect loyalty of the individual to the state. Social relations in authoritarian societies tend therefore to be as vertical as possible, with few horizontal ties. The less authoritarian a society, the more it is likely to have multiple horizontal ties and multiple organizational levels among its members.

The degree to which a society is differentiated is, to some degree, a measure of its strength. The recent collapse of the Soviet Union can be seen, to some measure, as the result of strains between a need for internal heterogeneity in human societies, and the imposed homogeneity of the communist state. Chaos theory, the basis on which the ideas of flexibility that are so important to adaptation are based, indicates that the greater the variation in a measure, the more likely it is to be plastic and adabtable (Gleick 1987). We should expect a society with a high degree of differentiation to be far more adaptable and flexible than one that measures lower on this index.

The degree to which a society allows (or encourages) a multitude of social tie topologies, is a measure of its freedoms. Note that this is unrelated to the question of primitiveness *unless* it is also the case that there are a small number of statuses in the society. A corollary is that the more authoritarian a society, the less it is likely to allow different voices to speak, placing humans in an invidious position.

10.5.4 Canonical Texts

A canon is a collection of texts. *Any* communicative civilization (which is an oxymoron: a civilization has some form of recording almost by definition) will accumulate many texts over the years, centuries, or millennia. If it uses some form of CMC, those texts will accumulate at a frightening rate. *Some* of those texts will acquire a canonical status: texts known to all or most of the civilization members, which are considered significant pointers to 'civilization', and which are preserved and possibly revered as repositories of standards. Others, perhaps because they are deemed important, or contain pointers towards the canon, might be deemed pre-canonical: they will be added to the canon, or fall into relative or complete obscurity. In some cases, such a canon can be relatively small: a central text with interpretations. The Indian *Ramayana* and *Mahabharata* are of this type. Others may be extensive and complex: the canon of human World Civilization contains poetry, books, buildings, painting, music, and praxes. Small canons suffer the same problem as authoritarian civilizational structures: they limit opportunity and variability. On the other hand, they provide stronger unity, as all members of the civilization must adhere to a limited set of texts. Large canons, on the other hand, suffer from fashion, as texts 'compete' (or rather, those who interpret and support them do) in the market of ideas. In other words, the canon is *generally* accepted, but specific texts have greater appeal to certain classes of or textual items. As external observers/potential members, we need to know, obviously, what the texts are, and what is the canon composed of. No less importantly, is the canon a relatively open one, subject to fashion (that is, some items are emphasized more than others at certain points in time) or is the canon a fixed, immutable text or series of texts, like the Jewish Bible or the Muslim Koran.

The number and variety of human canonical texts is vast, as is their heterogeneity of media and style. Since, as argued, human civilization is in the process of construction, we cannot say, for instance, that looking back from the year 2300, that Rembrandt's "Night Watch" will be included, whereas Hokusai's "The Great Wave" will not. What we can say, is that human civilizational texts appear to be heavily Western-oriented (at present, at least), display a great variety of media, and (again, at present) have few determinations or indications of standards, beyond the ability of a *cognoscento* (a person in the know) to display familiarity with these canonical texts.

CMC (Computer Mediated Communication), which is likely to be the medium of communication in a multi-star civilization, has an advantage and a disadvantage. On the plus side, electronic texts (even detailed and true replications on material objects using 3D printing) will ensure low-cost rapid access, and indexed availability to relevant texts, opinions about them, commentaries, and discussions. This is already the case in human science to some degree, and even traditional processes surrounding some canons—Jewish Talmud study is an example—are being transferred to electronic format to aid in dispersal and learning.[8]

The downside has an interesting implication. Electronic texts (notwithstanding the Canute-like efforts of interest holders ranging from the film associations to the US government) are infinitely replicable, *and easy to change*. For the cognoscenti of civilization that represents a true dilemma: are the texts in the canon authentic? Or are there several variations? This implies, on the technical side, that any civilization will invest heavily in some form of stamping and provenance to identify 'real' texts. It also suggests that, on the other hand, any civilization's texts will change over the course of time, and that the potential for splitting of a civilization is always inherent in the texts. This is not a question of quarreling over dogma or doctrine. These in-house squabbles, such as the Catholic/Protestant split, or the arguments between the various schools of Islam and between schools of Buddhism, are 'in-the-family' quarrels. None of the disputants denied the importance of the central texts (Bible, Koran and Hadith, and the Sutras, respectively) they merely argued, within the same framework, about how they are to be interpreted. However, once a canon is completely replaced (which may well be a completely peaceful affair, brought upon by different experiences and circumstances) we can see that two separate civilizations could emerge.

10.5.5 Material

The material aspect of civilizations are a combination of technological *and* social factors. From our perspective, it is unhelpful to say that ETI civilizations will have X technology, unless we can also argue the social implications of that technology.

[8]Talmudic study, which is one of the cores of Jewish culture, traditionally involves studying an original text such as a section of the Mishnah, as well as all interpolations, additions, commentaries, responsa, and objections. Since this has been going on for at least 2100 years, this implies an enormous canon of work, which put in-depth study beyond the reach of all but a few. The transfer of this enormous body of work to computer aids in indexing, searching and collating data.

The control of energy is of course the key. The more energy available, the more potential any social association has. A primitive society has little access to energy beyond satisfying individual energetic needs. A developed society has sufficient surpluses to develop complexity, where specialists live on the primary production of a portion of the population. An advanced society distributes this energy fairly (not necessarily equally, since some occupations demand more energy than others), and a transcendent society has sufficient energy to allow for individual wishes to come true.

How energy is *distributed* in the society must always be considered. We cannot make any a priori assumption that an ETI society is less primitive unless we are able to examine the energy distribution. We may (but should not) assume that the construction of large radio-telescopes is an indication of social development, but we cannot know what the cost has been in terms of distribution of resources.

That is, for any ETI civilization (or culture) we need to ask two closely related questions. First, where is it located on the Primitive <—> Advanced scale? The closely related question is how is that energy distributed, or, Does everyone have access to this energy? Clearly, a 'primitive' society (one where there are few or no surpluses, can, if the demographics are good enough, nonetheless accumulate large surpluses *for a small minority*, which could, in theory, use those surpluses to create a large radio telescope (something quite similar to 1950–2000s humanity). At the other extreme are cultures and civilizations in which energy surplus, in great quantities, is the norm. Nevertheless, we *still* need to ask how this energetic surplus is distributed. In general, we need to keep in mind, one form of social control by elites is conspicuous consumption (Veblen 1953), creation of expensive mass spectacles for the masses (Lane 1981) and practices that distinguish elite praxis and standards from those of the masses.

Approaching the material aspects of civilization in this way also encompasses two popular ways of viewing civilizations in the SETI literature. First, we can see that Kardashev civilizations, of whatever type, are simply 'local' expressions of a generalized rule, and the mere technological solution does not help in any way in understanding the ETI civilizations involved in any of the Kardashev types. Second, the only way to reach the type of civilization that might benevolently accept or isolate 'primitive' humans is if there were great surpluses of energy available to all members of the civilization: sufficient to ensure no-one preys on the primitives. Not impossible, but difficult and something to consider.

Another common approach has been to define advanced civilizations through the lens of information and control thereof, which is arguably again, a specific case of a general rule. Thus, the level of development is defined as "…the (common) logarithm of the total information content (in bits) of the ecosystem. This information resides in genes, brains, and extrasomatic memories such as books and magnetic disks." (Ball 1980: 658). While agreeing with Ball, it is also apparent that once again a measure of sheer mass is insufficient without at least a measure of the complexity and important of the bit density and distribution.

10.5.6 Introducing Some Complexity: An Analytical Tool

Now that we have some more-or-less objective ways of examining an ETI civilization, it is worth playing with these concepts to see what kinds of complexities are inherent in the different parameters potentially possible in 'normal', that is, more or less surrounding the 'assumption of mediocrity' position we have placed human (embryonic though it may be) civilization.

We've located several civilizational problematics, each on its own continuum. There are probably others, and perhaps, in practice, it will be found that the continua suggested here are unnecessary. So the best that can be said is that what is suggested here is a *method* by which we humans can begin to assess what it would mean dealing with ETI civilization X.

In the realm of standards:

1. *Ethical*. Negative "Thou shalt not" (minimalist) <—> Universal positive "Thou shalt" (maximalist) by command ethical standard.
2. *Aesthetic*. Fixed aesthetic standards <—> mutable aesthetic standards.
3. *Aesthetic*. Pervasive (cover all fields of activity) <—> Punctuated (cover only some aspects of life).
4. *Performative*. Pervasive (all activities) <—> Punctuated (only certain activities).
5. *Sources of discourse*. Single (rationality, religious, …) <—> Varied (rational/emotional/religious/…).
6. *Expansion*. E.g. a 'civilizing mandate' <—> Stability (no expansion nor missionary drive).

The praxis dimensions:

7. *Membership*. Open to all <—> Limited to those with certain inherent qualities
8. *Status differentiation*. Few statuses that have civilizational value <—> Many statuses for civilization bearers.
9. *Formal authority structure*. Authoritarian (hierarchical system of cognoscenti) <—> Flexible self-achievement.
10. *Violence*. Violence greatly circumscribed by civilizational standards <—> Low/no standards for violent praxis.
11. *Aggression*. Civilization standards keep violence to a minimum < — > Low/no standards limiting violence.
12. *Communication meta-rules*. General standards and praxis <—> Detailed standards and praxis.

The canon:

13. *Canon mutability*. Strongly fixed < —> Mutable.
14. *Canonical medium*. Single medium (e.g. written text) <—> Multi-media (written, dance, music, visual, etc.)

A multi-dimensional civilization-measuring tool in the making When looking at an ETI civilization from our position as human observers, it seems obvious that we cannot know all the parameters that we need to know to ensure that we can not only communicate, but also, potentially (for that is what a civilization is: a potential and framework for life) join. To do so, that is, to be able to adhere to an acceptable limit to the standards, praxis, and canons of the civilization, we need to be able to assess what the implications are. Many discussions of civilizations in SETI have been prone to assume that a civilization is a single dimension characterized by a particular technology or ethic, or whatever list the writer felt closest to. Here, instead, it is argued that civilizations must be characterized by a set of dimensions or continua. It is also argued that those presented above, may be only a part of the important dimensions we need to look at (and, indeed, readers are hereby invited to suggest other crucial dimensions). Alternatively, perhaps some of those suggested here are unnecessary.

More importantly methodologically, is that, until demonstrated or proved otherwise, the fourteen dimensions presented above, are independent. That is, there is no, or very scant evidence, that the location of any civilization on dimension X correlates logically, or, more importantly, empirically, with the same civilization's position on dimension Y. We may as well forget causation completely.

Attempting a multivariate analysis of a 13-dimension matrix is beyond this author's skill, though presumably it is feasible. In the absence of real empirical data outside the roster of human civilizations however, an analysis would seem to be facetious. Nevertheless, some *simple* (two variable) logical correlations are shown below, largely to give a taste of what is (or hopefully will be) required if we are to understand an ETI civilization.

Highly authoritarian (e.g. Parameters 5, 9 and possibly 12) + high status differentiation This seems to be a logical possibility. However, there is possibly a negative correlation between high degrees of differentiation and high degrees of authoritarianism. The high degree of differentiation allows individuals too many possibilities for creating horizontal ties. It seems reasonable to suppose that this sort of civilization is possible only when some technological device exists for keeping the authoritarianism growing. What this might well mean, is an unstable civilization with a much social tension, possible instability, and, if the human experience is any guide, a displacement of members dissatisfaction into aggressive channels.

Low authoritarian (5, 9, 12) + high status differentiation This seems to be very likely, given the negative correlation between the two factors. Such civilizations are likely to replicate our current world pattern to some degree, in that wealth, knowledge are widely and unevenly distributed. Such a civilization is likely to be stable, if the distribution of rewards is perceived to be equitable (a human example is post WW II Japanese society) or unstable if not. At the very least, this is a pattern that is familiar to residents of World civilization as it is currently constructed. It also seems to fit the rule of social entropy: complex social structures are hard to maintain for lengthy periods of time. Societies which have a greater degree of permitted variation may well be better adapted. It also seems likely that this combination will be correlated with a wide canon, minimum ethical standards to allow for flexibility, and variability in discourse rules. The danger of course for this type of civilization is that the ties between the elements—cultures and societies—that make up the civilization would have a centrifugal effect, bringing about a calving of the civilization into daughter civilizations.

High authoritarianism + low differentiation This seems to be a likely pattern. It may well be determined biologically, that is a hive species will exhibit this framework. However, it may well be arrived at through social mechanisms. It has not been very successful a pattern among humans. Nonetheless, it has been attempted as a social form again and again (e.g. PRC, Soviet Union, North Korea). Since it is an attractive pattern for a ruling elite, it may be expected to occur elsewhere also.

Low authoritarianism + low differentiation We would expect the possibility of a low intensity of ties between members of the society, and possibly a very high technology indeed, which would allow members of the society to live and function as social isolates. This could come about through war or some other possible decimation of the population, but also might be the consequence of rational, perhaps even moral consideration. A society like this would be within the realm of a deeply moral society, and may well also be indifferent to contact with other ETI and possibly even incurious about them.

As we have seen from this brief analysis, the construction of logical matrixes can at least structure the inquiry in such a way that we can make some (tentative in the absence of data) predictions about the ways in which civilizations might develop, isolated to some degree from questions that derive exclusively from the human experience. Furthermore, this allows some preliminary estimates of possible strategies to follow in communication with ETI. To take but one example, it would be incongruous to find a society exhibiting low authoritarianism and differentiation and high energetic distribution. It would imply one of three things: (a) a very small population base. How small is not measurable, but certainly something to think about. It would also imply very sophisticated, perhaps self-maintaining technology. (b) It might mean that there is a hidden factor that must be searched for. In other words, our information about a very crucial aspect of this civilization is not visible. Finally, (c) it might mean that the information we have received about them is untrue. In other words, we have been misled (if they supplied the information, lied to), in which case some conclusions should be drawn about continuing communication.

The weakness of this entire type of analysis is of course the ultimate bane of all xenological discussions: we lack empirical data. Thus we are, inevitably drawn to extrapolate from our own experience, no matter how much one struggles to disentangle the sticky webs of human experience. On the other hand, developing at least an embryonic, rudimentary tool for assessing the kinds of issues that could crop up, is worth some consideration, provided we accept that this tool absolutely needs to be modified and sharpened in the light of reason, and, hopefully, further data.

10.5.7 Summary: Parameters of ILF Civilizations

We have looked at *some* of the possible parameters of ETI civilizations to indicate the difficulties we will face *ab initio* when dealing with any ETI. These are far more complex than the rather simplistic pictures painted by e.g. Kardashev or Sagan, who tended to use a single parameter—technological prowess—as a measure of communicative possibility.

Where one looks at only *one* dimension of another civilization, one is almost always led to a very small array of solutions (i.e., in the cases of Sagan and of Kardashev "We can/cannot communicate with them" and "They will/will not be benevolent"). Looking at multiple dimensions has a positive and a negative effect. On the negative effect, identifying these dimensions, and which of them is relevant to a particular exchange, is difficult at best. On the positive side, because there are so many dimensions involved, there are also so many shades of gray. No more "We can communicate, yes/no?" and more of "Under what circumstances, with which partners in that civilization can we communicate what types categories and subjects of messages?".

10.6 Aggression and Hostility

The degree to which we (the human race and all our works) are potentially exposed to hostile action by ETI from outside (and inside) our Solar System has been a repeated theme for a couple of centuries, in both popular discourse and some scientific thinking. In contrast, many SETI writers have argued that a sufficiently developed ETI culture/civilization could not possibly be anything but benevolent. The extreme case argument is the "Zoo hypothesis" solution to the Fermi paradox, which is that out of benevolence, we are embargoed so as not to suffer culture shock from higher civilizations (Ball 1973). Curiously, and unsurprisingly, the same question can also have a completely opposite solution: they are predatory, and hiding not from us, but from each other (Brin 1983a, b).

The 'friendly-rational sophont' argument (Bracewell 1960; Learned et al. 2009; Sagan 1973a, b; Sterns 2004) in its manifold forms roughly asserts that any ETI species that had reached an interstellar age (whether it exploits that potential or not) is likely to be non-aggressive, and, indeed, positively benevolent. At the extreme, one argument is that the advanced levels of technology necessary for interstellar transportation imply complete physical integration of brains and information processing systems, and therefore competition and warfare are unknown in interstellar society (McGowan 1966: 366). The 'friendly-rational sophont' argument is based on two premises. First, the economic and material difficulties of physical space travel, which make aggressive moves prohibitive (Sagan 1966, 1973a, b). Second, the assumption that any species that has made its way and succeeded in maintaining a civilization for sufficient time to expand or communicate in the interstellar medium, will have also improved its social sciences to the point that it would learn to be nonaggressive (Geller 2013; Tough 1986). This latter assumption is also based on a further assumption that evolutionary paths will *necessarily* be the same as our own, with constant intra-species aggression as one prime motivation for cultural development. It is worthwhile examining this argument against possible logical outcomes.

The assumption of *rational* sophont that underlies this argument can be easily falsified. It is an elementary truism, one that many prefer to ignore, that 'reasonable' for you and 'reasonable' for me may well be different. 'Rationality' is entirely a culturally-biased idea. This has been demonstrated even within the human species. Thus the "rational man" used as a yardstick in English law, is quite different from that used in law in other cultures (cf. Gluckman 1972). Moreover, it could also be postulated that psychological, physiological, and physical levels of existence will affect "rationality" as well. Rationality is culture-biased above the physiological level.

Mere *distance*, critical as it may be for humans in human culture—Iran and Argentina are unlikely to come to war simply because they are too far, but they still can come into conflict—may not be the same for non-humans. Patience, paranoia, longevity, cultural form, may all permit an ETI culture to conceive of conquest, aggression, destruction, colonization where *humans* would rationally discount such efforts. Nor is it reasonable to suppose that inter-species aggression would be rarer than intra-species aggression, as Maruyama (1976) supposes. This is largely (though not wholly) true in the animal (non-sophont) kingdom. Where relations of predator-prey do not obtain, there is *rarely* (but not never) inter-species violence. This is not necessarily the case when one ILF species deals with another, whether because they may well be competitors for broadly defined ecological niches (solar systems, for example), because they might be inclined to predict, accurately or inaccurately, future aggressive actions and decide to make things happen quicker, or simply because one has an agenda that might affect the other; principally, however, because genetic predispositions to nonaggression or to violence are tempered by considerations of ultimate gain and loss, and, in civilizations, of issues such as standards and style.

Moreover, it is also conceivable—'rational' under certain circumstances—that an ETI civilization will display no aggression whatsoever *within* the species, and yet may be completely xenophobic. To suggest just one such source for the outcome, if more than one intelligent species have had to coexist, and compete, in the same planetary environment. This is not so far-fetched. It is virtually certain that several species of proto-intelligent hominids coexisted in Africa and Eurasia in prehistory (Falk et al. 1995), and it is not unlikely that more than one of them were tool users, yet only one hominin species survives today. Perhaps not because *Homo sapiens* destroyed them, but conceivably because we did.

One problem with discussing aggressive ETIs rationally is the human cultural propensity for scare stories. What is important is the fact that popular culture (whether we like it or not) in the form of TV shows, films, books, journals, dominates the world. And in most cases, "aliens" or "beings from outer space" are displayed as irrational, ravening creatures bent on unjustified and incomprehensible slaughter. Perhaps as a result, many such as Sagan and Drake (1975) and Papagiannis (1983), have gone to the opposite extreme, portraying any possible ETI as necessarily friendly, even benevolent. As the preceding chapters have shown, we must understand ETI in terms of their own background. That is to say, there is no reason to assume that ETI are innately benevolent, nor inherently aggressive: they have their own agendas after all. It thus becomes necessary to recognize the *possibility* of ETI who do not share the World Civilization's Western cultures' lip-service to 'humanism' nor Eastern cultures' lip-service to 'benevolence'.

To make things a bit clearer, let us distinguish between a number of dogmatic positions. 'Friendly' implies that the ETI concerned is positively inclined towards mutually beneficial contact. 'Antagonistic' implies that the ETI is, for whatever reasons, intent on attacking or otherwise dealing forcefully with other ETI. The aggression of a civilization as a whole towards other civilizations, and the degree to which violence is permitted (or encouraged) *within* a given civilization are distinct phenomena (though probably interrelated in practice).[9] Note that 'friendly' or 'antagonistic' could be effectively genetic and unalterable. Perhaps some arrangements might be worked out for aggression-prone species, such as Phoenician trading (see following chapter), but the basic condition is unalterable. More probably ETI will, at least initially, be somewhere in between with the potential (the same one *Homo sapiens* has) of being aggressive or not, depending on psychological, economic, cultural, and possibly political circumstances.

The language we use presupposes discontinuous states between, say 'aggressive' and 'friendly' which seems unlikely. "Antagonistic" ('bent on doing harm or perceiving others as harmful') is not the same as 'aggressive' ('having a high capacity for forcing themselves or their views on others'). Thus I would suggest that one continuum for examining ETI is *the degree to which they perceive their actions to be motivated by the desire to do harm/be beneficial in their own terms* (antagonistic-friendly) and another continuum *the degree to which they are interested in forcing their own positions on others* (aggressive-peaceful).

On the aggressive side, we might distinguish between different kinds of aggression, inasmuch as different kinds are likely to have different consequences for those outside the civilization. There are, even among humans, a vast range of motivations for aggression. Several examples, all from different human societies, will illustrate briefly the problem.

Human colonization has, at all times, involved slaughter of the indigenous population. This has been extensively documented particularly in recent Western cultures, which have brought a number of populations to extinction. In social Darwinian terms, this merely demonstrates the adaptive success of the colonizing population (no matter what our moral views of genocide may be). This form of aggression has been a feature of all human societies, whenever they have engaged

---

[9]Pinker (2002, 2011) at least argues convincingly that humans are inherently violent, mediated only by cultural practice.

in colonization: American and Australasian natives were victims of this phenomenon, as were Pueblo-cultures the victims of Athabascan speakers, and cultures that shared Chinese and Japanese geographical space.

Ideological aggression is another example. Here the sources of aggression express themselves in what amounts to an intellectual definition of some "other" and the translation of an aggressive intent into action. Human history is replete with examples, ranging from the treatment of Jews and Gypsies by the Germans to more recent genocides practiced by Cambodians against themselves, and Muslims towards Yazidis, to cite but some in a long and gloomy list. Since the rationality of ideological aggression is doubtful, a "Trojan Horse" invasion is *not* impossible (not that it is likely either). Convince another species to build something they think is for their benefit (say a computer program of great power) which is, in fact, an AI that will "save their souls." Conceptually, at least, this is not impossible, however unlikely or bizarre it may seem.

Aggression may also take place as a form of entertainment. One example is the so-called "flower wars" which the Aztecs engaged in against client and enemy states alike. In a flower war, the Aztec state would declare a period of war for a limited period of time, The victim state would be forced to send an army into the field, usually at a tactical and numerical disadvantage. The Aztec army would fight with the aim of taking prisoners for sacrifice. The entire exercise was considered, by the Aztecs at least (the victims were of course less entertained by the affair) a valuable form of entertainment cum worship (Moriarty 1969).

Finally, aggression may also derive from genetic requirements. Dogs, chickens, baboons, and humans, to cite but a few social animals, engage in dominance displays and controlled aggression *not* merely due to lack of resources (food, sex) but because of a drive for dominance. The drive is probably part of a selection mechanism, but its results are an almost constant jockeying for position among the participants. At an individual level it can be seen in non-technological societies such as the Yanomami (Chagnon 1968) as well as among technological peoples (Pinker 2011).

Aggression between ILF races forces consideration of the economics of the situation. Aggression, of whatever form, requires an investment of resources, which, in some instances (e.g. great internal efficiency and low population pressure) may not be worth fighting about. However, that assumes that races make decisions at some "rational economic" level, something that in the case of our species, at least, is not the case. True, a highly intelligent race, as we have noted before, will likely try to make rational decisions, and a civilization may well be advanced enough (it has large surpluses and sufficient energy resources to supply its members with what they wish to have) not to need violence to supply surpluses to its population. But that does *not* stop aggression, it merely restricts it into acceptable, civilized channels.

What has been said immediately above about causes for aggression in human society implies that there are different forms, or at least different modalities of aggression. To point out one difference, a society that is aggressive because it sees the social world in terms of a pecking order will be far more easy to deal with than a society that has an ideology that supports aggression against 'deviance': see orthodox Islam. In the first case the victim need only display expected signs of submission, and will be left alone thereafter.[10] In the second, case, nothing will convince the aggressor of the good intentions or lack of guilt of the victim: a much more difficult problem.

A rough continuum can be presupposed. *Peaceable* ETI have no, or very few motivations or capacities for aggressive action. This may be because they are genetically non-combative, or because they have evolved mechanisms for dealing cooperatively with others, or because they have had no experience of aggression. *Agreeable* ETI are predisposed to be friendly, but include a capacity, under some circumstances, to turn aggressive. *Assertive* ETI will be predisposed towards aggression, but will have mechanisms and the capacity to control that aggression (human beings, of course, fit into this part of the continuum). Antagonistic ETI are essentially xenophobic. They will react aggressively to any contact, and perhaps even search for other species with malevolent intent.

Peaceable <——Agreeable — Assertive——> Antagonistic

While we would all like to consider only the pleasanter possibilities, all of those suggested here are equally possible (it is, of course, hard to judge likelihood). While many have assumed a cosmic rationality and pacifism, this does not necessarily stand the test of careful consideration. Brin's far more nuanced classification of this sort is much more realistic from a social scientist's point of view, whether or not we accept his conclusion that hiding from aggressor species explains the Fermi Paradox (Brin 1983a, b).

To illustrate the varied possibilities inherent in aggression, it is possible to derive forms of aggression from one trait that is exhibited by humankind, and may well be a feature of many technological societies: consumption.

1. Aggressive consumers.

'Aggressive consumers' are antagonistic ETI whose motivations and actions derive, in some measure, from unrestrained consumption. This may be an inherent psycho-physiological orientation, such as a carnivorous species (e.g. Niven's fictional Kzin species), or the result of historical "Darwinism" in a species that has consistently overused the carrying capacity of its home and has thus evolved an expansionist drive. Sagan and Newman (1985) assume that such aggressively-colonizing societies would be self-limiting by reason of conflict. This is however not necessary. It is possible that such a civilization would destroy itself as it emerged from its planet, but once entering into the depths of space, it would more likely prey on others—planets, animal species, or ILFs—weaker than itself.

[10]Or more-or-less alone. Aztec armies repeatedly targeted past victims as objects of flower wars.

2. Opportunistic consumers.

Pure energy exchange considerations may motivate a piracy and raiding culture. That is, a species may evolve in a relatively anarchistic mode in which, originally preying on themselves, they come to accept preying on weaker members of any species as a legitimate and acceptable action. This is likely to be a cultural, rather than a physiologically based pattern. What specific form that would take would of course depend on the cultural aspects of that particular ETI. Slavery, for practical reasons, for use as food, for display and luxury, because of a dominance instinct, or for other reasons should not be discounted. However, again, given energy considerations, this is highly unlikely to be an interstellar pattern, unlike aggressive consumers who, by instinct and physiology, are oriented towards aggressive behavior. This could be translated into civilizational standards, detailing who, where, when and how to engage in this predatory behavior without penalties or loss of status, or accusation of standards being broken.

The difference between aggressive and opportunistic consumers is the difference between inherent, essentially uncontrolled consumption, and a learned behavior. The first may well be quite friendly, but unlikely to control predation if an opportunity or a signal 'I am prey' is present or displayed. The latter may well be 'teachable' in the broader sense of the word, that is, mechanisms of exchange can be worked out, and changes can be expected, unlike in the first case. This of course has implications for the ability of the human species to deal with such a civilization. To turn the picture on its head, it is likely that the human species, with its (natural, apparently) consumption increases should be classified as aggressive consumers: while we may *try* to control our consumption as a species, we have, in effect, not been able to do so.

Finally, it is important to consider the possible mutual effects of the *interactions* between ETI civilizations. If, as Bracewell (1974), Ball (1973), Geller (2013) and many other seem to think, ETI societies are inherently benevolent and rule bound, then there is unlikely to be much of a problem. But suppose, as Brin (1983a, b) has noted, that for every 100, or 1000, or even 10,000 peaceful civilizations, there is one xenophobic, even paranoid one? Such a species would virtually force many others to react to it in some defensive (and thus inherently aggressive) manner, even if many others would choose merely to hide from the aggressor.

3. 'Sharing' consumers.

Some ETI may be 'sharing' consumers: willing to consume up to the boundaries of interest of another ILF. This poses a number of problems for us. Defining what is actually being shared for consumption is one. The degree of exchange inherent in such a sharing enterprise is another (that is, setting the rules of the exchange). Given that civilizational parameters are likely to merge on the boundaries, what would probably be worked out are mutually acceptable standards of practice, that would allow such commensality.

4. Self-limiting consumers

At some point it is possible that either shared, or complementary technologies would create a technological commonwealth. Such a social phenomena would, almost inevitably, bring about some blending of the two civilizations, which in effect means the establishment of a single civilization with more than one internal civilizations or cultures: the Great Tradition/little tradition phenomenon replicated in truly grand scale. So long as such a 'super civilization' was able to develop the necessities—standards, praxis, and canon—there is no reason to assume that it could not continue *even if the super civilization's constituent elements are not purely peaceful, but include standards of aggression and warfare*. Consumption could thus serve as a bridge, rather than a divide, between civilizations with different starting points.

It needs to be added that the form of consumption *alone* will not determine the degree of aggression displayed by a species. Consumption is, however, a universal requirement deriving from energy exchange, and can thus be assumed to play a part in motivating and shaping aggression. *However* it is far from the only cause for aggression, and this must be kept in mind. And when consumption and warfare are constrained and framed by a set of standards and practices, there is no reason that, as in human history, agonistic displays (Gardner and Heider 1968) and conspicuous consumption (Veblen 1953) would not both be an important element of the civilization.[11]

10.7 Types of Communicative Civilization

Contact by electromagnetic means appears to be the most likely means of communication between star systems. The corollary is that physical contact, with the difficulties it implies is improbable. Nonetheless, because of the differences between species (at least) in the use of the term 'time' physical contact—the transmission, exchange, and return of physical objects (people or artefacts) is not impossible. Here we address the issue of physical contact in one major dimension: the set of relations between ILF civilizations.

The debate surrounding the Fermi paradox has continued unabated for several decades. Hart (1982) and most forcefully Tipler (1981), have argued that the absence of ETI on Earth is *prima facia* evidence that $N = 1$, that is, that ETI do not exist. A long list of logical, physical, and social reasons have been produced to show why this is not necessarily the case notwithstanding the absence of evidence. The argument need not be settled here. Hypothetically, the existence of *colonizing* ETI civilizations is a possibility. If it is a real phenomenon (and we have no *evidence* against that), it must be considered because it is likely to be a major factor in any consideration of contact. This is true whether or not we are invited to join the

[11]Consider agonistic systems such as code duello (Kingston and Wright 2010).

'Cosmic club' (Bracewell 1981) or whether we need only consider the mutual effects of different civilizations. As Bracewell notes, continuous interstellar contact makes *any* assumption regarding contact highly speculative since either ETI technology, or social imperatives might hide them from our sight. In this section I therefore intend to point out some of the more prominent social features that might emerge *if* colonizing space-faring species exist in the universe.

The possibility of having only a single expanding civilization throughout the galaxy, besides ourselves (that is, N = 2) cannot be entertained seriously. This is for two reasons. First, if two (us and any ETI) can exist, any other number is just as likely. Second, and perhaps more importantly, a very large social system is very vulnerable to change, splintering, and in effect, to becoming more than one civilization, and, considering the time scale, perhaps more than one species.

The basic premise here is that *if* physical expansion into the galaxy or universe is possible, then there are going to be several co-existing such civilizations. True, the factor L (longevity of a technological civilization) is important. For the moment, however, I intend to ignore it and assume, for the sake of analysis, that more than one expanding civilization will exist simultaneously at some point in time. If so, what kinds of ETI communicative strategies might affect human communication with ETI civilizations?

The letters between the Oranda king and the Tairo Surprisingly, we do have several examples of inter-cultural communications that took place notwithstanding long message delays. One example is the exchange of letters between the King of Oranda (Holland) and the *Tairô* (the government of the Japanese Shogun: the civili-military dictator under the emperor).

At the beginning of the seventeenth century, Japan—a nominal empire under the real control of a civil dictator who was also the military commander—decided to cut itself off from the rest of the world. The reasons were multiple. The Portuguese, who had been the first to contact Japan early in the sixteenth century, and the Spanish were importing religion into the country, threatening the status quo. Worrying news had reached the Japanese government that Christianity was almost always used as a Trojan horse on the way to colonization by Europeans. Portuguese and Franciscan friar arrogance were a problem, as well as the disaffection of peasants, encouraged by Christian ideas.

By 1637 the shogun—the military ruler of Japan—started enforcing a ban on Christianity enacted the previous century and banished all Portuguese from the kingdom, emphasizing the point by killing some 30,000 Japanese Christians and a handful of Portuguese resisters in Southern Japan. A few years later, the only foreigners allowed in were a few score Dutch traders, and a few hundred Chinese merchants, each imprisoned for the length of their visit on small islands in the far southern city of Nagasaki. Once a year, the Dutch chief trader was required to visit the capital in Edo (now Tokyo) a trip of about one or two months (depending on the weather and other factors).

Throughout the period 1637–1856, after which Japan was opened to traffic with the world, Japan and Europe exchanged messages of various kinds (they also traded goods, though that part spoils the analogy). Virtually no Europeans could read or understand Japanese, inasmuch as the Japanese government had forbidden teaching the language to foreigners. A few hereditary translator-interpreters in Japan opened up a new branch of learning: *Rangaku* ("Hollandology"). Poring over every fragment of information from fabulous Europe, they tried to piece together all knowledge they could from fragments of information including letters, speaking to foreigners (usually under the beady and reproving eye of a government spy), illustrations, books, and observation. European interest was far less systematic, confined to trade matters (the Japanese produced the best copper alloy and steel, and copious amounts of gold), and the occasional interest of a European savant (science was just getting started under the guise of 'natural philosophy').

The mode of transmission of official letters is of particular interest here. A carefully composed, and often edited letter was crafted in Japanese in Edo. The letter was translated by Japanese translators (who only had a partial understanding of the language, and almost no understanding of the social conventions and practices) into Dutch. The letter was handed to the Dutch chief trader on his annual visit. Several months later, the letter was handed to the captain of the single Dutch ship (in earlier periods, a flotilla of two ships) permitted annually to anchor at Nagasaki. Several months later, the letter might arrive at Batavia, capital of the Dutch East Indies. A year later, possibly being delayed in the Cape colony in south Africa, it would arrive at the Netherlands in the hands of the officers of the VOC (the Dutch East Indies Trading company) who may, or may not, pass the letter on to its recipient—the King of the Oranda people. The letter would be discussed, pored over, and perhaps published. Perhaps an answer would be composed (the Dutch, and the Europeans in general expected other people to be able to read European languages, so not much effort was wasted on translation). Some time later (a few months usually) when a Dutch ship was headed to Batavia, it would take the letter for the six months voyage to Indonesia, where it would wait for the annual ship to Nagasaki, where it would wait for the annual visit to Edo. All this, of course, if the ship bearing the message over any leg did not sink on the way, or the letter get soggy and unreadable due to poor storage conditions.

The exchanges between Japan and Europe simulate to a limited degree the possibilities of communication between humans and ETI. Let us therefore try to tease some of the issues out more specifically. We'll look at the message *protocol*, *format*, *content*, and it's *effect*. First, however, a word about the societies concerned.

At the point of cultural exchange—the mid-seventeenth to the early nineteenth centuries—Europe and Japan were on a par scientifically and technologically [we rely heavily here on three eminent scientists Kaempfer (1929), Thunberg (1795) and Von Siebold (1973)]. Unsurprisingly, in some areas (medicine, natural science) Europe was more advanced, though still superstition-ridden and non-scientific in many of their conclusions. Japan was more advanced in steel and metallurgy, agriculture, possibly in manufacturing. One thing that struck all observers was the quality of human management and social order, which they attributed to a better understanding of these things in Japan.

Message protocol Long time frame. Like interstellar radio communication, messages between Japan and Europe took *time*. By the standards of the twentieth century, extraordinary lengths of time. This meant that the modern habit of instant gratification in the news was completely absent, and could have created a crescendo of anticipation, but for other features of the message system.

Message priority It is important to note that the messages had very low priority. For both civilizations, the messages were swallowed up in other concerns: part of the general noise created by government, concerned as they were with internal conflicts, and for the Dutch, continuous struggles with the English and others and with colonization. When the messages *did* arrive, no one was in a rush to send them to the top political authorities.

Message Contents. Well, first, and significantly, no one sent an encyclopedia. There were several reasons for that. The Japanese, on their side, did not want information about Japan to leak out, for fear of assisting a potential enemy (and this, of course, sounds familiar to the SETI discussion). They also wanted to censor all incoming information, for fear of religious and political propaganda from the European side. The Europeans on their side did not care enough to send anything in particular, and it was up to the men of science (Kaempfer, Thunberg, Siebold and their like) to bring along whatever they could *which they often sold at a profit*. Both sides had more important things on their minds, and assumed they knew all they needed to know from previous sources. The Dutch king Willem at one point had one advantage: his adviser was Franz von Siebold, a physician and naturalist who had actually been to Japan. The shogun had another: a corps of trained *rangakusha* whose professional life was bound with studying foreigners in general (*yougaku*) and the Dutch in particular (*rangaku*).

Contents What did they write one another about? Basically, the one letter available said "We've been friends for a long time. Be careful, the British and Americans are coming and they want to swallow you up. Better open up the country to foreigners before they force the door open." The Japanese response was no less succinct "Thanks for your letter. We're not going to do what you suggest."

Effects What changes did these exchanges bring about? The Japanese, who saw themselves as the weaker party, consulted among themselves. For political reasons, the discussion was kept a secret, and only the Commissioner in Charge of Foreign Affairs, members of the Senior Council, and the Shogun himself were privy to the contents and debate. When the foreigners *did* arrive in force (the Americans in 1853–4), it was less the messages they bore than their artifacts (guns, warships, railways, food, clothes and books) which turned Japanese society upside down.

The effects of the letter exchange on European and Dutch societies were even less. The Japanese letter was stored in the Netherlands national archives, where it can still be read today. European society continued to mutate into modernity. There was a brief fad for Japanese art and aesthetics (Van Gogh was particularly affected), but that was fueled more by the opening of Japan and the arrival of Japanese artifacts than by the message exchange.

Some Lessons Messages, as we have seen, don't cut it. The likelihood of *Encyclopedia Galactica* being transmitted seems to be doubtful, unless someone (individual or corporate: think Gideon's Bible) fixated on spreading knowledge is behind that. And foreign entities need to be regarded with great caution: the Spaniards demanded free propagation of Christianity in Japan before they would issue a trade license. The Japanese who knew that Christian propagation was the road to subjugation, wisely refused.

Crucially, information control and benign neglect structured the one exchange we know much about. The Japanese government ensured that *all* communication with the barbarians was as controlled as possible. In the case of the King Willem message, no hint of the issues involved was spread beyond the need-to-know list. Given the physical reality—no material exchange, but a wide and uncontrollable radio medium—the same degree of control is unlikely in case of an ETI message, making a joke of the attempts to create a universally agreed on message from Earth (see "Consequences of Contact" below).

Above all, the length of time of a message circuit (message + response), the different interests of the governments concerned, their minimal understanding of one another, would seem to indicate that radio messages are unlikely to have much of an impact beyond a sort of reverse Tasaday seven-day-wonder: for a period of time an ETI message would be in the headlines, displaced in due course (hours? Days? Weeks at most) by the news of the newest Bollywood/Hollywood scandal, volcano, or war. True, a thousand or so religious fanatics will spring from the woodwork demanding we pay obeisance to the powerful strangers in their far off place. Another thousand will come forth demanding we close down all communications, and a further thousand will start building air-raid shelters or (in the US at least) stock up their survival supplies and load up on high-velocity ammunition. Joe Bloggs will continue doing what he was doing before. The degree of social and cultural change will depend to a minor degree on the message and its contents, but to a very great degree on the cultural, political, economic, and ideological situation (including religion and other ideologies) the message is received *into*.

10.7.1 Interested <–> Indifferent

Even in human society, private individuals and public figures have expressed reservations about the possibility of contact with ETI. Some of these have been founded apparently on simple xenophobia. Others on real fears of domination by ETI with greater power. Others on the fear that new ideas: religious or otherwise might imply for human society. All of these have occurred in the long-line communication between Japan and Europe. The same considerations must occur to ETI, and it is necessary to try and assess how this might affect the form and content of contact.

At a schematic level, it would be well to divide the possibility of ETI contact into three groupings for purpose of assessing how to deal with them.

- *Indifferent* ETI are those that have no interest in species-external communication. They may even resist it, perhaps forcefully, as do, for example, Andaman Islanders.
- *Interested* ETI are positively inclined towards communication. It is likely, though not necessary, that given the technological capability, they attempt to create the means (technical and intellectual) for initiating, or at least responding to contact. *Homo sapiens* belongs to this group (though if all SETI efforts are cut off, it may not belong here for long).
- *Occasional* communicators may have a threshold or trigger situation or event. Under some conditions, but not others, the ETI may not be interested in contact but once a certain threshold is reached, it may "change its mind".

The degree of intereste may well imply a number of things in terms of the consequence of that interest. Those ETI that have an interest in contacting us also, presumably, have either had experience in doing so in the past, or have thought about ways to go about it. This means, particularly if this is a first contact for humanity, that we are likely to be at a major disadvantage in negotiating with them: the only experience we have to draw on is the relationship between dominant cultures (largely Europe) and subaltern ones (largely everyone else), none of which is particularly pleasant or instructive.

There may also be indifferent ETI that we are *unable* to communicate with. Not for physical reasons, though there are many of those, but simply because they are so inflexible in some dimension of their culture and genes, that they cannot engage in exchanges *except* under conditions that other species are *unable* to accept. The problems inherent in contacting a species that is unable for some reason to contact other species cannot really be assessed, but must be recognized. The obvious possibility is to assume that some bypass means might be found: through simple signaling, by creating a shared artificial language such as LINCOS, or by using some intermediary. In other words, 'unable to communicate' might simply be a special case of the general problem of *how* to communicate. That is, some hidden variable or problem is to be solved first.

Far more problematic, in terms of this analysis, is communication with species who are completely indifferent to external communication. Broadly speaking, indifference may be strategic or tactical. By strategic indifference I mean that the species as a whole has no interest in achieving contact. This may well be because they are technologically so far ahead of humanity that they have no interest, or perhaps because their experience is that ILF of our particular pattern or extremely common in the universe, and thus uninteresting. By tactical, I mean that not communicating may be the beginning ploy in a larger game, or may simply be the result of current conditions (e.g. political) among these particular ETI.

Indifferent ILF's may be looked at from a different perspective as well. I have noted previously that the possibility of predator species must not be excluded. Predator ETI are likely to engage in communication under one condition: if they are unable, at the time of communication (that is, the potential still remains) to engage in predation on the species that they are contacting. They are also, of course, likely to engage in communication if they feel that the contact will yield information that will turn a potential victim of predation into an actual victim. It is probable that a predator species will regard all contact as the opening movement in an act of predation, and will therefore engage in communication as a matter of course.

Another ETI communicative class may well be composed of species 'in hiding' (Webb 2002) for fear of either 'killer probes' or predatory species in general (Brin 1983a, b). These will actively resist contact, usually by hiding their existence, or refusing to respond if detected. Admittedly a more sophisticated strategy may serve better, but there may also be circumstances unforeseen by us demanding a reply of 'no response'.

Tactical indifference may also be a factor. One such example has been discussed: dissemblance by a predator species not yet in a position to engage in predation. Alternatively, a species may be contact indifferent because it is, at the time of contact, engaged in some other activity taking up much of its efforts, i.e. warfare or social transformation. Again, the Europe-Japan case is instructive here: during the period of the Dutch factory in Deshima (Nagasaki City), the shogunate's major concerns were the potential of revolt among the 200-odd *daimyô* (lords) many with questionable loyalties, and the repeated rebellions by starving and unhappy peasants, often accompanied by millenarian religious organizations (e.g. the Buddhist Ikkō-ikki movements. Christianity was a minor element, virtually eliminated as a political factor in the massacre of Shimabara in 1637, but even the Christian uprising was more an economic issue than a political one). The Dutch major concern was colonization of profitable areas such as the Spice Islands in Indonesia, and wars at home with the English and Spaniards. Neither government had much time or energy to spare for establishing firm relations with some far off potentate who could not be a factor in their problems.

One conclusion to be drawn is that indifference to contact must be considered within a broader context. Given that species will engage in image management in most cases, indifference to communication had better be considered in its tactical aspect, rather than its strategic one. We should assume that given the means, *some* members of any species will be interested in communication, whether they are merchants, philosophers, scientists, pirates, missionaries, or any other related role. We should also know that such communication would be for the ETI's *own* interest, not necessarily either our benefit, nor for some transcendental, inter-species moral values.

There is a very important lesson that must not be ignored, something touched upon when we discussed personal parameters in civilizations. *All* social assemblages —groups, parties, cultures, and civilizations and everyone in between—are essentially composed of *individuals*. For reasons of coherence, we tend to 'zoom back' when making statements about 'civilizations', ignoring the personal element.

This, as Bar-Yam (2002) says, is an analytical convenience, one with a price. At different 'resolutions' we see different things, with different degrees and types of complexity. Now a *civilization* might have an anti-communicative stance (or the reverse), but, given what we know of civilizations—complex structures of commonly accepted standards and behaviors—we *cannot* and *must not* assume that every subordinate unit (whether individuals or some type or level of collective) will adhere to those standards. This has two implications. First, even anti-communication civilizations (or even cultures) might 'leak' communication with others, depending on the interests of different actors (again, single or collective) within the framework. Second, when humans approach an ETI civilization knowing little about it, if that civilization's standards are anti-communicative or indifferent to communication, if we do make contact, we are more than likely to be in communication with (and thus form our ideas of the civilization through) contact with fringe, possible outlaw or outcast members of that civilization.

10.7.2 Expanding and Non-expanding Civilizations

For some human societies, expansion was a mode of life. Western societies exemplify the case. Other societies, even when they were able to do so, expanded only selectively, through part of their potential range. The sixteenth century Chinese, though they had the most advanced technology for their time *and* growing population pressure, did not expand beyond the borders of mainland China (Finney and Jones 1985; Menzies 2003). The Japanese did not expand into the relatively empty spaces of the northern island of Hokkaido until the late nineteenth century. In both cases, political and ideological motivations limited what we sometimes see as a 'natural' tendency to expand a civilization.

Two sets of questions must be asked: what types of expansion are there, and what factors, if any, are likely to bring about a change of state from one to the other forms of expansion. Human history provides examples of a number of diverse modes of cultures expanding.

Expansion Intentional expansion operates through a motivated program of some sort, often ideological. Western expansions, including Spain's invasion and settlement of the New World, the US expansion into the West, Chinese settlement of Taiwan, and the Israelite settlement of Canaan were all of this type. The reasons and methods of expansion may differ substantially, from a major organized and planned national effort, to the simple encouragement of individual entrepreneurs. But some objective is assumed for the expansion.

A second possibility that must be considered is expansion that is neither intentional nor, in essence controllable: the expansion of technological innovation has, in effect, expanded and is bringing into being a World civilization on Earth. Physical and economic limits of growth would serve as a limitation even to such a 'cancerous' scenario. Or, possibly, with the emergence and potential dominance of another type of world civilization, such as that inherent in ISIS polemics, this will come about through politico-military rather than technical means.

For our purposes, the crucial question is whether, within the *civilization* concerned, there are pushes towards expansion, whether from a desire to promote the values of civilization, or because there is some such element within the canon. Note that in essence the expansion of a *culture* or a *state* (world or star-system) is a highly unstable proposition. As noted above, political systems would find it prohibitive to cross stellar distances, and cultures, dependent as they are on particular ecologies, would be transformed very quickly into different cultures, with perhaps retained influences at best from the home world. It is of course not impossible that some biological quirk, such a genetically dominant strains, would make either political dominance a possibility. Nonetheless, the daughter culture will almost immediately begin to change, and political change is likely to follow.

Above and beyond all hangs the value C. The cheapest way of expanding is by the expansion of ideas. This may or may not appeal to a particular civilization as a whole, or to actors (individual or collective) within it. But electromagnetic radiation, however convincing and pervasive the message, will alter the receiving civilization or society, but not necessarily be regarded as 'expansion'. So, with one exception—slow boats crewed by an aestivating species, a long-lived species, or a species capable of crewing generation ships—we really need not fear some galactic version of Cortez appearing on our shores as a matter of course.

Stability models The physical expansion of any species implies, as a basic parameter, the ability of that species to grow numerically. "Growth" in this context implies several dimensions. Simplistically, it means growth in absolute population size. At a more complex level, it also implies what is usually listed under development, that is, growth of economic and technological product, measured both in terms of access to and use of energy and information and in complexity of the technology (which in essence are two sides of the same coin). At a still further remove 'growth' can also imply growth of sophisticated and overarching moral and philosophical guidance for a civilization or species. All three of these are related in obvious and non-obvious ways. The growth of one may mean stagnation in another, or they may, under other conditions, heterodyne.

All this is to say that although the *gross* features of growth (which is the basis for physical expansion) have been identified (for terrestrial species at least), not enough is known about the interaction between the levels to make any more than a rough estimate for non-human species (particularly those, which includes all ETI, about whom we have no data). One feature, though, may be worth examining: the limits of growth.

A number of theses have been advanced largely in the context of disputing the Fermi paradox, about the limits of growth of ETI. Each thesis about limits to growth essentially presupposes that one or another form of growth will come at the expense of the two others. Papagiannis (1983) for example, feels that the limitations on both population growth and technological/economic growth will bring *any* long-lived intelligent species to the point of expanding in the direction of intellectual growth.

This is the point of what Brin (1983a, b) calls 'transcendence'. Another alternative is Dyson's 'cancerous technological growth' (Dyson 1966), whereas still another is continued numerical expansion through the universe (Haqq-Misra and Baum 2009). Clearly, each of these is only a partial solution. The universe, whether headed for a heat death (as it expands and entropy increases), or collapse (depending on which cosmological theory proves to be the correct one) is the ultimate limit to growth in whatever thesis is developed.

A second major limit to the expansion of ILF society must be considered as well: other civilizations. Spreading through the galaxy or cosmos, these will provide limiters to *physical* possibilities of expansion for one another. Granted, that major engineering efforts such as Kardashev civilizations or the exploitation of non-solar orbits will increase the actual volume of space available for expansion.

In examining the possibilities for expansion (and thus the types of civilizations one might encounter in the cosmos), two factors seem prominent. On the one hand, expansion of any particular species or civilization may be constrained externally. That might be because they run out space to expand (that is, they have occupied their entire galaxy) which seems unlikely, if only because in our Galaxy, they would have encountered humans, and we have no record nor evidence of that happening, or because, as Brin says, the ecology of spacefaring species is essentially a natural one composed of prey and predators, and both hide their presence (Brin 1983a, b), which seems unlikely: confident powerful prey cultures would not hesitate to broadcast their presence if only in the search for allies.

On the other hand, expansion may be limited by *internal* constraints. Conceivably because some inherent flaw brings about downfall (Gibbon 1846), or possibly because they pass beyond some event horizon as their technology and culture bring about new intellectual horizons, and physical expansion is no longer an issue (Papagiannis 1983; Vaas 2009).

Whatever proves to be the case (inasmuch expansion and stability are by no means mutually exclusive as there may be more than one interstellar civilization), expansion does not appear to be something to fear, and stability means more than one variable.

10.8 Change in ETI Civilizations and L Redux

The length of time in which a civilization may persist as an identifiable structure (intellectual, social, and perhaps material) L is an absolutely crucial variable for SETI, and is of general interest for us here as well. In effect, L relates to the ability of an ETI civilization to adjust to change. The issue of change takes up a great deal of space in the behavioral sciences (Barth 1981; Inkeles and Smith 1974; Parsons 1964; Steward 1977). Roughly speaking, it implies (a) for the individual, different access to opportunities, both expected and allowed. (b) For a society, the emergence of new institutions or their curtailment. (c) For a civilization, additions or deletions in the canon, different standards, and evaluations of praxis. At all levels, the introduction of new concepts or the abandonment of old ones or their modification.

The most significant aspect is the expectation that change is acceptable, perhaps necessary. Many human societies do not expect change as an inevitable part of their lives. This is true of most technologically (and thus socially) simple societies. One day, one year, one lifetime is expected to be more or less like any other. Such static societies and their world views are neither pathological nor necessarily bad, but they do have a disadvantage when exposed to societies that are more dynamic and thus more open to adapt to new circumstances and pressures.

A culture may also have a retrograde view of change. That is, change may be a part of the world view, but change for the worse. The Meso-American view, for example, was that the world was getting progressively worse, rather than better (Fagan 1984). In classical Chinese thinking, the current age reflected badly on the ancient age of the sages, their simplicity, and sharing, which later ages had lost. There was no expectation of improvement, only a hope for keeping things going by emulating previous, better, ages (Schirokauer 1989: Part I). This may sound like a profoundly pessimistic viewpoint, but in point of fact it is not. To the contrary, it seems to have offered hopes for betterment *with a defined goal in mind*: the restoration of a supposedly golden age.

What is generally termed a "modern" or a "Western" viewpoint is a *prograde* dynamic view of change: things will change in a positive direction for ever and ever. This poses a number of problems: run-away change, with no defined objective in sight can be destructive and disruptive, and in the case of the modern human world often is. This world view is not unnaturally seductive to many, but may not be materially more sensible than the other two. It also has a major weakness, unacknowledged, but very risky: change as envisioned superficially by Western thinkers (e.g. Fukuyama 2006) is also a major threat to people's stability, continuity, and social position, which they will resist with all their might (Ashkenazi 2012).

Finally, there is a third dynamic possibility: the possibility of continuing cyclical but climax-less changes. Though it is hard to think of a human society that tends to include this position in its viewpoint, it is an attractive ideological viewpoint which has been proposed to some extent as a political philosophy.[12] Anarchism is founded upon it, and Mao Zedong has been accused of fomenting it.

A static world view will probably reject changes brought about by communication with ETI, but might be powerless to resist them. A retrograde dynamic world view may reject the changes, and be well able to resist them, as it appeals to past successes against which measure the changes fail. The prograde world view might be ready to adopt changes wholeheartedly, as, presumably, would the anarchistic viewpoint.

[12]Though it is common in religious ideology, in, for instance, Buddhism and Hinduism.

10.8.1 Change and Development in Practice

Rather than contemplating the issue of change as an intellectual structure, let us consider it in terms of a common word used (and misused) in the international and local spheres: development. Development implies a change towards a more complex, or more fit, state from some ground state. In the human experience, development is closely related to changes in the ability to procure and share calories. Often in the literature development is confused with the availability of tools and technology. Leading to many kinds of bizarre, not to say dangerous, materials, tools and processes forced upon the 'underdeveloped' by the 'developed' through agencies such as the UNDP and USAID. Parallel to the three views about change described above, we can also how the process of development can affect the variable L.

Terminated/static development is a process in which a culture stops developing. That is, there are no 'more efficient/effective" artifacts, mentifacts, or sociofacts emerging in this culture. This may occur as result of natural or artificial disaster. There are no (or very few) examples of such societies on Earth. What are plentiful are societies in which development of any sort is very slow. Prehistorical societies we know of, as well as some more primitive societies on Earth are of this kind. There is no evidence of a culture having "given up." Evidence cited to the contrary —Australian and Tasmanian aborigines, American Indian tribes, the Hottentots in Southern Africa, who have died off through supposed contact with a more developed society (usually European, in the cases cited)—has usually proven to be more than a bit biased. Most of these have been destroyed from without, usually by having been hunted or starved to death. Nevertheless, it is certainly possible that a culture will view development as detrimental, and will covertly or overtly discourage development practices, as the Chinese empire did in its later reiteration (the Ch'ing period, 1600–1910). In such a case, as with the Chinese empire, the introduction of ideas from other cultures such as Earth will be unwelcome, whether ideas, social practices, or technologies.

Cyclical development has been cited by some historians, notably Toynbee (1964) as an inevitable, natural process in civilizations. To what degree this is indeed the case—a natural law—is debatable. Some civilizations—the Chinese are a notable example—have gone through what appear to be developmental cycles, in which periods of development in all or many fields have been followed by periods of retreat and quiescence (Eberhard 1971). Which may account for late Ch'ing resistance to development. The oft-cited retreat of the Chinese from overseas expansion during the Ming period (1368–1600) is one result of this process (Menziesz 2003). Nevertheless, there is clear evidence that though the Chinese imperial *polity* underwent cycles of boom and bust—a Chinese saying is that no family retains its social position for five generations, and this is certainly applicable to imperial dynasties—society, the economy, and technology in China did develop, albeit not as systematically nor in as straight a line as Europe (Needham 1969; Wolf 1978).

Continued development is a situation in which technology and science continue to expand their knowledge, certainties, and practices. This is the situation today, even if some studies have argued that there is an end in sight (Kahn 1978; Meadows et al. 1992; Hirsch 2005). This has also been on strand of argument about the Fermi paradox (Hetesi 2010; Tinsley 1980). Resource exhaustion, overpopulation, and waste-heat generation may well spell the physical end to planet-bound cultures. Interplanetary cultures have much higher limits, since energy is effectively free, and resources far more common in a multi-planet environment. Still, it is possible that even these limits will be exceeded, with consequent disappearance of the culture and its ability to sustain radio telescopes.

The term "development" was purposely used here in as broad a sense as possible. it can mean better living conditions, improved technology, structured organization, or other things (or all of them). For the purposes of discussing L, it means a combination of technological development, and sustained use of technology. This of course raises a very important question, which is sometimes taken as a given by e.g. Ball (1980): do moral and social development necessarily have to parallel technological development? In other words, could a society of rampant violence, self-interest, and dog-eat-dog morality evolve a sophisticated and complex technology? Well, Ayn Rand certainly seems to think so. Certainly 'pure' capitalism is based on that idea. In practice, while such a society, unpleasant as it may be to contemplate, *might* encounter some form of instability, there may be no reason to stop such a society from functioning for a very long time. In fact, if versatility and adaptability are intrinsically desirable (Ball 1980), then a capitalist form constitutes the perfectly adaptable society. We may not *like* such a society, and an ETI society of such a type would constitute a major threat to human existence, but as a Darwinian organism, it could be perfectly adaptable to meet any and all eventualities, and there is no *intrinsic* reason we can think of at this stage, that would cause it to collapse entire.

Is there a limit to development? Levels of development can be defined in terms of complexity, versatility and ability to control the environment. One development definition is "… [T]he fraction of the free-energy content of a system contained in internal structural information. The key word is *information*" (Ball 1980: 657) Civilizations 'develop' to the extent that they accumulate new information and knowledge. When maximum efficiency is approached, societies, like ants, stop developing. Development at that stage can only be counterproductive. *What* such a society could look like is speculative. What such a society would *not* have, of course is much choice, inasmuch as everything is known. This constitutes a logical paradox, which weakens the idea of no-change considerably: while an ETI society could *conceivably* attain perfect knowledge of themselves (a precondition for victory, as Sun Tzu noted) they would still lack true perfect knowledge unless all other societies (at least) were incorporated: including, of course, ourselves.

Most discussions of L have conflated civilization with culture. As we have seen, there is a great deal of difference in practice. When we come to L, the question therefore breaks down into two separate questions: How long would a civilization, composed of N cultures continue? And how long could a culture X sustain itself? To restrict the question even more, we can accept the standard SETI definition of continuity to the presence of large radio-telescopes able to receive and transmit interstellar messages.

This immediately suggests and answer at the civilization level. Civilizations require some amount of exchange (frequency and amplitude unknown) of communication between its nodes: the cognoscenti or literati who are invested in it and sustain it. If such communication—exchanges of ideas, standards, and styles—does not exist, the separate 'lobes' of the civilization will drift apart over the centuries and millennia. There is a *technological* solution: the development of some FTL communicative system would sustain a civilization for a long time, even if other species (e.g. Humans) were unable to sense the exchange. Given reality, however, FTL communication may be possible but is unlikely. This means that there will likely be a tendency for civilizations to fission over the long term. However, this tells us nothing about the variable L. One can assume that the daughter civilizations will continue to maintain a communicative ability even if that is not sufficiently frequent or repetitive to sustain the original civilization. There are other reasons a civilization might not survive: renunciation of standards in favor of some other ideological standpoint, diminution of practice, perhaps because of the demise of all authoritative practitioners,[13] or because of changes in recording practices, secrecy, or some other feature, extirpation of the canon. The crucial issue, however, is that technology is not imbued in civilization. Civilizations, particularly interstellar civilizations are *embedded* in technology: it allows the civilization to flourish through lines of communication, record maintenance, and so on. But the absence of a civilization does not imply the death of the ability to create and maintain whatever measure we use for L.

The death of a *culture*, on the other hand, may spell the end of interstellar communication. This can come about through war and the destruction of technological capability, or some other natural or unnatural disaster. It may come about because the culture loses interest, rises to some sort of transcendence where electronic means are no longer important, or because of some major change in the institutions that make up the culture. How long such a process would take is anyone's guess. We know that physical destruction of multiple cultures is possible —we have been living under the threat of ABC [Atomic, Bacteriological, Chemical] death for the past half century and more. We know that ideological changes in civilization—if, for example, the current nascent World Civilization is replaced by an Islamist one—may bring about cultural stagnation in SETI research (at least until a new Islamic science develops on the ruins of the Western-based one).

[13]Which has happened with some human societies, notably those where the practices are a secret. See for example (Barth 1975).

10.8.2 Experience with Culture Contact: Cultural Diversity and L

Humanity, as noted, exhibits great cultural diversity. Some three thousand separate cultures have been identified (Murdock 1961) and the number is probably very conservative. Many other cultures have existed in the past. Very few people in the modern world have not come into contact with members of other societies. The few marginal examples cited, e.g. the Tasaday (Marshall 1989; Vine 1989) and some isolated jungle [e.g. Yanomamo (Chagnon 1968) and Mbuti (Turnbull 1961)], island (e.g. Andamanese (Fry and Söderberg 2013)] societies, have already come into contact with other cultures since their initial contact with outsiders (and the presence of an anthropologists constitutes contact, even if limited) or have turned out to be fakes. Inevitably, cultural diversity is an accepted fact. We are therefore more ready (even though not without doubts and opposition) to accept the fact of cultural diversity, and even (for some of us) to see it in a positive way.

Is this a universal? Not necessarily. It is easy to assume, and there are obvious reasons to suppose, that highly authoritarian societies will display as little diversity as possible. As a consequence, we are likely to find that they have a low tolerance for different world views, to the point that they might feel them threatening. Whether they then attempt to destroy, modify, or ignore those 'aberrations' depends on other factors.

As a general rule, the more external contact any given culture has, the more it is able to adjust to (not necessarily welcome) contact with others. What they then do —anything from reject forcefully to welcome without restrictions—depends on the ideological and other cultural factors of the culture concerned. And the outcome, too, depends not only on the base start of the receiving culture, but of the encroaching culture as well.

There does not appear to be any way to assess what posture any given culture will adopt towards strangers. In human experience, simple societies have been welcoming and aggressive. So have complex societies. And the same is true of more developed societies. Aside from primitive societies that *generally* tolerate, even encourage strangers, but that can turn aggressive when their interests are threatened (consider many Amerind societies) or when met with aggression themselves [e.g. the Andaman Islanders (Fry and Söderberg 2013; Cooper 1989)], complex societies display the same variability, and so, on the whole, do developed societies.

From a xenological perspective, we must accept the possibility that perhaps *some* ETI societies may be extreme xenophobes. How this will actually play out, given the distances involved is an important question. Brin's (1983a, b) extreme position aside, we simply cannot predict that ETI will, or will not, be aggressive or violence prone, nor what shape this outlook will take. Nor, for the same reasons, can we accept safely the other extreme position, that because other ETI civilizations will be more advanced technologically (and perhaps socially) than we are, they will be inherently benevolent.

What can we say in conclusion to this section?

The *degree* to which information about an ETI society will impact on another, if it is relatively stable, is unknown and depends heavily upon the complexity of the society in question. Chinese society—complex, multi-statused, wealthy—was able to resist changes engendered by the West for a long time, until technological superiority forced it to reassess its technological priorities (though much of the social and cultural substrate remained unchanged until relatively recently, and Chinese culture and society continued to hum merrily along). In contrast, another part of Sinitic civilization, the Japanese, embraced change as a principle, which is still held today. The Hottentots, with a simple society, were unable to resist change to the point of decimation. We could accept, as an unproven axiom, that the more complex a society/culture is, and the more heterogeneous its institutions and statuses, the longer is L likely to be, since at the very least *some* elements will adjust to major, even catastrophic changes.

10.9 And What About Space?

Reference has been made to the possibility that humans and ETI may be more similar to one another in many ways, than 21st century humans are to our ancestors. Certainly, the Flynn effect (Flynn 2007; Wicherts 2008) would seem to indicate that modernization causes intelligence to change to a measurable degree. In the human experience, outer space is a reasonable step in the process of development, modernization, globalization, and capitalism. "Clearly, we will not create a major infrastructure outside of Earth's atmosphere with interstellar migration foremost in mind. Rather, the primary goal will be wealth. We will occupy the Solar System partly for adventure, partly for science, but mostly to get rich" (Brin 1985: 43). This also means better radio and optical telescopes (Beckwith 2008), and the development of a very different sub-set of human cultures and outlooks (cf. Crawford 2010; Maruyama 1976). This is likely to be true of ETI as well: the differences between planet-based and space-based societies may be small or large, but they will be. Among other things, space-based societies will have better resources to find and possibly contact other ILFs.

Living in space is bound to bring about changes. Some of them may well be physiological, others will certainly be social and psychological. Simple physical changes among human populations do not take long, surprisingly. They may be the result of changes in diet, or even the result of natural selection as the Dutch example shows: taller men marrying taller women and having more offspring than shorter people (Fredriks et al. 2000). Space, which represents a vastly different environment, will have effects (currently unforeseen) on human populations that spend their lives there (Launius 2010; Maruyama and Harkins 1975).

What about ETI *civilizations*, and how does this affect the issue of communication? Almost by definition, ETI civilizations are likely to be space travelers. Beyond having population centers on planets, there may also develop purely space-based colonies such as those proposed for humanity by O'Neil (1975) and O'Neill and Gump (2000). Such colonies, whose physical confines are likely to be small, whose ability to communicate will depend on things such as orbital periodicity, solar events, and orbital deliveries, will represent a new ecosystem to all who participate. As a result, notably in those colonies that have very distant orbits (for humans, Jovian orbits and beyond to the Oort cloud) are very likely, over the generations, to develop their own cultures, with all that implies: they may well be within the orbit of the home planet's civilization, but without a great deal of expensive coercion, are likely to be separate polities, economies, and social systems, in other words, different culturally as modern humans are from ancient Sumerians.

This also has implications for L. Initially, space colonies are likely to be heavily dependent on one or more populous planetary ecologies. Many, however, will eventually not only become autonomous, but independent (in the proper sense of the word, not the political). The demise of the culture or population of planets will not affect them much at that stage, and because of their nature, they will continue to develop and use space-based radio telescopes as an essential communication and navigation service. This means that 'punctuated' L is a very real possibility: the decline, transcendence, or senescence of a particular culture is unlikely to stop radiation of radio telescopes by space colonies.

References

Aijmer, G., & Abbink, G. J. (Eds.). (2000). *Meanings of violence: A cross cultural perspective*. Oxford: Berg.

Ashkenazi, M. (1992a). Not the sons of adam: Religious response to ETI. *Space Policy, 8*(4), 341–350.

Ashkenazi, M. (1992b). Some alternatives in the sociology of space colonization: The Kibbutz as a space colony. *Acta Astronautica, 26*(5), 367–375.

Ashkenazi, M. (1995). Will ETI be space explorers? Some cultural considerations. In G. S. Shostak (Ed.), *Progress in the search for extraterrestrial life* (pp. 507–520). San Francisco: Astronomical Society of the Pacific Conference series #74.

Ashkenazi, M. (2003). The fats of the land: How sumo wrestlers, tempura and viking became foodtainment in Japan. In H. Walker (Ed.), *The fat of the land* (pp. 22–30). Bristol: Footwork. 22–30.

Ashkenazi, M. (2012). Development is destruction, and other things you weren't told at school. In V. Franke & R. Dorff (Eds.), *Conflict management and 'whole of government': Useful tools for U.S. national security strategy?* (pp. 91–125). Carlisle, PA: Strategic Studies Institute.

Ball, J. A. (1973). The zoo hypothesis. *Icarus, 19*, 347–349.

Ball, J. A. (1980). Extraterrestrial intelligence: Where is everyone? *American Scientist 68*, 56–63.

Baron, N. S. (1998). Letters by phone or speech by other means: The linguistics of email. *Language & Communication, 18*(2), 133–170.

Barth, F. (1981). *Process and form in social life*. London: Routledge & Kegan Paul.

Barth, F. (1975). *Ritual and knowledge among the baktaman of New Guinea*. Oslo: Universitetsforlaget.

Bar-Yam, Y. (2002). Complexity rising: From human beings to human civilization, a complexity profile. In *Encyclopedia of Life Support Systems (EOLSS)*. Oxford: UNESCO, EOLSS Publishers.

Beckwith, S. V. W. (2008). Detecting life-bearing extrasolar planets with space telescopes. *The Astrophysical Journal, 684*(2), 1404–1415.

Bracewell, R. L. (1960). Communications from superior galactic communities. *Nature, 186*(4726), 670–671.

Bracewell, R. L. (1974). *The galactic club: intelligent life in outer space*. Stanford: Stanford Alumni Association.

Bracewell, R. L. (1981). Manifestations of advanced civilizations. In J. Billingham (Ed.), *Life in the universe* (pp. 343–350). Boston: MIT Press.

Brin, G. D. (1983). The 'great silence': The controversy concerning extraterrestrial life. *Quarterly Journal of the Royal Astronomical Society 24*, 283–309.

Brin, G. D. (1985). "Rocs' eggs and spider webs: The first steps toward building starships. In B. R. Finney, & E. M. Jones (Eds), *Interstellar Migration and the Human Experience* (pp. 42–49). Berkeley: University of California.

Caplan, J. (2000). *Written on the body*. Princeton: Princeton University Press.

Casal, U. A. (1966). Japanese cosmetics and teeth-blackening. *Transactions of the Asiatic Society of Japan Series, 3*(9), 5–27.

Chagnon, N. A. (1968). *The fierce people*. New York: Holt, Rinehart & Winston.

Cheng, C. -M. (1973). On Bunjin: The literati of the Edo period. In S. Ota, & R. Fukuda (Eds.), *Studies on Japanese culture* (Vol. 1, pp. 3–10). Tokyo: The Japan P.E.N. Club.

Cooper, Z. (1989). Analysis of the nature of contacts with the Andaman Islands during the last two millennia. *South Asian Studies, 5*(1), 134–147.

Crawford, I. A. (2010). Astrobiological benefits of human space exploration. *Astrobiology, 10*(6), 577–587.

Crihfield, L. D. (1978). *The Institution of the Geisha in modern Japanese Society*. Stanford University.

Curtis, P., & Nichols, D. (1993). *MUDs grow up: Social virtual reality in the real world*. Palo Alto: Xerox PARC.

Dalby, L. C. (1983). *Geisha*. New York: Vintage Books.

Dawkins, R. (2009). *The god delusion*. New York: Random House.

Deardorff, J. W. (1986). Possible extraterrestrial strategy for earth. *Quarterly Journal of the Royal Astronomical Society, 27*, 94–101.

Dyson, F. (1966). The search for extraterrestrial technology. In R. E. Marshak (Ed.), *Perspectives in modern physics* (pp. 641–656). New York: Wiley.

Eberhard, W. (1971). *A history of China*. Berkeley: University of California Press.

Ehricke, K. A. (1979). The extraterrestrial imperative: Part 1; Evolutionary logic. *Journal of the British Interplanetary Society, 32*, 311–317.

Fagan, B. M. (1984). *The Aztecs*. New York: W.H. Freeman & Co.

Falk, D., Gage, T. B., et al. (1995). Did more than one species of hominid coexist before 3·0 Ma?: Evidence from blood and teeth. *Journal of Human Evolution, 29*(6), 591–600.

Finney, B. R., & Jones, E. M. (1985). *Interstellar migration and the human experience* (pp. 299–276). Berkeley: University of California.

Fischer, A. (1966). Flexibility in an expressive institution: Sumo. *Southwestern Journal of Anthropology, 2*(1), 31–43.

Flynn, J. R. (2007). *What is Intelligence?*. Cambridge: Cambridge University Press.

Frake, C. O. (1964). How to ask for a drink in Subanun. *American Anthropologist, 66*(6), 127–132.

Fredriks, A. M., Van Buuren, S., et al. (2000). Continuing positive secular growth change in the Netherlands 1955–1997. *Pediatric Research, 47*(3), 316–323.

Friese, H. (1963). Qualification for office: Expertise or character. In J. M. Menzel (Ed.), *The Chinese civil service* (pp. 77–83). Lexington, Mass: D.C. Heath and Co.

Fry, D. P., & Söderberg, P. (2013). Lethal aggression in mobile forager bands and implications for the origins of war. *Science, 341*(6143), 270–273.

Fukuyama, F. (2006). *The end of history and the last man*. New York: Simon and Schuster.

Gardner, R., & Heider, K. G. (1968). *Gardens of war: Life and death in the New Guinea stone age*. New York: Random House.

Gat, A. (2013). Is war declining, and why? *Journal of Peace Research, 50*(2), 149–157.

Geller, H. A. (2013). Harmful ETI hypothesis denied: Visiting ETIs likely altruists. In D. A. Vakoch (Ed.), Extraterrestrial altruism: Evolution and ethics in the cosmos (the Frontiers collection) (pp. 65–78). Berlin: Springer.

Gibbon, E. (1846). *The history of the decline and fall of the Roman Empire*. London: J. Murray.

Gleick, J. (1987). *Chaos: Making a new science*. London: Heinemann.

Gluckman, M. (1972). *The ideas in Barotse Jurisprudence*. Manchester: Manchester University Press.

Grady, S. M., & Reality, V. (2003). *Simulating and enhancing the world with computers*. New York: Facts on File.

Guimaraes, L. (1918). *Samurais e Mandarins*. Rio de Janeiro: Libraria Francisco Alves.

Guthrie, S. (1980). A cognitive theory of religion. *Current Anthropology, 21*(2), 181–204.

Haqq-Misra, J. D., & Baum, S. D. (2009). The sustainability solution to the Fermi paradox. arXiv: 0906.0568.

Harris, M. (1988). *Good to eat: Riddles of food and culture*. New York: Simon and Schuster.

Hart, M. H. (1982). An explanation for the absence of extraterrestrials on earth. In M. H. Hart & B. Zuckerman (Eds.), *Extraterrestrials: Where are they?* (pp. 1–8). New York: Pergamon.

Haugh, M. (2010). When is an email really offensive?: Argumentativity and variability in evaluations of impoliteness. *Journal of Politeness Research. Language, Behaviour Culture, 6* (1), 7–31.

Havens, T. R. H. (1982). *Artist and patron in postwar Japan*. Princeton: Princeton University Press.

Heppenheimer, T. A. (1978). *Colonies in space*. New York: Warner Books.

Hetesi, Z. (2010). Energy use, entropy and extra-terrestrial civilizations. *Journal of Physics: Conference Series 218*.

Hiltz, S. R., & Turoff, M. (1978). *The network nation: Human communication via computer*. Boston: MIT Press.

Hirsch, F. (2005). *Social limits to growth*. London: Routledge.

Hobsbawm, E. J., & Ranger, T. O. (Eds.). (1992). *The invention of tradition*. Cambridge: Cambridge University Press.

Hogan, B. (2010). The presentation of self in the age of social media: Distinguishing performances and exhibitions online. *Bulletin of Science, Technology & Society, 30*(6), 377–386.

Hsu, F. L. K. (1975). *Iemoto: The Heart of Japan*. New York: Wiley.

Inkeles, A., Smith, D. H. (1974). *Becoming modern: Individual change in six developing countries*. Cambridge, MA: Harvard University Press.

Jones, E. M. (1981). Discrete calculations of interstellar migration and settlement. *Icarus, 46*, 328–336.

Jones, E. M. (1982). Estimates of expansion time scales. In M. H. Hart, & B. Zuckerman (Eds.), *Extraterrestrials Where are they?* (pp. 66–76). NY: Pergamon.

Kaempfer, E. (1929). *The history of Japan, giving an account of the ancient and present state and government of that empire. Together with a description of the Kingdom of Siam*. London: No Publisher.

Kahn, J. (1978). Marxist anthropology and peasant economics: A study of the social structures of underdevelopment. In J. Clammer (Ed.), *The new economic anthropology* (pp. 110–137). St. Martin's Press.

Kashdan, T. B., & Silvia, P. J. (2009). Curiosity and interest: The benefits of thriving on novelty and challenge. *Oxford handbook of positive psychology, 2*, 367–374.

Kingston, C. G., & Wright, R. E. (2010). The deadliest of games: The Institution of Dueling. *Southern Economic Journal, 76*(4), 1094–1106.

Kracke, E. A., Jr. (1957). Region, family, and individual in the Chinese examination system. J. K. Fairbank (Ed.), *Chinese thought and institutions* (pp. 251–268). Chicago: The University of Chicago Press.

Lane, C. (1981). *The rites of rulers: Ritual in industrial society*. Cambridge: Cambridge University Press.

Launius, R. D. (2010). Can we colonize the solar system? Human biology and survival in the extreme space environment. *Endeavour, 34*(3), 122–129.

Launius, R. D., & McCurdy, H. E. (2007). Robots and humans in space flight: Technology, evolution, and interplanetary travel. *Technology in Society, 29*(3), 271–282.

Leach, E. (1954). *Political systems of Highland Burma*. Cambridge MA: Harvard University Press.

Learned, J. G., Pakvasa, S., & Zee, A. (2009). Galactic neutrino communication. *International Journal of Astrobiology, 6*(03), 15–19.

Lewis, M. E. (1990). *Sanctioned violence in early China*. Stonybrook, NY: SUNY Press.

Liang, Q., & Xiangsui, W. (1999). *Unrestricted warfare*. Beijing: PLA Literature and Arts Publishing House.

Lipnack, J., & Stamps, J. (2000). *Virtual teams: People working across boundaries with technology*. New York: John Wiley & Sons.

Lusted, H., & Knapp, B. R. (1992). Biocontrollers: A direct link from the nervous system to computer. In *Proceedings: Medicin Meets Virtual Reality. La Jolla: University of California at La Jolla.*

Marshall, E. (1989). Anthropologists debate Tasaday hoax evidence. *Science, 246*(4934), 1113–1114.

Maruyama, M. (1976). Designing a space community. *Futurist, 10*(5), 273–281.

Maruyama, M., & Harkins, A. (Eds.). (1975). *Cultures beyond the earth: The role of anthropology in outer space*. New York: Vintage books.

McGinnis, A. J. (1990). Counting coup and cutting horses: Intertribal warfare on the northern plains, 1738–1889. Cordillera, CO: Johnson Books.

McGowan, R. A., & Ordway, F. I. (1966). *Intelligence in the universe*. New York: Prentice-Hall.

Meadows, D. H., Meadows, D. L., & Randers, J. (1992). *Beyond the limits: Global collapse or a sustainable future*. London: Earthscan Publications Ltd.

Menzies, G. (2003). *1421: The year China discovered the world*. New York: Random House.

Moriarty, J. R. (1969). Ritual combat: A comparison of the Aztec "war of flowers" and the medieval "melée", presented at a meeting of the Southwestern Monuments Association in 1968.

Morris, D. (1967). *The naked ape*. New York: Jonathan Cape/Bantam.

Murdock, G. P. (1961). World ethnographic sample. In F. W. Moore (Ed.), *Readings in cross-cultural methodology* (195–220). New Haven, CT: HRAF Press.

Murphy, M., & Levy, M. (2006). Politeness in intercultural email communication. *Journal of Intercultural Communication, 12*, 1–11.

Needham, J. (1969). *Science and civilisation in China* (Vol. 1–7). Cambridge: Cambridge University Press.

Newman, W. J., & Sagan, C. (1981). Galactic civilizations: Population dynamics and interstellar diffusion. *Icarus, 46*, 293–327.

O'Neill, G. K., & Gump, D. (2000). *The high frontier: Human colonies in space*. Burlington, Ontario: Apogee Books.

O'Neill, G. K. (1975). Space colonies and energy supply to the earth. *Science, 190*, 943–947.

Okakura, K. (1964). *The book of tea*. New York: Dover Publications.

Papagiannis, M. D. (1983). Natural selection of stellar civilizations by the limits of growth. IAA-83-272.

Parsons, T. (1964). A functional theory of change. *Social Change*, 83–97.

Pinker, S. (2002). *The blank slate: The modern denial of human nature*. New York: Viking.

Pinker, S. (2011). *The better angels of our nature: The decline of violence in history and its causes*. Harmondsworth, UK: Penguin.

Sagan, C. (1973a). *The cosmic connection*. New York: Doubleday.

Sagan, C. (Ed.). (1973b). *Communicating with extraterrestrial intelligence*. Cambridge: MIT Press.
Sagan, C., & Drake, F. (1975). The search for extraterrestrial intelligence. *Scientific American, 232*, 80–89.
Sagan, C., & Newman, W. I. (1985). The solipsist approach to extraterrestrial intelligence. In E. Regis. (Ed.), *Extraterrestrials: Science and alien intelligence* (pp.151–61). Cambridge: Cambridge University Press
Schirokauer, C. (1989). *A brief history of Chinese and Japanese civilizations*. New York: Harcourt, Brace, Jovanovic.
Siebold, V., & Franz, P. (1973). *Manners and customs of the Japanese*. Rutland, VT: Charles E. Tuttle Co.
Shahar, M. (2008). *The Shaolin monastery: History, religion, and the Chinese martial arts*. Honolulu: University of Hawaii Press.
Smith, J. M. (1974). The theory of games and the evolution of animal conflicts. *Journal of Theoretical Biology, 47*, 209–221.
Stephenson, D. G. (1981). A classification of extraterrestrial cultures. *Journal of the British Interplanetary Society 34*, 486–490.
Sterns, P. M. (2004). Metalaw and relations with intelligent beings revisited. *Space Policy, 20*(2), 123–130.
Steward, J. H. (1977). In J. Steward, & R. Murphy (Eds.), *Evolution and ecology: Essays on social transformation by Julian H. Steward*. Urbana: University of Illinois Press.
Steward, J. H. (1986). Levels of sociocultural integration: An operational concept. *Journal of Anthropological Research, 42*(3), 337–353.
Thunberg, C. P. (1795). *Travels in Europe, Africa, and Asia, made between the years 1770 and 1779* (Vol. 4). Volume IV: Containing travels in the empire of Japan, and in the Islands of Java and Ceylon, Together With the Voyage Home. London: F. and C. Rivington.
Tinsley, B. A. (1980). Technical development and colonization as factors in long-term variation in limits to growth. *Cosmic Search, 2*(4), 10–12.
Tipler, F. J. (1981). Extraterrestrial intelligent beings do not exist. *Physics Today, 34*, 9–81.
Tough, A. (1986). What role will extraterrestrials play in humanity's future? *Journal of the British Interplanetary Society, 39*, 491–498.
Toynbee, A. J. (1964). The nature of the growth of civilizations. In A. Etzioni & E. Etzioni (Eds.), *Social Change* (pp. 27–39). New York: Basic Books, Inc.
Turnbull, C. M. (1961). *The forest people*. New York: Doubleday.
Turner, E. (1984). Galactic colonization and competition in a young galactic disk. *IAU symposium 112: The search for extraterrestrial life: Recent developments*.
Vaas, R. (2009). *Life, the universe, and almost everything: Signs of cosmic design?* arXiv:0910.5579
Veblen, T. (1953). *The study of the leisure class*. New York: The New American Library.
Vine, I. (1989). Myth, truth, and method. The Tasaday tribe. *Ethology and Sociobiology, 10*(5), 391–392.
von Hoerner, S. (1975). Population explosion and interstellar expansion. *Journal of the British Interplanetary Society 28*, 691–712.
Waldvogel, J. (2007). Greetings and closings in workplace email. *Journal of Computer-Mediated Communication, 12*(2), 456–477.
Walters, C., Hoover, R., & Kotra, R. K. (1980). Interstellar colonization: A new parameter for the Drake equation? *Icarus, 41*, 193–197.
Webb, S. (2002). *If the universe is teeming with aliens—Where is everybody?: Fifty solutions to the Fermi paradox and the problem of extraterrestrial life* (p. 2010). Berlin: Springer.
Wicherts, J. (2008). What is intelligence? Beyond the Flynn effect. *Netherlands Journal of Psychology, 64*(1), 41–43.
Wolf, A. (Ed.). (1978). *Studies in Chinese society*. Stanford: Stanford University Press.

Chapter 11
Frameworks of Communication and Exchange

Obviously, one of the most difficult, and yet the single most important issue in xenology is to make clear and overt the underlying cultural assumptions that will frame the exchanges—intellectual and material—with another intelligent race. Basic concepts such as 'trade', 'peace', 'trust', even 'war', which are essential to any form of enduring worthwhile communication, are not so simple when we consider the invisible baggage. In this and the following chapters we shall examine these issues within various frameworks we are familiar with from our human experience. This is not to say that these are the only frameworks possible; they are simply the ones *we* as humans, are familiar with, and within which we are likely to interpret these types of exchanges. Our ETI correspondents may well view them in other ways.

11.1 Introduction to the Science and Art of Intercultural Contact

The interface *between* human and non-human cultures can be discussed because some examples of cultural contact are familiar to all of us. Even if all take place within a single race, we can extrapolate to some degree from human experience. It also is not necessary, in this particular case, to distinguish between ETI from different physical surroundings, since we can know little about that. We can afford to be ethnocentric and say that what concerns us is the way in which these ETIs, whatever they breath, look like, or eat, will interact with other ILFs. The specific *causes* for a set of behaviors to emerge, while interesting, and even very important for actual communication, are not as significant at this point, as they have been discussed earlier. Obviously we cannot guess what individual reactions to contact are likely to be, thus we have to examine all possible *general* reactions, and then discuss the likelihood of one or another, against our earlier fuzzy measures of the

M. Ashkenazi, *What We Know About Extraterrestrial Intelligence*,
Space and Society, DOI 10.1007/978-3-319-44456-7_11

sorts of organisms possible. Once again, we must take refuge in that most attractive, insidious and simple answer: the analogy.

How humans conduct intercultural conduct: Paying attention to clues

The fundamental unit of intercultural communication is the individual. "Cultures" come into contact through individual products of these cultures, multiplied by the number of interactions. Therefore, in order to understand the process, it is a good idea to reiterate how humans communicate, and how this communication, which has an interface and a hinterland, takes place with strangers.

Three theoretical perspectives inform micro-communication between strangers. Goffman's concept of the presentation of self (Goffman 1956, 1959, 1967; Goffman and Best 2005) helps show how the actors in a conversation projects a series of signals to ensure that each one has autonomy, and displays respect to the other while protecting their own social position. Role theory (Biddle 1986, 2013; Dahrendorf 1968; Jaekson 1972; Solomon et al. 1985) is the branch of sociology that develops insights into interactive micro-behaviors of individuals based on a set of principles first articulated broadly by Dahrendorf. Finally, Skinner and his followers' behaviorism theories, while often challenged, provide supporting explanation for *why* roles develop they way they do (Bushell 1969; Lyng 1988; Skinner 1953, 2011, 2014; Skinner and Ferster 2015).

A number of principles are relevant here. First, each one of the actors in an exchange comes to the exchange with a set of communicative behaviors based on background role definition and culture. Second, exchanges, particularly regularized continued exchanges bring about mutual shaping of behaviors. Each actor provides, consciously and unconsciously, cues that help reinforce the behaviors of the other. This occurs whether the actors are of the same or different cultures, though not necessarily to the same degree. Intentions, incomprehension, cultural barriers and incompatible objectives affect the content of the exchange. With time, however, experienced parties bring to the table an understanding of the reactions of the other party.

Third, disagreements, disputes, and conflicts can arise within an exchange *either* because of misreading of the meta-rules the communication operates under (which are shaped by the roles played and their mutual shaping) *and/or* because of mutually incompatible objectives *and/or* because of poor definition of the terms being exchanged. In other words, communications between strangers have three series of difficulties to overcome: of *presentation* of self/roles, of semantic *content*, and of *compatibility* of objectives.

Taking this analysis into the realm of communication with ETI, we need to add some potential difficulties to overcome. Regis's (1985: 238–9) argument that we could not learn from totally alien ETI seems to be a terminological trap: at what stage does an ETI become totally 'alien' to us? Given the nature of the universe, while it is *possible* that we would find some exotic intelligent life form (ignoring the question of how would we know it is intelligent) it is unlikely that we would find no way of at least initial communication, which, in turn, if regularized, would imply some common shaping, and better understanding. However, that being said, it is useful to keep two points in mind, both articulated several times in the xenological literature. First, we must avoid the 'similar in parts, therefore similar in whole' fallacy. That is, just because we are able to communicate on one level, does not necessarily ensure we are able to on other levels (or for other topics). First, the only thing we can be certain of is that ETI, like us, will have at least some form of 'what's in it for me?' (Molton 1981). We cannot assume that they are necessarily more moral than we are *in the way we define morality*.

Second, we cannot assume much about ETI background, without a great deal of exchange. After all, while we *assume* that humans will contact ETI through a broad based national or international effort, it may be a single successful backyard astronomer who gets the prize. Does that astronomer represent humanity? Even if he is a raving neo-Nazi? Contrariwise, suppose our respondent on the other side is a member of a fringe lunatic religious group?

At the base of our human communicative assumptions, we need to keep in mind a list of differences, following Narveson:

1. Physical differences: Physiology, anatomy, composition which would mean the ETI responds to external stimuli we don't or the reverse.
2. Sensorium: Different physical effects reacted to.
3. Information processing: (a) The ETI might be better/worse at recognizable similar operations such as logic; (b) perhaps it will perform operations unfamiliar and not performable by us.
4. Affectivity: It might not have, or have amplified, or have different emotive capabilities.
5. Values: Its life goals may be so different as to be unrecognizable.
6. Personnel Organization: The question of the degree of individuality of units or their coherence.

(Narveson 1985: 250–1)

Differences in 1–3 would alter the communication and require careful assessment to identify the problem. Differences in 4–5 would alter the content of the moral relations we might have and affect an entire slew of issues, as we shall discuss below.

11.1.1 Sociological Variables: Presentations of Self

Any form of communication can be reduced to a set of exchanges between two individuals, labeled *ego* and *alter*. The rules by which such exchanges occur in all human societies have been explored by Dahrendorff (1968) as a set of role expectations. Homans (1969) and Barth (1966, 1981) have studied the ways in which the aggregation of such dyadic exchanges go towards making up what we call 'society'. And, in every such exchange there is an element of playing a role, what Goffman (1959) has called "presentation of self". We have explored some of the implications in Chap. 7, here we need to go into the issue of roles in more depth.

I would suggest that the basics of the rules by which humans carry out exchanges are likely to be present in exchanges with other sophont species. Moreover, these rules undoubtedly apply to contact between ETI civilizations. First, it should be possible to examine contact in terms of role theory. That is, in terms of the emissions and expectations that each side has from the other. Second, these emissions and expectations carry an element of *exchange* about them. Exchange covers the entire range of *possible* benefits that each side may get or expects to get. That is, it is used in the social-symbolic sense (Barth 1981), and not the economic. Third, each side to an exchange is going to go through a managed or partly managed process of self-presentation. That is, based on the information it has of the other, and its analysis of the implications, it is going to strive to provide a picture of *itself* that fits it's goals: projecting power, attraction, submission, dominance, … to the other. This principle will hold well when civilizations of approximately, or supposedly, equal qualities approach one another. Imbalanced contacts between a very powerful (scientifically, politically, or economically) civilization and a much less powerful one will be different. The more powerful is *less* likely to try and manage its presentation: it simply would not care *unless* there was some overriding reason (such as an internal morality) to present itself in a particular way.

The mode of communication—radio or physical contact by relativistic travel—will be important. The issue of FTL communication, while a logical option, is not, of course, supported by current physical evidence and can basically be relegated to the back burner. Radiant communication has an obviously long turnaround time. Physical contact even longer. Recipients and potential respondents to such a message are going to take their time about responding (even if such considerations are only a small fraction of total message turnaround time). Radiant communication also differs from physical communication in an important way. It is both controllable and 'packeted', by which is meant that information content arrives as a complete unit. In contrast, physical contact means that communication is variable, the content difficult to control, and there is a constant process of reexamination.

In discussing contact, therefore, we must make reference to the various implications of these sociological rules. We should also consider whether the *contact mode* will affect these rules. If we use slow radio waves, or slower relativistic flight, how will these contact principles be affected?

11.1.2 Generalizations About Communicative Content, Diplomacy, and Phone Calls

To simplify the discussion, we shall follow the schemes suggested by the trio of ideas we have presented before.[1] I assume it will put certain sociologist noses out of

[1]Not that this is the only possible scheme, but it has the advantage of being simple, concrete, and replicable.

joint, but this is not an essay on sociological theory. Role theory, reinforcement, and presentation of self represent a good, parsimonious set of analytics for the needs here.

An interstellar communication between ETI and humans will consist of four elements, each of which has internal variables we have already discussed before. We will have a *Sender*, who emits a message. A *Recipient*, who is the intended receiver of the message or an eavesdropper. There will be *message content*, ranging from 0 (no content, merely a beacon or blip) to 100 (cf. Learned et al. 2008) which would be a Galactowiki. A set of *meta rules* might be inferred from the sender and content of the message (and yes, the less we know about them, the poorer our understanding of the meta rules).

1. Who is the sender?

There are three significant variables for any sender.

First is this a message from an individual (however defined) or a collective? In the first instance the value of the message must be weighed against the question of where the individual in question is in relation to the association of which it is part. Is this individual normative (most others will be like it) or peripheral (at either end of some bell-curve of 'standard' individual for it's society). In the second instance, it is possible (since the group might be completely homogenous, or authoritarian) that the message represents some form of consensus, the elements of which might be teased out of the text.

The second, and more important variable is the question of social location and nature of the sender. At one extreme, the message could be emitted from some form of central authority. It could thus be viewed as representing a political (and possibly social) consensus, and an authority with which agreements can be reached (very important for our own central authorities, i.e. governments). In the second instance, the message might be emitted from some institutional source. In the human case it might be the IAU/IAA, or even the World Trade Organization: not a central authority, but nonetheless, a significant institutional expression of the culture. Finally, it is possible that the followers of the Great God Ghu have gotten together to build a radio telescope/a Von Neumann machine, and sent off a radio/material message. That is, a group/individual representing no-one but themselves.

Third, and perhaps most significantly, we want to know whether we are receiving a message from a culture (which implies a political authority at some point) or a civilization. In the first case, one of the implications is the potential for diplomatic and other formal relationships (including conflict). In the second, the primary question they are likely to be interested in is simply "Are you civilized?" In other words, do the receivers adhere to certain standards. And, since civilizations are usually not political bodies 'formal' relations would be out of the question: communication in the receiving association is best left in the hands of whatever cognoscenti are relevant to the sending civilization, so as to work out the rules.

2. Who is the recipient?

The same set of questions are relevant to the recipient, notably when the issue of developing a reply comes up.

Are individuals or groups the recipient? And are there differences between the ways in which individuals (in different roles, we should add) will react to the message? "Take us to your leader" will mean different courses of action for a citizen of authoritarian North Korea, presidential US, or near anarchic Palestine.

Crucially, how the recipient will interpret the message? "I bring good news" can have multiple implications, depending on the nature/role/world view of the recipient.

Finally, what will the individual/group do with the message? And, if a response is to be made, who shall formulate/execute the response: any who choose to do so? A central authority? If so, which one? In the human case, much ink has already been spilled over the topic, so the issue will be dealt with in greater detail in a later chapter.

3. What is the message content?

The content variable is a moving target. As a matter of course, the more text the message contains (including pictures, sounds, and material goods if any, from a physical probe) the easier the decryption. And of course, if it is a regularized exchange of messages, the easier still. As a generalization, we can say that the content of a message ranges from 0 (the content is completely incomprehensible or content-less) to 100 (the content includes the Galactowiki or some equivalent, designed to be understood by the meanest intelligence, and chock full of useful knowledge about the sender, among other topics). Neither extreme seems likely. The nature of the message will yield, at the minimum, some clues such as the location of the sender. Galactowiki seems unlikely *except* if the sender represent an ETI civilization (which is interested in ensuring that its standards proliferate), or, conceivably a religio-philosophical missionary movement, in which case the information is likely to be somewhat skewed. Govenment Wikis are more likely to be along the lines of "Import fees at x % will be levied …." and "If you pass the following line you will be shot on sight …."

4. What are the meta rules?

Another crucial variable will be the implied (and possibly overt) meta rules hidden in and around the message. What are the protocols? Is "Dear Sir:", "Your Excellency:" or "Hey, Bro:" addressed? How is the sender's self presented: "Your humble servant" or "Yours in working solidarity"? Are there expressions of esteem towards the recipient, and what do they say about the sender? Understanding these and other emissions within the message will help in identifying the type of ETI *structure* we are dealing with, and how they see the world.

11.1.3 Intentions, Barriers, Incompatible Objectives and Misunderstandings

The process of communicative exchange among humans is fraught with complex dance of shifting intentions, barriers of the unspeakable, incompatible objectives, and plain misunderstanding. We need to keep those in mind when considering the process of communicating with other ILFs, which we might do with greater caution, but which is still open to such problems, careful as we may be.

Intentions: The interplay of strategy and tactics

Underlying every communicative act is a set of (sometimes incompatible) intentions. These may be purely tactical and short term, or strategic, involving a lengthy effort and process. They may also be conscious—the angle of a Japanese businessman's bow is the result of complex calculations about relative individual and group status—or unconscious (few people give much thought to the everyday term "God be with you" in English).

The importance of knowing what one wants to communicate, and how that could be perceived by the other, cannot be overestimated. The first is imbued in one of the classics of human civilization (Griffith 1963): know yourself. The second is a major premise in any discussion of strategy, including Sun Tzu and others. So communicating with strangers, whether they are perceived of as a danger or a blessing or even divine (McLemore 1970; Simmel 1950; Wood 1934; Yoshida 1977) is always a dance of intentions, revealing and hiding oneself in an effort to strategize the relationship.

Barriers: that which cannot be thought of, let alone talked about

Here is a thought experiment.[2] Suppose an ETI we come into contact with offers us a bargain. Their culture is composed of gourmands who wish to taste meat from all possible sources. They would like to license the human genome, in order to breed humans for the table. They are far enough away and powerful enough that we cannot stop them if they decide to do so without our permission. They are polite enough to ask first. Should we accept the offer? Should we decline it? Should we forbid the process (even though we cannot enforce the ban)?

There are a great many aspects of human life that are outside the pale of 'proper' discourse. Toilet habits (except within scientific discourse, slang, and certain extreme sexual fetishes), cannibalism (even though it has been a human practice among simple and even complex cultures), incest (though practiced among some very low and very high status humans). We need to keep in mind that none of these taboos automatically translate into meaning or even great interest for an ETI whether due to different physiology, culture, or both. The obverse is also true: aspects of behavior that we find quite normal in public, may be taboos for the other.

[2] I owe the genesis of this experiment to Larry Niven's *The Draco Tavern* (2006).

The point we must keep in mind is that cultural taboos are moving targets: they evolve and change along with other aspects of the culture. To take one case, the US taboo on body hair, notably axillary hair for women, and for men showing their genitals to other men is as recent as the 1950s. Before that no one thought anything of either as public issues. Crucially, we must be aware that for any given ETI, these features of our culture/lives/bodies may have little more than academic interest. What *is* likely to arouse interest (for either side of a communication) is the attempts to hide or obfuscate these issues. *Why* 'these strangers" are hiding a feature that to us is perfectly unremarkable, is *very* remarkable. And it ought to be so. Certainly when we are exchanging information with a civilization, we ought to be are of those areas that are considered a taboo, a standard not to be breached, since it provides us with hints not only about how to behave, but also about how *not* to behave, however irrational the premise.

Secrets, incompatible objectives and misunderstandings

Secrets and lies are the commerce of most communication. Consciously or unconsciously, we shield ourselves from giving away information that *we* consider prejudicial to our interests, whether because it presents ourselves in the wrong light, or because it conveys an advantage (more on that when we discuss responses to an ETI message). We must be aware of the fact (and are likely to engage in it ourselves) that communications may contain various degrees of prevarication. We cannot assume that every piece of information we receive is going to be truthful. This is so whether we are communicating with a civilization at the same technical level as our own (confined to a single planet) or a transcendental one. It is manifestly true whether we are communicating with an individual or some corporate group. In the absence of an ability of establishing unambiguous truth, we must assume that any detail we get, *could* in theory, be false.

Why is that possible? Many authors, as noted, have assumed that our ETI peers are necessarily benevolent and idealistic. This may or may not be the case. Even if it is the case, this does not mean that they will be providing us with truthful information all the time. It is conceivable, and ought to be kept in mind, that for reasons of benevolence and idealism, they may hide data from us, for reasons we cannot conceive of.

11.2 Disagreements, Disputes, Conflicts and Wars

Much ink (or, these days, electrons) have been spilt over the issue of conflict. Conflict is endemic in our society, however defined. An extensive intellectual and practical effort is devoted to "Conflict Resolution", whether interpersonal or inter-societal (cf. e.g. Burton 1990; Glozman et al. 2015; Melin et al. 2013; Powers and Kirkpatrick 2013). While not inevitable, the consequence of what has been said earlier is that conflict is possible. In order to understand how this plays in relations between ILFs, it is useful to parse the term 'conflict'

Consider the realm of conflict within human societies. We can identify four ascending types of agonistic displays:

- *Disagreement.* Fundamentally an expression of intellectual opposition. It may be expressed [in facial or bodily gestures (cf. Watson 1970; Morris 1982), sounds, or words], but it may well be expressed textually: long accounts of the disagreement by one or more sides. Nevertheless, even in those cases, these texts are intended to express the realm of intellectual, rather than active, opposition.
- *Dispute* constitutes a disagreement in which the *expressive* elements begin to predominate. This may take the form of spoken acts *presented theatrically* and often aimed at onlookers as an inducement to create allies or establish lines of offense and retreat (Bailey 1968; Swartz 1968). It may be a ritual in the form of a lawsuit (cf. Maine 1906; Hoebel 1967; Moore 1978), or even of a religious ritual which is intended to express the disagreement (Gluckman 1954). Typically, too, disputes display a *time* dimension: they take place over periods of expression and response, sometimes very lengthy cycles of these (consider a lengthy law suit). The expressions and responses are, in many societies, typically regulated *ritually*.
- *Conflict* introduces a *physical* dimension. Very often this is in the form of violence (including the threat of violence). A conflict, therefore, has all of the qualities of the lower intensity agonistic exchanges and displays, with the addition of a violent or potentially violent element in which the positions of the contestants *appear* to be mutually exclusive and irreconcilable (Avruch 2001; Balikci 1973; Brahm 2003; Nudler 1990; Ross 1986).
- *War* is a subset of conflict in which actual violent acts are perpetrated by one side or more, and in which ongoing violence is the expected mode of exchange. The entire social unit of each side may participate, passively or actively, in the conflict. Some wars are bound by rules (e.g. intra-European warfare between thirteenth and twentieth centuries) others are no-holds-barred struggles where the entire society is considered an opponent, and thus subject to violence, e.g. nuclear warfare and wars of extermination and subjugation such as the Indian wars of the US, and similar practices elsewhere (Anderson 2011; Chadefaux 2014; Grandi 2013; Ross 1986).

It is helpful to keep in mind that while members of the group can understand with a great deal of clarity where and when an agonistic display can be located in this hierarchy, the same is not true for observers from other societies. Disagreements among Mediterranean people (Italians, Greeks, Spaniards, Israelis, Arabs) are typically loud, shouted events, hands waving, faces glaring, which can terrify members of other societies who may assume the disagreement is turning into a conflict, whereas in actuality the protagonists are merely pressing a point. Disputes among Japanese appear calm, reasoned, concerned with long silences and minimal gestures even if the topic is deadly: Middle Easterners and American observers will often assume there is nothing more than mild disagreement at the worse.

ETI would be in much the same situation, magnified, as a Japanese in a Mediterranean dispute. Different interest positions *may* be reconcilable or not; but agonistic displays from one culture to another—whether making faces, writing diplomatic notes, or parading the army—are easily misunderstood by someone from another culture. The actual issue at the heart of the disagreement, given the difficulties of intellectual understanding between intelligent species, may not even be comprehensible without a great deal of negotiating about the fundamentals, themselves possibly either not apparent or incomprehensible to the other side.

On top of this, is a level of how the agonism is *presented* to the other side, that is, the meta-rules of presentation used by either side. Rituals are partly inherent in our phylogeny; all ape species play the intimidation game very loudly, with glaring eyes and open jaws, often waving their arms and sticks in the air, thumping the ground, bouncing about, and hurling branches, earth, and other objects: surely this sounds familiar (Goodall 1986; Symons 1974)? This is anchored partly in our physiology, as our speech and handedness brain centers are intimately linked, so unsurprisingly, we talk using our hands, (Beattie 1983) and partly in our learned culture. Understanding the meaning of a ritual is difficult enough for a human outsider who at least may know or have similar referents herself, but a non-human? This is not to say that all 'odd' and misunderstood proxemics lead to conflict or war, but that very different modes and expressions of communication can start a disagreement, that *can* lead to conflict, and from there to serious physical violence unless both parties are very conservative in their interpretations, and are slow on the trigger.

The fact of (potential, if highly unlikely) physical presence is an additional complication. Human beings communicate through a variety of modes, not all of them on a conscious level. In addition to speech, we use facial gestures to embellish our speech (Morris 1982), as well as to emphasize the truth value of our statements (Watson 1970). Size and spacing of discussants is important to us, albeit unconsciously (Hall 1959). The role of olfactory values has not been explored fully, but that might be important as well (Kohl and Francoeur 1995). Thus Molton's (1981) observation—that we will tend to apply our aesthetic/cultural criteria to prominent (to us) features of ETI regardless of conflicting signals they may be emitting purposely—is very important. *Inevitably* we are going to apply certain subjective criteria to any physical contact.

For many of those involved in SETI, interstellar conflict is the sum of all fears. After all, the reasoning goes, anyone able to contact us is also going to be technologically far in advance of us, and therefore the conflict would be a no contest: humanity could not possibly survive, never mind win, such a clash.

It needs to be kept in mind that conflict, like any other activity, follows certain rules which are, if not well understood, than at least understandable. Conflict does not mean that some actor X launches a full scale attack on actor Y until victory or death. Even violent conflict is more subtle.

In the human experience, conflict (that is, violent exchanges, including war) follow a bell-curve of intensity, which notionally divide any conflict into three unequal stages: initial hostilities, ongoing full conflict, and die down (Augsburger

1992; Avruch 1998; Bercovitch and Jackson 2009; Burton 1990). Only rarely does a conflict start with an all-out assault. In most cases, there is a great deal of posturing and ritual presentations during which one or more sides try to (a) estimate the willingness to fight of the opponent, and (b) intimidate the opponent without going into full fighting mode. It is also an opportunity (sometimes intended by one or the other side) to allow a third party to intervene, either as a mediator, or as an ally.

Posturing can include anything from strong messages, through demonstrations, to raids and 'controlled' local violence on the borders or hinterland. Not all conflict mutates into war, and even wars are fought for a variety of reasons, following a variety of meta-rules. In feudal societies (in Middle-ages Europe, feudal-period China, Medieval Japan, and ancient India) 'knights': specialists in warfare challenged one another individually to battles. Among the Cheyenne Plains Indians, individual combat, while violent, was subsidiary to counting coup: touching an enemy with a special coup stick as a form of personal humiliation. Many rural communities ranging from New Guinea through Japan to the Andes engage in ritualized, seasonal battles, during which small numbers of individuals are injured and sometimes killed. In contrast, 'complex operations' (Schneider et al. 2001) and 'unrestricted warfare' (Qiao and Wang 1999) are forms of war in the twenty-first century in which no restrictions are acknowledged. The point here being that there are different *modes* of warfare.

So we really need to ask several questions. First, is interstellar conflict (and possibly war) possible? Second, would intelligent, sophisticated, and technologically savvy ETI engage in war, and if so, against whom? What modes of war might we expect, if any, and do these modes encompass all potential cases? Third, is it possible to generate a set of 'rules of war' such as those imbued in the modern world by the Geneva Convention and other regulations? And finally, given the difficulties of setting a warfare mode, how can one bring upon a cessation of combat and the emergence of peace?

All of these questions have arisen at one time or another in human discussions of warfare, conflict, and their varying permutations and possibilities. Given the narrow scope of this book, it is impossible to deal with all of these issues with the thoroughness they deserve. A few points need to be made from the start. First, it does not seem to be inevitable that ETI do *not* engage in conflict (or war). None of the moral objections seems to hold universally, given the wide variety in phylogenetic and social forms discussed above. Even the existence of a civilization is not, in and of itself, a guarantee that polities that share a civilization will not engage in war. As we have seen from our own history, people who share the same civilizational ideas can happily enough engage in very savage warfare, as the histories of India, Europe, and China demonstrate. Second, we would expect interstellar war to be uncommon, if only because of the velocity limiter C, and the likelihood that it would be cheaper for any society to acquire resources from astronomical bodies, than to go looking for trouble. So interstellar war is hugely unlikely: the rewards would be small, and the effort enormous.

Nevertheless, war over inherent ideas certainly seems possible. That is, given sufficient ethical or ideological justification, including extreme xenophobia, religious impulse and a missionary motive of an extreme sort, it is conceivable that even the most morally directed ETI will engage in wars. And, as we know to our cost, wars of religion can sometimes demand, and receive, extreme sacrifice from those who engage it. What perhaps limits these sorts of wars is simple economics: even extreme cases of religious wars, ranging from Islamic crusades[3] from the seventh to the twenty-first centuries, the Taiping rebellion, and various European crusades, were fundamentally, from the perspective of many of those in charge, economic opportunities. And the economics of war across space are so onerous, the returns will likely never recoup the costs.

The 'Rules of War' we have today are the result of centuries long intra-*European* experience of war. They were never intended to apply to those people and combatants outside the 'civilized' (that is, European) countries. They only became a universal ideal with the evolution of the Hague Conventions (1899 and 1907) and later the Geneva conventions. Even so, these conventions were not adopted by all potential combatants until after World War II, and have been disputed, and in practice repudiated by legal maneuvering by many signatories. Certainly the evolution of weapons and the rise of guerrilla and terror tactics they allowed, has brought about the rise of many groups that do not adhere, in fact repudiate, these conventions.

For our purposes, yes, it is possible that laws of war will emerge if war is an existent form among ETI. There are three problems. The first is that planets and other potential biospheres are extremely vulnerable and fragile. Toss enough high speed mass in the direction of a known orbital location, and nuclear winter arrives (Firestone 2011). If we are speaking of space habitats, they are even more vulnerable. The second is that all sufficiently advanced ILF, including ourselves, are likely to be extremely reliant on electronic digital means to keep themselves going, whether as a culture, a polity, or a civilization. Disrupt these, perhaps because their beliefs are anathema, and one can take over, mould, and change, the entire intellectual working of the target, and of course their actual physical existence as a communicating civilization (Carrigan 2006; Shuch and Almar 2007). Third, one form of space colonization that has been discussed quite extensively in the SETI literature is colonization by micro robot, notably self-replicating Von Neumann machines (Merkle 1992). One, or a small swarm of these, reproducing in, for sake of argument, the orbit of a giant planet or rich asteroid belt—with sufficient materials and energy available—could, in theory wreak simple havoc on any space-traveling or even planet-bound society.

Whether rules to limit that sort of total warfare exist in the cosmos is possible, but, again, unlikely. The limits to warfare seem to be, once again, the limits of distance, and the question of *cui bono*? What kind of foreseeable benefit could there

[3]Yes, an oxymoron, I know, but do not have a better words for a focused, aggressive, religiously motivated war, notably given the multiple glosses of the Muslim word *jihad*.

be? Taking over communications *might* benefit a crusader, but may also backfire, or have the targets develop their own mutated version of the original message. Destroying an entire race for some unforeseen reason, whether by space bombardment or Von Neumann death, does not make any more sense than the creation of a universal, enforceable set of benevolent rules of law: both are possible, neither is a certainty.

War comes in many modes among humans. At one extreme are wars fought under severe, sometimes self-imposed restrictions. 'Knightly' battles fought in various feudal periods (Marshall 1999), Aztec "Flower Wars" (Hicks 1979) and other ritual combats (Moriarty 1969; Gardner and Heider 1968) are examples. At the other end are wars fought for complete subjugation and/or extermination. These range from American blood feuds, through Tamerlane's wars, to the modern age's Balkan conflicts. Is this likely to be true among ETI as well? As noted, interstellar war is unlikely, but that does not exclude inter-planetary warfare from coming about within the frame of a single solar culture.

It is possible that *if* a civilization has emerged among the (possibly) numerous cultures and polities of a race that occupies it's entire solar system, that civilization will also include some standards and behavioral demands during periods of war. Some forms of war, e.g. at the subjugation/destroy end of the spectrum, will likely be considered improper and forbidden. It is also potentially possible that as the civilization spreads outside its home system, or comes into contact with other civilizations, that these principles might become more-or-less universal, though we should not bet on that occurring.

11.3 Contact

When considering the possibility of contact (that is, initial indication of presence, whether the content marker is set at 0, 100 or anywhere in between) between two ILFs, there are a number of general features that must be borne in mind. For example, the problems inherent in contact are simplified the more often it occurs. That is, if any particular ILF has had contact with others, then it is likely to be better equipped to handle contact (all other things being equal). Molton (1981) notes that this is true *provided* either side has sufficient interest to study the other. There are two corollaries to this. First, it is likely that an older culture will have had more experience, and thus is better equipped to deal with the issues than a younger one (because they have had other contacts) and, if we find upon contacting any ETI culture that the contact is going smoothly, we should consider the likelihood that they have had prior experience (whether they tell us of it or not). And though many (e.g. Tough 1986) feel that an older society will necessarily be benevolent, the only common idea that might be driving ETI is "What's in it for me?" (Molton 1981).

A general model for the parameters of contact has been suggested by Maruyama (1976). His model is explicitly a model of relationships between human space colonies, but some of the issues are relevant to a discussion of contact between ILFs as well. In practice, Maruyama suggests there appear to be four possible types of 'contact environment':

1. A generally hostile and mutually exploitative environment possibly due to different evolutionary and development paths and the distances between the parties.
2. Amicable competition in which free enterprise reigns between the parties (the different ILFs) which could bring about disputes at the borderlines, should those come into being.
3. Isolationism, where independent forms, not dependent on one another will continue to exist side-by-side and self-sufficiency is the rule, and possibly the goal.
4. Organic network model, where the strength of one ILF contributes to the weakness of others, and contact brings about social and economic replicability, and mutual safety notwithstanding (Kuttner 1981) who argues that technology is always convergent and that peer cultures will have little to offer one another.

One advantage of this scheme (which may not be comprehensive) is that it is scalable: at a lower scale, a section of a given ILF may be involved in a network with other races, while the rest of that particular race may be isolationist. At a larger scale, islands of hostility may be surrounded, or in contact with larger networked associations. An understanding of these patterns of contact, if possible, allows us to sort out, and establish our own position, where N = 2 or any greater number.

The contact environment has a number of pitfalls one should be wary of. At one extreme is the supposition that since *any* contact will primarily be with a race that is far in advance of humanity (technologically, and possibly socially) they will be benevolent sharers, providing us with information and technology that will allow us a quantum leap in understanding (see e.g. Arbib 1979). The other extreme is that we come into contact with a race that is inherently and irrevocably xenophobic. With the former, there are three problems. The first is why would benevolence, rather than any other posture *necessarily* be the one advanced societies adapt (though I am prepared to concede the *possibility*). The second is that gifts of that sort have two effects. They shape the less advanced society *in the image* of the advanced society. Our own experience of that—US policy insistence on replicating the US wherever it can—is a very sorry one, as it breeds resentment and resistance (Ashkenazi 2012). The third is that such benevolence, since it shortcuts whole processes in the receiving culture, is likely to create a crippled version of the donor culture: dependent, and unable to deal with its own problems.

The xenophobic extreme also is problematic. First, given interstellar distances, a xenophobic culture will be unlikely to advertise its presence, if avoidable, and if unavoidable, might simply hang out a sign saying something like "Stay out. Trespassers will be shot": A message that has a content value slightly above 0. We need also consider that 'xenophobia' and 'aggression' are not necessarily synonymous. If the xenophobic ETI is strong enough, it might simply isolate itself and ensure there is no contact with others, as the Japanese *sakkoku* (closed country:

1639–1853) and its Korean parallel exemplify. If the xenophobia is somehow inherent (genetic or otherwise) and the culture sufficiently powerful and technology savvy, they could enforce this for a long time. If it is a matter of policy, at some stage, when conditions change, we could expect a change in the stance. If, on the other hand, the race is actively aggressive (food, slaves, treasures, flower wars: take your pick), then other ILFs need to be on their guard when dealing with the, whether these aggressives adhere to a "Catch more flies with honey" philosophy, or merely intend to wreck other cultures for some purpose of their own.

Finally, back to the issue of transcendence. A transcendent civilization—one in which the rules of energy are in abeyance, and possibly social ones as well—would likely come about in one of two forms. The 'simple' form is one in which the biological and digital technology sides of a culture blend—a wetware society—to such an extent, that most of our assumptions about social living fade away. In such a case, a society that is no longer dependent on external stimuli (beyond energy input) may well become uncommunicative with others, since all social variety and other stimuli are virtual (Curtis and Nichols 1993; Grady 2003) IT based. If one adds the potential of artificial intelligence to the mix (Barbrook 2007; Bobrow and Brady 1998; Dick 2010; Kurzweil 2010; Venkatesan 2010), the results in terms of interest in contact with other cultures may well dictate complete isolationism. Of the 'complex' form of transcendence (Smart 2012)—when a culture supersedes physical reality in some non-mechanical fashion—there is little to be said. We simply have no data to work with, whether they would be amenable or opposed to contact, though the second would be likelier: what do they benefit from contact with the worms?

Summary. The contact environment is more complex than merely an exchange of messages. The nature of the contact environment, as well as the two parties to the event need to be examined carefully. This section deals only with the ETI side; the human side will be dealt with in Chap. 13, where the entire issue of reaction to contact among humans is explored. But we need to add one warning. We will tend to apply our aesthetic/cultural criteria to prominent (to us) features of ETI regardless of conflicting signals they may be emitting (Molton 1981). Therefore the possibility of errors based on preconceptions may well be high and we need to make quite sure that however we react, it is on the basis of reason, not emotion. This is true whether we are engaged in contact through the electromagnetic spectrum or other radiation contact, or physical contact, unlikely as the latter might be.

11.3.1 Electromagnetic and Physical Contact

Radiation contact encompasses all possible or potential forms of contact via a radiation medium, whether that be electromagnetic radiation, or hypothetical, and so far unproven tachyon or gravity wave communication. Physical contact includes both manned missions and autonomic or semi-autonomic probes. Contact by radiation—the identification, interception, and interpretation of some form of electromagnetic signal—has been the majority assumption of most searches for evidence of ETI (Fridman 2011; Morrison 1977; Shostak 2010; Traphagan 2014).

There have also been some arguments that physical contact has occurred in the past in the solar system (Forgan and Elvis 2011). This also includes observational evidence of the existence in the past of ETI by their artifacts (Carrigan 2010). Many physical scientists would deny the possibility of ever having any form of physical communication between ILFs. This derives from good solid physical and economic calculations that have been discussed earlier. Physical contact, as argued previously, is at this point, merely a theoretical possibility. Nonetheless, the possibility does exist, if only as a low-probability hypothesis, and thus must be explored.

Contact is always time dependent: the period from emission to reception, even at light speed of radiation contact takes hundreds to hundred thousands of years. Therefore, we may assume that a society that engages in intentional contact, particularly one that does so regularly, assumes that it has great durability, otherwise, it could not survive to benefit from its efforts (Sullivan 1964: 264). This is of course even more true of physical contact.

Since we can assume that any society engaging in contact is going to try and present itself in the best way possible, can we make any second order generalizations about that society? Given that the contact is likely via radiation, we might not be concerned with any forms of physical aggression nor expansion (though more on this later). The objective of contact appear to lie in one of two realms: scientific curiosity and the exchange of information, and spreading ideas and information. Certainly given the first case, the contacting ETI is going to be expecting an exchange, that is, a return on their investment.

Since contact is the first move in an exchange, it is likely that two factors will underlie the initial signal. First, the presentation of self of any ILF will be based on *its* analysis of desirable characteristics in a partner. This, in itself is revealing. Let us examine, briefly, the original Pioneer 10 spacecraft message. For example, it makes much of human form, arithmetic, and chemistry. It says nothing about exchanging quid-pro-quo. It would seem therefore that the ILF who sent the message (the human race as represented by NASA, and thus largely, the USA) is not concerned with some sort of intellectual exchange *or would like to present itself as such*. There is no mention of exchange, either because the ILF is not concerned with exchange, because exchange is risky, depreciated in that society, or perhaps because they are insincere. We (the receiving ILF society) do not know which of these is true. We do know, based on the simplicity of the message, that the technology is similar to ours. Therefore they are likely to engage in presentation of self, therefore they are *interested* in portraying themselves as disinterested in exchange. We have, perhaps, gained additional data from what is *not* in the message and can shape our own reply accordingly.

This latter brings us to the second factor underlying contact. Every reply will be shaped, to some extent, by the perception the receiving ILF will have of its alter. This can be represented schematically. Each party to the interaction will present itself in some particular way. Alter will interpret the presentation in light of its own background. It will then modify its own presentation so that the effect it wishes to achieve will be more likely, given its understanding of the other sides presentation.

Since this sort of exchange will take place over the span of centuries, millennia or longer (assuming no exotic ansible[4] communication such as gravity waves or tachyon communication occur), there will of course be plenty of time to script, rehearse, and orchestrate the play. This is true assuming incidental broadcasts are not received by either side. As a matter of general principle, the more varied information is received by ego, the better ego will understand alter, and be able to script a play to alter's liking. This will occur much faster at electromagnetic ranges: the range at which Earth's incidental broadcasts could be received is currently circ. 119 ly and growing. Any ETI civilization listening at that distance will have a huge amount of data that will enable it to narrowly focus its script to our liking.

11.3.2 Radiation Contact

Radiation contact differs from physical contact in a number of ways (that is, beside the physics of the two sorts of event). Most importantly, it implies that each side is better able to control its presentation of self. Michaud (1998); Goldsmith (1990); International Academy of Astronautics (ND) and others have argued that humanity must answer any communication from ETI with a single unified message. This particular position comes from a position of relative weakness, that is, from an assumption that any contacting ETI is going to be more powerful (because technologically more advanced) than we are. A direct consequence of this is the (misguided, in my opinion) attempt to manage our self-presentation, by sending off a unified message.

Thus the *advantage* of non-physical contact is the fact that since it must pass through a chokepoint—the transmitter—it is amenable to manipulation, and to 'putting the best foot forward' to the transmitting society. However, two issues affect this feature. In the human case (and we can assume this is true for many societies new in space), radiation contact is by way of electromagnetic waves propagated through space. Our own species started its process with the first Marconi broadcasts (circ. 1905), so the electronic leakage is a bubble about 129 ly in radius at 2016 (though extremely limited for the first decade, thus in practice 119 ly). And this bubble is chock-a-block full of information about the human

[4]Instantaneous universal communicative device proposed by LeGuin (*Rocannon's World*, 1966; *The left hand of darkness*, 1969), here used as a thought experiment in instantaneous communication.

species, ranging from its wars and triumphs, to how it wipes its nose and other areas: that is, information both valuable and puerile, though which is which, depends on the observer. A second issue is that contact can be inadvertent, as for example identification of industrial residues in the atmosphere as an indication of a technological culture.

11.3.3 Physical Contact

We have to take for granted that the likelihood of physical contact with other intelligent life forms in the universe is low. Very low. This does not mean that the odds are nonexistent. And we need to keep in mind that there are many implications for the (low probability) of physical contact. Given the fact that sending a relativistic starship is extraordinarily expensive (about 50 years current total world energy output for a single one-generation trip to Alpha Centauri and back), anyone who actually does so is fabulously wealthy and and/or must have an overriding urgency to perform the feat. Another, far more likely possibility is intelligent Bracewell probes of some sort. These too imply technology, albeit possibly of a lesser (that is, more reachable for us) order.

The possibility of physical contact lowers the ability of either side to engage in controlled presentation. Controlled presentation in physical contact would be extremely difficult, since all participants are likely to at least be giving off unconscious and unscripted clues. The process of evaluation and counter-evaluation will 'speed-up' to the rate of normal processes of that sort in any society. Part of the that show is the genetic and learned habits ingrained in any culture.

To take but one example, in almost all human cultures, sexual acts are also private. This may not be the case for an ETI who might regard it as a practice that one engages in much as we drink a glass of water. Overtly, humans might be able to cope with such a factor, but it will inevitably affect the public perception of these ETI, as well as the private, perhaps unconscious feelings of human negotiators towards the ETI. To take it one step further, suppose the ETI engage in *rishathra* [public inter-species sex as a formal means of establishing confidence. Larry Niven *Ringworld*, (1970)] or some other activity that we tend to regard as naturally private or opprobrious, such as defecating while talking?

Obviously, when a species decides to engage in contact, it must make allowances for 'strange customs'. As Dyson noted, we cannot and must not force our taste on others. The problem is that all humans are conditioned from childhood, in their various cultures, to behave in certain ways, and, more importantly, to regard certain acts that are physically possible, as highly repugnant. It is certainly conceivable that humans might *never* adapt to the presence of certain activities which might seem perfectly reasonable, even required, by another species. A solution might be strong conditioning of the negotiators, or even raising negotiators in an artificial environment which would allow them neutral reactions to many practices, In effect, raising a class of morally neutral human beings. The potential answers raised here as examples are far less important than an understanding of how deeply our instinctual and learned reactions go beyond 'sweet reason'. That is, our cultural conditioning is *so* strong, that we must be aware of the sad possibility that we might never be able to tolerate the direct presence of a particular ETI species. We might turn this around as well. We may very well find an ETI species who are so conditioned to react negatively to certain practices that we, or other ILFs may feel are natural and even necessary, that they are unable or unwilling to participate in contact. Since we have no data, we may be anywhere on the continuum ranging from completely flexible in our reactions to strange behavior patterns and appearances, to completely inflexible.

11.4 Interstellar Relations: The Diplomatic Relations of ETI

"A diplomat," says Ambrose Bierce, "is a man sent abroad to tell lies for his country." Human polities have long employed professional and semi-professional individuals to advance their interests with other polities. Along the way, particularly since the fall of Napoleon, certain understanding have evolved (being cynical, largely by diplomats to protect themselves) about diplomatic activities. In the absence of physical diplomats roaming the universe "Ensuring a state of tension short of war," as Laumer puts it (Keith Laumer, *Retief*), what kinds of diplomatic issues need we consider?

Terminologically, we need to be aware of the concept that *civilizations* (as the concept is used in this book) do not conduct diplomatic missions. Diplomacy is the province of states: a cultural form, which may or may not exist among ETI. The concept of a state—a polity that controls a defined territory, within which its law runs supreme, which is sovereign and free of legitimate interference by other states, and which relies ultimately on the sole legitimate use of violence within its borders—is a European concept. Developed initially during the Peace of Westphalia which ended the Thirty Years War in Europe in the seventeenth century, it is currently the dominant concept of the state throughout the nascent global political system.

The state concept is often conflated with the *nation* concept: a political system in which a particular culture (language, society, and learned behaviors, possibly religion) is coterminous with the borders of the state. In practice, and notwithstanding the hot air of repeated rhetoric about 'self-determination' and 'the right of all people to express themselves' by governments and international bodies, this position is an out-and-out lie. Cultures cross borders (consider the Basques straddling France and Spain, and the Kurds straddling Iran, Iraq, Syria, and Turkey) and the borders of the nation-state always include one or more majority culture(s) that dominate(s) the state (peacefully or by force), and one or more cultural minorities that are kept in the state, often by force and threats (consider the fairly recent American Civil war, and Chinese domination of Tibet and the Uighurs, the UK and Scotland, Wales, Ireland, and Cornwall). 'Minorities' that might number in the millions (consider some of the nations that make up the "Federation of States" AKA the Russian empire: certainly sufficient for the establishment of nation-states of their own). In practice, all nation-states today are *empires*: a state in which one culture (and its preferred form of cultural institutions) dominates and controls, sometimes, but not always, sharing power with other cultures within the state's boundaries.[5] It is worth examining whether these conditions hold water outside humanity's narrow perspective. So let us get back to the nation definition as it is commonly understood internationally.

Polity. A polity is a political form, an institution within which decisions are made about the allocation of resources. As we have seen, for most human states, this means a particular culture (which is often a numerical majority, but could be a numerical minority) defines the rules, and thus the allocation. Would the same be true of a polity that consists of numerous different ecologies (planet[s], moon[s], asteroid colonies, space stations) orbiting at different frequencies, having different gravities and ecological systems? How would they agree on how to distribute resources, or even what the rules of the game are?

Controls a defined territory.

Consider a culture that has reached into space.

- First, instead of an area, there is a 3 dimensional volume, most of it—over 99 % —vacuum. Whether one can control such a volume or not, is it *worth* controlling? It may be argued that minimum orbits or trajectories are the real 'territory' of a solar or interstellar state, but how is this control to be exercised, and at what cost?
- Second is the problem of interpentrability. Nation-state A, largely the creation and product of Race x, is interested in F-type planetary systems. Nation-state B, dominated by Race y, only interested in K-type solar systems. They may interpenetrate for hundreds of years, not realizing that some parts of State A are surrounded by parts of State B. In other words, boundaries cannot be determined by lines or even surfaces: the universe does not allow that. The same might be true if there are more than one solar state, where interests and interpenetrability must also include shifting locations as objects in far-solar orbit will have more in common with others in similar orbits for perhaps lengthy, but nonetheless finite, periods of time.
- Third, as noted, mechanics of star systems mean that everything is in motion, much of it predictable, of course, but even in an average-sized (we assume) solar system such as ours, time lag between the home planet (where we assume the capital lies) and a colony in the Oort cloud could be more than an hour at C. So how do we define 'control', notably if the colony in the Oort cloud has other interests, feels abandoned by the uninterested and uncaring mother planet, and want to run its own affairs? Do we send in the space marines? Assuming our space marines are on board a ship capable of 1 g acceleration, they would still

[5]The only non-empire state I can think of offhand is Iceland.

take weeks and days. And do we want to resort to force every time any of those colonies refuses to pay its taxes? So control over even a solar state-empire is going to be a tricky thing. Moreover, how do we define territory? Orbits intersect, areas of interest interpenetrate for a phase, and then the configuration changes. This means that the 'territory' is constantly mutating and changing throughout the solar year and beyond.

Within which its law runs supreme. It is of course conceivable, that a small colony way out in the Oort cloud is simply not *interested* in investing in the whole legal mechanism, and simply agrees to delegate this to the capital. This is the case in the British Commonwealth, where many small states, notably those in the Caribbean, allow the British Crown to assume the burden of e.g. a supreme court. But the legal issues are not as simple, when we compare a space colony to a nice and cozy planet. I am aware of no law in any country that forbids lock-step marching. This may be a life or death issue in a space station, where the lock step could heterodyne and vibrate to the point of breaking the structure. So what does one do if the supreme court simply has no law forbidding such an event, and a culprit is brought before it, blamed for initiating a lock step?

Sovereign and free of interference by other states. Is demanding access to solar resources, such as starlight (for energy) water asteroids (for water and oxygen) or methane from heavy planets a given? Is it part of the sovereignty, and if one uses them freely, is one not exposing someone else to starvation or death from oxygen depletion? Parenthetically, even in the modern Westphalian regime, the concept of absolute sovereignty is cracking, with the assumption by many (powerful) states, of the duty to protect [unarmed civilians from terror, and war crimes. Tan (2006)], as well, of course, as clandestine and even overt 'regime change' doctrines (cf. e.g. Litwak 2007; Rice 2008).

Sole legitimate use of violence. We need to ask again, how does one deploy violence, even legitimate violence, in a sparsely populated 'territory' with varied speeds of communication and transport? And the use of force is not so simple, given the inherent fragility of both space colonies and planetary habitats: high speed rocks anyone? More importantly, if the total population of an Oort cloud colony has different interests, different environment, and possibly different physiological requirements than the home planet's, is the law of the home planet 'legitimate' for the colonists?

All this is to say is that the Westphalian regime, which in any case is a badly frayed, patched quilt, may not be the most applicable in a human-run solar system. And, if our ETI are indeed in advance of us, then they may well be aware of the problem, and evolved a different solar political regime. Which brings us back to diplomacy: is diplomacy possible with a non-state? And what would the limitations and advantages be?

11.4.1 Diplomacy as a Practice

With the problematical position of the state in mind, we can, nevertheless, try to examine the role of diplomacy, most importantly considering the physical and other issues of interstellar communication. Jönnson and Hall (2003) have made a case that there are a number of communicative dimensions of diplomacy [which, they argue, is essentially a form of communication. See also Avenhaus and Zartman (2007: IX)] that need to be kept in mind:

Verbal versus non-verbal communication

Diplomats provide both verbal communication, and non-verbal, including what is now termed 'soft power', that is, the ability to project an image of their state through cultural, media, and other presentations.

Public versus private communication

Diplomats communicate their state's interests through private communication between themselves and the government (officials) of the host country. Increasingly, however, due to the rise of formal and informal news networks, their work is also done in public and in the light of public opinion: the host public, the home public, and kibitzers from the world.

Ritualization of diplomatic communication

In order to ensure degrees of acceptance, obligations, and, a cynical social scientist would say, their own safety and privileged access to power, diplomatic communications have been ritualized since the post-Napoleonic era. Diplomatic bodies and baggage are sacrosanct, communications are formalized, and status hierarchies dominate all discourses.

Technological development

Technological development, notably in the communication industry, has revolutionized diplomatic exchanges. First, due to widespread literacy and education, far more states and individuals can play the diplomatic game. This has led to what is called "second channel" diplomacy(secret negotiations between unidentified, informal parties while formal negotiations are taking place) and even "third channel" diplomacy: negotiations and communication between non-diplomats, including from the industrial sphere, NGOs, and virtually anyone else who wants to play and can garner support from some constituency for their activities (cf. Cooper 2007). Finally, there is 'public diplomacy': "…the way in which both government and private individuals and groups influence directly or indirectly those public attitudes and opinions which bear directly on another government's foreign policy decisions" (Gilboa 2008: 57). The availability of world-wide fast transport, telephony, and the internet adds to these changes, as they do to such activities as spying out the negotiating position of the party(ies).

It is also useful to keep in mind that diplomats may be the tool that states employ to engage in so-called 'soft power' that is, activities designed to highlight the cultural and other achievements of the diplomat's state, as well as to open third channel access to those in power (or out of it, but hopeful), to provide bait for implanting positive views of the diplomat's state, to hint at the degree of power the state is willing and able to bring to bear, and to deepen economic opportunities in the host country by providing examples of the visiting state's competence, industry, etc.

How does this play in the context of communicating with ETI? We need to recognize that only *some* of these forms are applicable to ETI. We cannot (or at least, are unlikely) to send a diplomat to appear personally in the political centers of an ETI civilization.[6] Of the four communicative dimensions proposed above, we can note several items that could be important for diplomacy of the sort humans are familiar with.

Given the nature of the media involved, it is likely that much of the diplomacy would need a mix of mediaranging from the verbal (written or auditory, which in turn may depend on the communicative forms of the ETI in question) to the visual, auditory, and even tactile (assuming a handshake can be ensured between human programming and ETI 3D printing).

We can specify some basic attainments we ought to expect of a human diplomat to ETI *in addition* to the normal expected of a diplomat (some legal and political background, chicanery, keeping a straight face, presentation of self, etc., but not the ability to consume large quantities of liquor, which is unlikely to come into play)

- Familiarity with and ability in IT/CMC technologies to the level of expert;
- An ability to be versatile and facile in adapting to the performative demands of the host culture, in other words, an actor able to observe and replicate the proxemic behavior of another race: no small task when we consider potential physiological differences;
- An ability to interpret the legal and other meaning(s) of texts and presentations made by non-humans, into human, or at least humanly-comprehensible idioms.

Whether any human diplomat would be able to play all of these roles still needs to be assessed. Certainly, at some stage, we might be faced with the necessity of developing such experts.

11.4.2 Diplomacy and the State: "Maintaining a State of Tension Short of War"

Keith Laumer, the *Retief* series

If we accept that any ETI we come into contact with are likely to be technically in advance of us, their dispersion through space greater than ours, and diplomatically more experienced than Earth in inter-species contact, we need to ask "How

[6]That having been said, it is conceivable that a sufficiently sophisticated AI or near-AI, modeled carefully on a human diplomat, could, in theory, be transmitted at light speed to a receptive audience somewhere in interstellar space. This provides some interesting problems, but really does not change the paradigm much.

can we handle ETI?" We might also ask whether there is room in this matrix for professional diplomats? After all, it is unlikely that Earth (or any faction of its population) would be able to send a human diplomat to an ETI entity. ETI *might* have better inter-stellar propulsion, or the biology to accept such a posting, but that too is moot.

There are two potential diplomatic roles for diplomats per se. One is to negotiate intra-human disputes about the uses of information retrieved from an ETI culture/civilization. The other is to help in developing responses (though as we shall see subsequently, a unified human response is neither likely nor may it be desirable).

More importantly however, is whether humanity could negotiate a relationship with an ETI polity of whatever form, given Earth and humanity's weaknesses vis-a-vis ETI. This is a problem that we humans still face. The current world political picture is of world-ranging powers (the US, Russia, and China hurrying to catch up), regionally dominant powers (Brazil, India, Nigeria, South Africa, for example) and a large number of small, internationally minor, and relatively powerless states: Barbuda, Antigua, Timor Leste, Tonga, etc. Nonetheless, as Cooper and Shaw (2009) point out, the smaller states do have and impact, and if they are able to leverage such advantages they do possess, they can fight, and even succeed, in gaining their diplomatic objectives "What small states lack in structural clout, they can make up through creative agency" (Cooper and Shaw 2009: 2). In other words, fight smart. If anywhere, this where the practices of human diplomacy *could* bear fruit, with the experience of chicanery, legalism, and formalization, delay, and other tactics that are the hallmark of human diplomacy over the ages. Using those, even smaller states can survive and even prosper in a diplomatic arena (Baldacchino 2009).

11.4.3 The Value of Civilizational Ambassadors

Civilizations are not states, and thus are in no position to send ambassadors in the formal political sense. However, we can conceive of individuals, and indeed, entire groups, whose purpose is to learn or to teach a civilization's standards, behavior and texts. There have been numerous examples of civilization ambassadors throughout human history. To some degree, the missionaries—of Christianity, Islam, and Buddhism—that spread out to convert the 'heathen' could be seen as ambassadors of a civilization, during the period when distant travel was difficult and time consuming.

So the possibility of cognoscenti from an interstellar civilization offering to teach members of another civilization about their own, should not be discounted. If, indeed we are dealing with a *civilization* then the use of diplomats—representing an Earth civilization as sort of embryonic *junzi*—might be an appealing option (Finn 2003). Preferably, of course, we would develop some *junzi* of our own. This however, presupposes that we are able to develop a true world civilization, one in which there are standards of behavior that are a cynosure, and where we can develop cognoscenti and teachers of those humanly universal standards.

11.5 Exchanges, Trade, and Credit

At this point it becomes necessary to discuss an issue dear to all our hearts: money. The following discussion is rather abstruse, and not all readers will enjoy following it. Moreover, the position adopted is not that of an economist (who might disagree with the basic premises), but of an anthropologist who believes (following Bourdieu 1986 and Firth 1977), that capitalist economics are merely a special case of social economy.

If there is one thing we do know about economic contact and trade from our own history, it is that very often one of the parties eats, the other gets eaten. Human experience of economic contact has in most cases been that where parties of very unequal powers meet, one tends to overwhelm the other. Trade is an exchange based on an assumption that each side strives to maximize its benefits from the exchange. A subliminal, but incorrect message is also often transmitted: that trade allows each party to withdraw from the exchange when it wills to. This may not necessarily be the case, as the history of colonialism on Earth makes clear (Reinhard and Sturge 2011). And, exchange depends very strongly on information. As a rough rule of thumb, the party with less information and lesser ability to disengage is going to get the worse of the exchange. Exchanging a worthless plot of land for rare and valuable glass beads might not have seemed such a good deal *if* the land owners had known to what use the land would ultimately be put and if they had understood the other party's concept of 'ownership'.

11.5.1 The Benefits of Trade and the Issues of Credit

First, a quick recap. Economic activity consists of three elements (Forde 1963; Polanyi 1968). *Production* is the creation of usable goods. *Consumption* is the use of such goods. *Exchange* mediates between the two when more than one producer/consumer are present. "Goods" from this point of view are not necessarily material goods. All of us would agree (particularly when required to shell out a chunk of our meagre salary to see a movie) that a performance has an economic aspect. In effect, the consumer exchanges a physical good for a non-material one. Well, what about paying by credit-card? One could argue (correctly, the bank would say) that the physical money is not necessary, it is merely a counter in the game, and anything, even a slip of paper (or an electronic code in a complex computer such as a mobile phone), would be adequate as counters. This is where exchange, and, more importantly for our discussion, credit, come in. Credit relies, in effect, on a series of trust building activities and institutions which symbolize and create a certain reality for those engaged in it. The minute the players cease believing in the credit, 'poof' it disappears like a soap bubble. Don't believe me? ask anyone who lived through the Great Depression, or ask anyone about the Wall Street crashes, 1980, 1984, 2009 and 2011/2012 economic crises, or 2016 Brexit.

Now, how does one establish credit (never mind funding for SETI) for an individual or corporate body twenty light years away, and, no less importantly, forty years away in time? After all, one central premise in SETI is that major benefits are likely to accrue to the human species from contact with ETI. If that is the case, can the ETI call on that credit? In view of the fact that messages will arrive at rare intervals, we should expect two issues. First, each sender will try to include as much information as possible, so that more trade options will be on the table. Second, due to the long time between exchanges, purchases of means of production, including bank deposits and stocks, have the potential for destabilizing local economies, notably if the locals engage in compound interest (Adams 1984).

To sharpen the issue, I make a distinction between two sorts of credit. Physical credit is the obligation of actor A to repay, under stated conditions, physical goods that have been provided by actor B (it is immaterial if some sort of exchange, i.e. goods to money to service takes place as well or not). *Social credit* (Cohen 1977) is a related, though different kettle of fish. Social credit is the *estimate* by one actor, of the ability and willingness of another actor to fulfill the first actor's expectations. I loan you money, and have thus, according to a banker, provided you with credit. More significantly, however, even before loaning you the money, I have assigned you a 'credit rating': the degree of certainty I have that you will return whatever is loaned to you. Social credit therefore *underlies* physical credit. It depends on *information*, and that is a composite. Roughly speaking, the more I am able to empathize with you, the more I am able to predict, or am prepared to take a risk, in granting you physical credit, the more likely I am to grant you physical credit. Moreover, and this is one of Cohen's main points, the more relations we have in common (that is, the more homologies there are between our social networks) the more I am able to trust you, since I have varied means of access to you, both positive and negative. The fewer the homologies (that is, points on my network that are also points on yours) the less we are able to trust one another, and must rely on past history with one another and/or on nebulous, possibly unreliable 'public' information.

All this, of course has bearing on interstellar economics. Basically, if we are to establish an economic relationship over time (and all interstellar exchanges are over time) we need *also* to establish trust, which means a *social* relationship: the more we know about one another, warts and all, the greater the trust (which may not need to be global: "I trust this person so far …').

11.5.2 What to Do on Arcturus When Your Credit Card Maxes Out? Generating Trust in Interstellar Relations

All exchange involves trust. An infant receiving a bottle from a parent *trusts* the bottle is full of milk, not acid. When I pay my money down at a shop, I *trust* the shopkeeper to acknowledge I have paid, just as the shopkeeper trusts me to pay once goods have been handed over. My banker trusts I will put money into the account on a regular basis. Trust, as we can see, springs from role expectations (see Chap. 7): as I emit behaviors, I am reinforced by *alter* whose emissions, in turn, are reinforced by my behaviors. Trust is merely a special case of this: habituation to a set of expectations. All of these role behaviors, however, no matter how short in duration (e.g., the offer of a milk bottle) have some initiation: a period in which the trust has not been established. And most are mediated, to some degree, by tangible rewards: milk, purchases, cash. Those that do *not* involve material exchanges are, in effect, 'modeled' on those that do: in other words, we are 'habituated' to behave with trust. And, because this is *behaviorally* conditioned, the degree of trust varies with experience, role, frequency of successful (or unsuccessful) exchanges, tempo (number of trust exchanges in a period), contingency (how important the exchanged things are to us) among other factors.

But how do you generate trust with someone who does not share many features of your emissions, nor have the same responses as you do? We do have historical precedent. The Phoenicians, Herodotus (iv: 198) tells us, when approaching an unknown shore, would need to generate trust with new clients. They would land, place a row of goods (presumably of low value?) on the beach, and retire. The natives would approach, examine the goods, and place items they considered fair exchange in a parallel row, then retire in turn. The Phoenicians would approach the two rows of goods, and if the local offering was good enough, would leave their own goods, and retire with the local offerings. The natives would then emerge and take their trade goods. Presumably (though Herodotus does not say so), cheating was discouraged by arrow shots from either side: each side could protect their goods, but not attack the other party.

From our perspective, Phoenician trading is both a good way of opening negotiations, and represents a problem. Information, once disclosed, is a flown crow: we cannot retrieve it. Physical goods are unlikely, give the structure of the universe as we now know it, to be exchanged (we're not patient enough to wait several thousand or hundreds of thousands of years to examine a pack of ETI granola and decide we don't want it). Crucially, however, the issue is *not* the items of exchange, but the non-material development of trust, which will allow us to exchange the *really* important things, such as valuable information.

How does one develop trust in another, notably when communication between them takes years or centuries? This is likely to be one of the major hurdles humanity —or any other ILF—will need to overcome. First, as a rule, no information, no trust. If I have had no experience, nor reliable knowledge of you, how could I possibly trust you enough to offer you credit? True, an advanced culture, one with lengthy experience stretching over millennia, of dealing with less developed cultures, may well have an acute insight into all the possibilities, and may thus have the equivalent of actuarial tables: "This culture is a biological omnivore, warlike and politically inclined, …. We therefore assign them a trust index of X, and a credit rating of BBB." Or something of the sort. Alternatively, and certainly for humans, a lengthy period of familiarity and exchanges will be needed before most people (and economic institutions) will trust, and provide credit. Of course, as in any economic situation there will be bulls and bears. Some people will be willing to take a gamble on credit in the hope of great gain, others will hold on to see how the bulls fare, before jumping in.

Crucially, the most important item in developing trust will be verifiable records of how culture X has dealt with other cultures over a period of time: whether members of other cultures have been given a fair shake formally in the practices of diplomacy, in the courts or their equivalents, in the formulation of formal agreements (AKA contracts and treaties). There is unlikely to be a 'trusted intermediary' in the first instance. Hard, varied, and holographic data would be the strongest 'trust' currency available, for, as we know, forty or a hundred or a thousand years away, there is no recourse if one's correspondent happens to be a swindler, much less so if the 'swindle' is the result of a cultural misapprehension on our part. The exchange of such information: internally consistent, from varied sources is absolutely vital if we are to reach a situation of trust, and thus exchange of items (artifacts, mentifacts, sociofacts as the case may be) of real value. Even so, of course, there are no guarantees. The nature of the universe makes the term *caveat emptor* a physical reality for which humans must be prepared.

11.5.3 Economic Modes Across Time and Space and Their Social Consequences

Several alternative modes of exchange in interstellar medium may be posited. On the one hand we may have a *tangential* economic system. A tangential economic system is one in which each economy benefits from the other's communicative output without regard to the other. Thus, if an ILF is broadcasting information *for its own purposes* (for example, we broadcast a lot of science programs), and another ILF benefits, there is no direct reciprocity, even though exchange is taking place. Their economies are thus tangential.

Alternatively, where one or both of the societies may well broadcast for the specific benefit of the other. In which case, it is conceivable some reciprocity will be expected. In such an *exchange* economic system, the issues of social and physical credit become paramount. If the economy of an interstellar culture is affected on a regular basis by the economy of another, if regular and continuing economic exchanges take place, then the economies are exchanging.

A further model is also possible. Many human economies have evolved into a what economic anthropologists call 'peasant economies' (Gamst 1974; Kahn 1978; Scott 1976). A peasant economy is one where a society, composed of subsistence farmers, is forced to participate in unequal terms in a metropolitan economy. In order to do so (whether because it is taxed in cash rather than kind, or because there are certain resources only available from the metropolitan economy for cash, e.g. health care), the peasant economy, in addition to subsistence crops, must also raise cash crops to sell on the market. This sort of exchange is highly unequal, since the peasant's cash crop is generally of low added value.

If the ideas of expanding ETI civilizations, and of their technological superiority over humans (or some other ILFs) are correct, it is likely that the interstellar economy will include 'peasant' economies. These will be economies in which large metropolitan cultures will enjoy good communications, regular and lengthy exchanges of credit, high technology and great wealth. Associated with them will be a mass of less technologically capable societies. Without actually engaging in predatory behavior, a metropolitan ETI culture will exploit whatever products are offered by isolated, perhaps non-interstellar cultures. Those will suffer from a great demand for metropolitan products (and, if we include *Homo sapiens*, styles), periodic economic slumps (as their products' values shift in metropolitan economies), poor communications, and lower technological and scientific benefits. It should be added that even if some of the interstellar civilizations are 'benevolent' in the sense used before, this is not likely to change the situation to any great degree on the economic front, unless the standards of the civilization include a principle of care for the less able or fortunate. The experience of development of nations on Earth demonstrates the difficulty of overcoming peasant problems (Agnew 1982; Barlett and Brown 1991; Landes 1998). If we are indeed lower on the technological and expansion scale, then we might also end up being the peasants of the galaxy: one small culture among many, overshadowed by larger cultures and the rules of economics.

11.6 Have Your Lawyer Call My Lawyer: Legal Niceties and Otherwise

And with trade, inevitably almost, come lawyers. That is a shortcut to the idea that regular smooth exchange requires an agreed upon framework within which to operate, and that exchanges can lead to confusion, and accidental or deliberate misunderstanding. As a result, the presence of a mediator or arbitrator (for the difference between the two roles, see Bailey 1969) is necessary, both to establish the specific case, and to establish precedents for further, similar cases. While amateur and occasional mediators work well in certain societies, usually the job of elders, in an inter and intra-cultural system, professionals may be best; hence lawyers.

And for lawyers, law of course is an essential. There is some traction to the idea we discussed in Chap. 7, of some form of universal trust in the negative Golden Rule: do not do unto others that which is hateful for yourself. Whether that would be elaborated as a metalaw for inter-cultural relations, is moot. Crucially, in the absence of some form of metalaw, precedent may be the only thing to go on, in which case, to cite Michaud (1979) contact is much more likely to demonstrate the politics of inequality, and the outcome of legal contestation may well depend on the ethos of the more powerful species and on its conceptions of the value of intelligence in the universe.

Law of the powerful requires that advocates for the less powerful would need to be knowledgeable in the law of the powerful. This is *not* equivalent to being at home as a member of a civilization, though there are likely to be some contiguities. But law is the province of communities, whether at the small community (AKA village, hamlet, gang, space-station) level as appropriate, or at the broader community level (state, nation, or other large political structure). The complicates the entire issue, notably, as with nation states on Earth, being a member of the polity concerned is a prerequisite for practicing law.

11.6.1 The Question of Personhood

Every society has some formal or informal definition of who, or what, a person is. We all make a rather clear distinction, at the very least conceptually, often semantically and verbally, between "people" and "objects" even though the boundary between the two varies from one culture to another. Persons are autonomous, within limits, and they have the quality of actions for which they are held, or may be held, responsible. This in turn implies that not even all members of the human species are "persons", at least in the legal sense. This is true of every society we know of. Children, and sometimes women, the mentally infirm, are not, or may not be, considered full "persons" for the purpose of the law. On the one hand special restrictions may exist on their movement (in the US in many places, there is a curfew imposed on the category 'minor' which forbids their movement after a certain hour of the night. In Saudi Arabia similar restrictions are placed on women). On the other hand, they may not be liable for all of their actions (thus, again, children in the US for whom a guardian is responsible, and women in Saudi Arabia, who require the permission of a male *wali* for all public activities) The rights of individuals under law are a direct result of their status as persons.

This works the other way as well. In most modern societies there are fictitious persons—corporate bodies such as limited firms, associations, and the government itself—which are considered persons for the practice of law. This is a legal nicety evolved from Roman law, but adopted slowly by most participants in the emerging global civilization as the result of pressure by these corporations. This is not, however, the only way to go. In traditional Chinese legal thinking, people were responsible for their actions. Period. Some statuses had preferential treatment.[7] Most did not, but personal responsibility remained even where collective action was the issue.

Another aspect of personhood is a legal question as well. Jurisprudence in the UK recognizes a concept under law called 'the reasonable man', which we have mentioned before (Epstein 1973). This is a legal construct, particularly important

[7]Those who passed the imperial examination, whether they acceded to government office or not, were excused judicial torture for many less serious crimes.

when judging people's actions and liabilities deriving from them, for instance in cases for damages, which purports to establish a standard of reasonable behavior. Did the person under discussion do whatever a reasonable person would have done under the circumstances? The assumption underlying this statement is that an individual cannot be required to do more—fence his property, curb his dog—than any other reasonable person can be expected to do. As Gluckman has shown (Gluckman 1972), the concept of the reasonable man exists in non-British jurisprudence systems as well. Thus we can amend the idea of the reasonable man to say "reasonable behavior in the society concerned." Now, in the case of ETI, we have two problems stemming from this concept. First, the nature of reasonable must be extended to non-humans as well *and sometimes behaviors forbidden in one society, may be absolutely, perhaps even physiologically required, of members of another species*. Second, the reasonable man is a *legal* defense in many cases. How then do we identify a "reasonable" ETI? And the reverse is even more important. If, as Michaud (1979) argues, interstellar legal relations are decided by the powerful, and if *we* humans, are not the powerful, the definition of reasonable man will require us to adapt to behaviors expected of a reasonable man under other laws. So if the basic thrust of morality in a culture we have come into contact with is *caveat emptor*, we are unlikely to find any legal relief when we complain that the shipment of brown gobblesmacks we ordered are rotten: under the law, it is our responsibility to behave reasonably and check we will not be cheated. Which we can't do, because the transaction took place forty light years away.

Moreover, if, as is possible, collectives of any sort are *not* recognized as legal persons, *our* entire econo-legal system could be under challenge. The existence in modern global emerging civilization of legal entities is *so* important, that some such legal entities—corporations—are more powerful than many states, are able to evade those laws aimed at natural persons, and underlie our entire technology and material existence. Would such corporations be able to function the way they do in an inter-ILF environment, if the 'natural law' there was that only real biological individuals were 'persons' and responsible for all actions undertaken by and for them? The directors of GE, Monsanto, Panasonic and other corporations shudder at the thought.

11.6.2 Strangers

The history of humanity is replete with the examples of the stranger, and strangers vary among several axes, the most important of which are *power* in relation to 'us', *propinquity* in relation to 'us', *numbers* in Relation to 'us', and *strangeness* in relation to 'us'. Each of these axes will be discussed briefly before looking at them as a whole.

The issue has received some attention in sociology, particularly since the seminal work of Simmel (1950) and Wood (1934), but it is far from exhausted. All human societies that we know of have some way of dealing with strangers. These ways, though particularly human, are universal in their possibilities. In order, however to understand how reactions to strangers occur, and what their implications are, we must understand what a stranger is. There are, of course, a number of different possibilities, some combinations of them in human society are worth exploring.

The question of how powerful a stranger is, technologically, politically, or economically shapes human lives, particularly in the modern era. Historically powerful strangers have made serious impacts on human societies. One prominent example is the arrival of Cortez in Mexico and Pizarro in Peru. The Conquistadores conquered as they did because they were militarily superior in many ways to those they attacked. However their initial success was at least partly due to the awe they brought upon, both because they fit into one or another of the mythologies of the people they conquered, and because of their strange appearance and technical prowess. The question of the fall of the Aztec "empire" has been debated at length. What is significant here is the strangeness that the Aztecs and their neighboring Mesoamerican nations reacted to, which, initially at least, lent the Spaniards an aura of sacrality which the Aztec nihilistic view in particular was unable to resist (Fagan 1984).

The fewer encounters one has with strangers, the more strange and possibly more mysterious they may be considered. In some Japanese communities, strangers, both because of their supposed power, and because of their rarity in rural Japan, were deified (perhaps a "just-in-case they really can harm us" view of the locals. Araki 1978; Sakurada 1980). Throughout pre-modern Japan, with some relicts existing to today, 'traveling deities' have been an important signifier of possible rewards and punishments (Ouwehand 1964).

We can argue—from the perspective of those who may well be on the weaker end of the stick—that the moral position of a society is measurable in its ability to relate to strangers. It is not that eating strangers or making love to them makes one morally superior or inferior. It is that they are morally different, and *that we are entitled to prefer one over the other*. The closer the stranger is to our ideal of 'human' in her presentation and demeanor, the more we are inclined to treat that stranger as similar to ourselves, and thus spread the protection of our practices over that person. Wrongly or rightly, we are more likely to accept LGM (Little Green Men) over BEM (Bug Eyed Monsters). Even if we accept Watson's claim that persons are conscious selves who can "envisage a future for themselves and that have desires about those future states of themselves" (Watson 1979) this does not necessarily mean that a stranger will be acceptable to all. However, the definition of what a person is, crucial to any system of human law. This is no small matter. Deprive an individual of personhood, and that individual is an object, manipulable, but not responsible.

Would therefore we/they be defined as strangers, with all the baggage that could imply? Or would we/they have automatic 'personhood' in the host society? Humanity's record in that regard is not a brilliant one, and it is certainly possible that the legal status of humans in an ETI society will be that of 'stranger' rather than that of person. Given our own penchant for depersonalizing others—as Indians (both eastern and western), Jews, Rom, Lapps and many others can attest—we may have great difficulty in dealing with ETI individuals as persons in all senses of the word, whether or not we and they are in physical contact.

In summary, the issue of law is complicated partly by our human conception of it, partly by the question of the universality or otherwise of morality, and what emerges from that, and partially by the realities of power and of distance. The issue of law is deeply troubling and the claims of such as Fasan (1970) that universal law is possible and likely will hold water if, and only if, the cultures we come into contact with follow the same logic as Fasan and others such as Watson (1979) who believe that benevolence springs from the universal brow, which we have no *valid* reason to suppose is true. If we do need lawyers, they must have an understanding that law in different places reflects different realities and environments. And, if we do need then, we shall need a lot of them, since we shall be playing an away game in a strange land.

Contrariwise, we would do well to ensure that we have a definition of 'personhood' that is immutable throughout the human world. Otherwise we might end up misidentifying some 'traveling deity'—a powerful ETI (or one with a powerful relative) that has wandered innocently into our ken—as a non-person and have to bear the penalty for mistreating a stranger.

11.7 Trading in Concepts: Saving of Bodies and Saving of Souls

Beyond trade, something intended to benefit both parties, human beings are also altruistic. Examples of that include the selfless devotion of Inquisitions everywhere trying to foisting their creed on the pagan and heretic, and the selflessness of people from Bakunin through Trotsky, Guevara, and Trump in preaching and practicing their political commitments through all odds. These are examples of attempts to demonstrate the manifest truth of certain concepts and ideas to people as-yet unconvinced of their necessity. Note that this is a given, almost universal quality of at least one ILF we do know quite a lot about: *Homo sapiens*. Note too, that questions of relative *expenditure* are not raised. Followers of the Church of Jesus Christ of the Latter Day Saints devote ten percent of their income to *their* convictions, followers of the Al Quaida and ISIS brands of Islam risk their lives for theirs.

There is no reason to suppose that things *must* be different for ETI. It is also, therefore possible, that one coin of exchange will be 'intrusive ideas'. Here we enter into a very murky realm indeed, having two rather unpleasant sides. Will we be willing to accept their preaching, and if so what will be the consequences? Will we (or some of us) be insistent that we preach at them, either in exchange, or as an act of piety, and what will be the result of that? Consider an ETI that practices cannibalism (that is, eating members of its own species) and burning selected offspring. Suppose, moreover, that they consider this highly moral behavior, *and will not deal with other ILFs on any other basis*. How do we react? Is this a case of contact refusal, and are we justified in working our way around it?

Parenthetically, we need to recognize that an electromagnetic message, even if it is no more than a Sunday school tract, will provide us with information about the bases of the communicating culture. Keep in mind that religions evolve from a particular social matrix, traces, at least, of which are retained in the culture concerned. Humans are *still* arguing about the relative places of males and females in society, and the seeds of inequality are inherent in the ways in which religions present themselves: "Our father who art in heaven" not 'our mother'. So it is worthwhile listening to intrusive ideas for at least two reasons. First, such transmissions include truths about the transmitting culture. Second, it does tell us something about *at least one* view among our opposite numbers. It also implies, at least, that the transmitting culture is likely to make efforts to clarify even those concepts that are absolutely basic, and absolutely clear (to them, at least). This means that a great deal of information, non-religious information, is going to be piggybacked on our Sunday school tract.

This returns us to electromagnetic contact: relatively cheap, the most common trade item born by it would be ideas.[8] Before discussing those ideas, some of the qualities of electromagnetic contact generally not considered by engineers and scientists must be dealt with. The most important one is of course that it *radiates*. As a general principle, *radiation contact is not exclusive*. One of the ideas implicit in trade is that each party *excludes* other parties from the transaction: if it were free, why should I repay? How can we 'trade' (excluding assumptions of benevolence on the other side) without encoding our message? We are forced to the conclusion that 'trade' (as your average capitalist or bureaucrat understands the term) is not possible unless one of two conditions obtain: interstellar economy is of a radically different kind than the exchange system common on Earth, and therefore credit is unnecessary, or there are only two trading partners in contact reach (or, a directed focused energetic source is used such as a laser). The first may very well be the case. Free market capitalist economics are a special case of general economic exchange on Earth. It should not surprise us (though it may disappoint many a free market partisan) that free market philosophy is not the dominant one elsewhere. The second too, is possible (notably the use of lasers by ETI, cf. Kingsley 2001), and then we must find ways to establish our credit rating with the other side. Here too, free-market rules hardly obtain if there are only two partners to the exchange.

To return now to the question of ideas. One common human meta-idea is that ideas have a power, a truth value, that goes beyond the need for empirical truth. This belief is so universal among humans that one is tempted to see it as a basic, perhaps genetic component of the human mind. The expressions of this meta-idea lie in the realm of morals and belief. Briefly stated, some statements, if they have an

[8]There is always the possibility of using some type of 3D printing by the receiving party. Notwithstanding, the actual "material" transmitted would be information for the printer, not the 3D object obtained.

apparent truth, are considered 'true' even if there is no empirical evidence for the proposition. Religious beliefs are one form of such ideas. Political ideologies another. Moral philosophies a third. The precise details are of no great concern here. What is of concern is an idea that is a corollary of meta-beliefs: that exposure to meta-beliefs makes them more 'true'. That is, the more people who exhibit such ideas, the more benefit is likely to accrue to all holders of the idea, and, the more 'truth' inherent in that idea.[9] The behavioral pattern associated with this corollary is proselytizing activity. Consequently we can assume that at least one effect of contact will be the direction of ideological initiatives by the human species (or parts of it) towards any ETI contacted. I have argued elsewhere (Ashkenazi 1991) that the nature of the message (religious, in the reference cited) and of the initiating and receiving religion are important for understanding their effects. We have no idea whether or not ETI will have religion at all. If we accept the convergence argument at its most extreme form (Bylinsky 1981; Raup 1985) then ETI are likely to have religion simply because it is a concomitant of the kinds of intelligence that are likely to embark on SETI. In contrast, Dawkin's (2009) claim that religious belief is genetic may have validity, in which case, ETI are unlikely to have religion since their genetic makeup would be different from our own.

Among other things, proselytizing attempts are cheap for the recipient. That is, the sender generally does not expect a return on his investment, at least not in the short run. In the long run, both churches and political/ideologicalparties often expect some return (tithe, party membership dues, power of increased membership) but that has less to do with their nature as ideological organizations than with the nature of organizations in general. So it is possible, *if* our ETI transmitters have something analogous to religion, and *if* that religion is not exclusivist and non-proselytizing (such as e.g. Hinduism, Shintô, and Judaism among humans) then we shall be faced with messages that might (even if we do not know that) be less factual than statements of supposed meta-truths.

The same is true in reverse. The broadcasts of televangelists are radiating through the universe at C. Faint as they are, a sufficiently sophisticated technology might be able to intercept and view them.[10] What will they make of the ranting of a televangelist? Is this a performance? A part of a debate? A meta-truth? A factual statement? In many ways such a broadcast is all of these and more, with different styles and details dominating different televangelists—American, African, Indian, and so on. And how are we/they know to which realm of discourse this particular message is to be assigned to?

---

[9]This concept, which is easily demonstrable in any public worship ritual, is a development of Guthrie's statement that "participating publicly in a ritual is equivalent to acquiescence to the theological, moral, and cosmic beliefs behind the ritual" (Guthrie 1980).

[10]That they are technologically more advanced than us is a basic SETI assumption, so rasterizing television signals should not provide an insurmountable technical challenge.

Missionary and evangelical activities of any stripe represent the truism that not all exchanges are market exchanges. This does not mean they do not benefit the transmitter, only that there is no direct exchange between the partners. Moreover we should be aware of the fact that not all transmissions, or exchanges of information, are based on what we might call 'truth' as an absolute.

11.8 Selfish and Altruistic Exchanges

The question of non-factual information raises a related question: what could we consider to be 'altruistic' trades of information?

11.8.1 Reciprocity and Non-reciprocity

Our universe appears to be constructed on the TANSTAAFL principle.[11] So, in effect, are most societies. In broader terms, it is a human social given that activities, goods, etc., must be reciprocated (Mauss 1966). That is, if I want an ice cream, I must reciprocate by paying some other resource. If I am disruptive, I may expect negative sanctions. If I love, I expect to be loved in return. Does this principle extend to ETI? On first glance, the answer should be 'yes'. The universe, it seems, operates under the same laws everywhere. Energy or matter cannot be created from nothing. Use implies deduction from the whole. The generalization to social relations seems obvious, but is it? Middle Easterners, from Morocco through Afghanistan, as well as other cultures elsewhere, share a trait of negative reciprocity: harm me or mine, and I will make it a priority to return the harm. The reverse is also true. The Ik, described by Turnbull, whose society had been utterly disrupted, had no concept of reciprocity: utter selfishness was the rule (Turnbull 1972). We must ask, therefore, to what degree, and in what form, does the idea of reciprocity exist in an ETI society, *if only because, in many ways, it is extremely prominent in ours.*

We need to take account of two points: range and scope of reciprocity.

- The *range* of reciprocity basically asks are reciprocal acts directed *only* at the one to whom ego is returning a favor/injury, or does it extend to e.g. members of the original donors group? Species? Kin? It also asks whether reciprocity is general (you did me a favor, I do a different one for you in return) or specific (you gave me an x, I return an x)
- The *scope* of reciprocity asks what things are reciprocated? Only money? Money and services? Goods? And under what circumstances?

---

[11]There Ain't No Such Thing As A Free Lunch. I gratefully acknowledge Larry Niven as the originator of this term.

Two human cases in point. Many Japanese keep a 'gift book': a notebook in which they record specific sums they receive from social others (friends, co-workers, neighbors) on formal occasions such as marriages, births, funerals etc. These sums, if one wants to continue the relationship, are expected to be returned, and failure is either a social solecism or an outright declaration of disdain (Befu 1974). In contrast, Americans often provide favors for significant others, in the expectation of having a favor—not necessarily similar—demanded in return (Cropanzano and Mitchell 2005; McGraw and Tetlock 2005). Virtually all other societies have some variant of those two options.

The proposition that each side in an exchange (or communication) will seek to maximize the benefits gained is an extremely powerful one. On the surface, at least, it seems intrinsically human. This seems to be the case because our own experience in the physical world tells us that nothing appears to be free. On examination, such a statement has two cases. The weak statement has it that an ETI in communication with us (or, by extension, another ILF) will require a commercial or commercializable return. In other words, what the weak form of the statement implies is that communication between ILFs will be at the level of "direct exchange" (Polanyi 1968; Sahlins 1972). The strong version of the same statement suggests that a basis for all ILF societies will be an expectation of individual or species benefit, even if such benefit may not accrue to the individual (or even the species) in question, otherwise 'redistributive' or 'delayed' exchange. Redistributive exchange assumes that some central office (in the sociological, not organizational sense) will accrue all benefits and redistribute them in some way to all members. Michaud's (1998) elegant solution to the 'final problem' (the heat death or collapse of the universe) is a redistributive exchange one: from according to his ability, to each an equal benefit. A sort of cosmic *kibbutz*. Reciprocal exchange assumes that exchanges go from status to status: a parent provides resources to a child, so that the child, in turn, will provide resources to its own child, when the time comes.

11.8.2 Other Forms of Exchange: Enter the Kula Ring

We tend to think of economics as a relatively straightforward form of exchange: A has something B wants and vice versa, they exchange to mutual benefit. However, this may well be a very limited viewpoint. We have noted that exchanges among humans typically follow three rather different identifiable patterns all of which are present *to some degree* in all societies. *Market* exchange is what the capitalist (and communist as well) regard as economic exchange: A exchanges some object X with B who provides object Y. *Delayed* exchange is where A supplies some good to B, who then supplies that good to C, in whom A has an interest. Doesn't exist in your hyper-capitalist environment? But almost everyone raises children with little thought of return, largely so that *they* will raise one's grandchildren. Finally, there is *redistributive* exchange: A receives (or, more often, demands) a good from X, X_1, X_2, … X_n, and then redistributes the good in some fashion to X_1, Y_2, Y_3, X_n.

Sounds familiar? Of course it is, the IRS does that every year. These are straightforward exchanges. But can such forms of exchange emerge in interstellar commerce, where market and redistributive exchanges are highly unlikely, material exchanges almost impossible, and even intellectual exchanges (which we discuss in the next chapter) would be problematic.

Humans, surprisingly enough, do have experience in lengthy delayed exchanges of *intellectual* and not just material value. The *kula* ring is a persistent form of exchange, still practiced today among islands in Melanesia (Malinowski 1961, 1968; Leach and Leach 1983). In the *kula* ring, shell ornaments of specific kinds circulate clockwise or counterclockwise, according to type, through a chain of islands called the Trobriands, northwest of New Guinea. The exchange of *kula* objects is a vector for the exchange of consumable surpluses such as fish and garden products, pots, and adzes, as well as part of the social and political mechanisms of the participating communities, as men acquire and transact business with their peers from other domains and their own. The trade is principally in aesthetics; the foodstuffs and other consumables are given as gifts to one's hosts and are not formally accounted for. The quality of the *mwali* (the traded armbands and necklaces), and their histories of ownership are bargained for very vigorously.

It is important to understand that the *kula* ring is *not* a market of objects. It is a market of ideas. The shell necklaces and armlets are *not* intrinsically valuable. The value *is inherent in the exchanges*: the 'story' each item carries with it. If at the same time some mundane presents are exchanged as well, that is, in the understanding of the people in the *kula*, merely a side issue (a capitalist would argue this is not the case, but I prefer to accept the local's opinion as, at the very least, a valid one). Now this is precisely what we want to exchange with ETI: ideas. Paradoxically, while we can exchange ideas, we cannot (probably) exchange the material objects.

'Trade' notably in ideas may be polysemic. It may have an economic values, a ritual value, social value, and so on. What is crucial is understanding this polysemy. As we exchange objects and ideas over time, so to we are acquiring/providing other things as well. Critically, as the Lenape Indians learned, one needs to know what is behind the munificent offer of hard-to-get glass beads. Not only that, but the concepts of 'ownership' and the use to which a particular gift or exchange will be put, are important parts of a trade agreement.

11.8.3 The Zoo Hypothesis Again

This leaves us with the various permutations of the Zoo Hypothesis. Assuming multiple, extremely powerful, technologically well developed civilizations in the galaxy, *someone* would have made contact with us, deliberately or not. Deliberately for much of the same reasons we would: curiosity, scientific knowledge, new perspectives, desire to conquer the unknown. Unintentionally, for simple behavioral reasons: the civilization itself, and individual components nested within itself—

groups, societies, corporate organisms, individuals—would show signs of their presence as they go about the normal organism processes of seeking positive reinforcement. The only way such a process could be curtailed is if either there is a very strong *negative* reinforcement for showing oneself (e.g. Brin's predator hypothesis) which of course does not solve the problem, or if there is some *social* transcendence: some way of getting away from the behavioral basics of the universe.

Consider the paradox of transcendence. If (and there is no evidentiary nor logical argument against it) reinforcement motivation is indeed fundamental, then the development of a highly sophisticated technology almost automatically means indications, at least a spoor, of other civilizations in space. For a civilization to emerge that would create a Galactic Zoo, or Galactic Club, it would have to transcend the behavioral rule. A transcendent ETI civilization would not therefore be defined in terms of the degree to which it controls energy, as Kardashev would have it, but in *fundamentally altering the nature underlying biological (and possibly physical) existence.*

To summarize, the Zoo Hypothesis in its many permutations imbues a major difficulty which is not, at its basis, a technological one. Insofar as we are able to predict, behavioral processes underlie much of what simple as well as complex organisms such as civilizations do. Logically, it can be argued that this fundamental rule set extends beyond biological organisms. It is also, in many ways, destructive to other organisms (individuals, groups, cultures), *even where there is some benevolent impulse generated in a society.* For a civilization to be able to maintain a Galactic zoo or club, it would have to be able to transcend its fundamental behavioral underpinnings, something much closer to godhead than any mere technical ability at controlling the energy output of a galaxy.

11.9 The Issue of Transcendence

Many of the answers to the Fermi Paradox—notably Ball's Zoo Hypothesis—can be characterized by reformulating the question as "Whether the development of civilization—any civilization—involves a phase transition from a 'normal' civilizational state, to a transcendent one?" To answer this question, it is necessary to clarify some basic terms.

Dawkins's (2006) characterization of genes as 'selfish' is unnecessarily limiting. *All* biological systems are inherently 'selfish' since all biological systems, including their derivatives—societies and cultures—are bound ultimately, by the desire for continuity and by resource (energy, materials, and room) considerations. In the social realm, these are expressed by what is known as the March of Dimes rule (Palisi 1970): an organization that has achieved its overt objectives, will define new objectives in order to continue its existence. Assembling sufficient energy to be useful *at any scale* against entropy, requires effort and a reward for one's activities. In a peculiar way, Dawkins's argument is reflected by the behaviorist argument formulated by Skinner (1957, 1969) and others about the behavior of animals. Skinner demonstrated that providing rewards to animals (pigeons, rats, and eventually, humans as well as SETI scientists) caused changes in their behavior.

The behaviorist idea can be expanded *mutatis mutandis*, to cultures as well. Basically, cultures conceived of as complex organisms (Chang 2010) will respond to behavioral stimuli. Positive reinforcement will encourage certain kinds of behavior, such as space exploration. The idea of rewards determining behaviors, I would hypothesize, is likely to underlie the behavior of life forms such as AI as well. Simply put, unrewarding behavior will be shunned, rewarding behavior will be encouraged at all levels of complexity. It also means that in the process of expansion (if that provides sufficient positive reinforcement) the presence or needs of weaker, less technologically developed ILF civilizations will be trumped by the behavioral processes. An analogy from the processes inherent in human civilizations is useful. In the 21st century world, which for the argument at this point consists of a single civilization (even if we know it is only a nascent civilization: civilization *in potentio* as it were), there is a more-or-less accepted value of preserving rare species, and even rare simple societies. This recognized value—held very strongly by many people—is trumped solely by the fact that for most humans, rewards are intrinsic. It is not that we have anything against pandas, but that the value of cutting down their forests is so much more rewarding on an individual level, for most individuals. As a cumulative result, no forests, and therefore no pandas.

The same mechanism will hold at the galactic scale and beyond. The behaviorist mechanism is so fundamental and so simple, that only on Occam's Razor grounds I would argue that it is a universal social principle. Critically, the principle holds true *even* for cooperative organisms: their 'altruistic' behavior is rewarding in behaviorist terms over the long run. Which brings us to Kardashev, Ball and Ćirković. Kardashev's argument is essentially a technological one: as civilizations evolve *technologically* they develop ways to retain as much local energy as possible. At stage I, the energy of their planet, at stage II, of their solar system, and at stage III of their galaxy. Doing so they essentially create something of an informational and energetic 'black hole' and thus disappear from the ken of lesser civilizations. Ball's hypothesis is much simpler: that very advanced galactic civilizations keep less advanced civilizations in a kind of informational zoo or preserve, to allow them to mature on their own, something like allowing cheeses to ripen. Ćirković's argument (Ćirković 2004; Ćirković's and Bradbury 2006; Cirkovic and Vukotić 2008) is more complex, but essentially boils down to a variation of Ball's zoo hypothesis.

The zoo hypothesis is problematic in general because one underlying assumption is that cultures or civilizations are monolithic entities. They are not; they are the aggregate of many types of relationships and inter-individual transactions. For a cosmic zoo to exist, the zookeepers must all be unitary, and have developed their social sciences, or their criminology to such a degree that they can control the activities of billions (perhaps trillions) of individuals in their culture. The idea of an interstellar race or races all marching in lock step, I find a more chilling than that of a freewheeling capitalists, as I assume do many of my readers.

Kardashev's argument is similarly problematic in this context. Such civilizations would *still* be subject to the strains imposed by behavioral rewards. For some time, perhaps a long time, they would have sufficient energy for their needs, but eventually entropy, which never forgets and never forgives, would catch up with them. Behavioral reality would also force internal stresses and strains which would, ahem, burst their bubble, as local needs outweigh pan-civlization ones (unless some other mechanism intervenes).

A different perspective on Ball and Ćirković's arguments may be needed. I take both (as well as others) to refer to the possibility of a race 'transcending'. What does transcension actually mean? Specifically, for Smart (2012) and perhaps others, transcension appears to be a better knowledge of the physics of the universe, and the ability to manipulate them. Presumably Kardashev's position would be similar. Civilizations leaving the universe through black holes or by other means (Smart 2012), or merging biological and machine life (Dror and Harnad 2008; Kurzweil 2010; Schank and Birnbaum 1994) are all examples of definition of transcendence by technological prowess.

Another example can be found in religion, transcendence means something that "…is a transcendence that is supposed, as such or at its limit, to outstrip the limitations that affect the human condition and the condition of natural things and of all things other than itself" (De Nys 2009: 17). Normally, to say that x "transcends" y is to say that x is beyond y, is not immanent in the condition of y, and/ or is not accessible, given that condition." (De Nys 2009: 18). So far, a good description of *Western* ideas of transcendence. The problem is that Western categorizations tend to be either/or, with few shades of gray. In contrast, in Chinese thinking, transcendence is one step in a chain of lower and higher existence, a continuity "[T]he Chinese spoke of gradations, hierarchical continuities, chains of being. In translating *xian* as 'transcendent(s)' … it denotes one or more transformed persons, as 'transcendence' when it names their exalted status, as 'transcend' when it names what they do… *xian* … have ascended to links in the chain higher than those occupied by even the best human beings … I intend "transcendence" … in a … concrete and metaphorically grounded sense of 'going beyond,' 'surpassing in status, excellence, and/or power,' or 'ascendance over'" (Campany 2002: 5).

Where are our hypothetical transcendentals located on he continuum between a De Nys transcendental (with whom by definition, we share no points of contiguity) and a technical transcendental, with whom we possibly might have points of congruence, if only those of a baby to an adult?

Principally, is there any way we can communicate with our transcendental? If not, and if that transcendental race affects our existence in some fashion, we must make accommodations for those effects, whether we designate them as acts of nature or of god. If we can communicate with them, however remotely or with difficulty, *and* we have some mutual interests (for example, the possibility that we join them in a hundred or a million years) then we must treat them with caution and respect as due any other ETI, if we are to survive, prosper, and maintain some standards.

11.9.1 Exotic Life-Forms

To this point we have barely touched upon 'exotic' life forms, that is, the potential for truly alien life: non-oxygen-based life (Rampelotto 2010), life in different types of matter such as black holes (Lefebvre and Efremov 2000) or ball lightning (Sanduloviciu and Lozneanu 2011), with truly different sensoria (Censor 2005), or having different makeups than we are familiar with as 'life' (Schneider 1977; Molton 1978).

The crucial questions are of course whether (a) if such life exists, can it also give rise to intelligence and perhaps culture, and (b) how common could such life/intelligence be, on a basic scale of none to many? Given that the perversity of the universe extends to infinity, it would seem at least *conceivable* that exotic life as described might exist, and that intelligence could emerge from that matrix. *If* that is indeed the case, then I would suggest that, on the whole, such races would *in general* follow the rules proposed throughout this book. They would need systems of physical continuity, thus offspring (of some sort) and a means of socialization, they may develop some, all, or a transcendent set of the institutions we have discussed throughout this book, albeit both identifying these institutions and getting information about them would be much more difficult.

We must also entertain another possibility: that exotic life is truly transcendent, in the sense articulated by (De Nys 2009: 18). For example, if black holes were a form of intelligent life their existence would not allow the sending of messages outside the event horizon, and even if that were possible, living at different time scales than do biological organisms, such messages might not impact at all on familiar life consciousness. In other words, again, in the pragmatic sense described by De Nys, they would be truly 'transcendental'. This poses something of a problem: if such life is truly transcendental, it may be pervasive throughout the universe, the Galaxy, the Solar System, our planet, or ourselves, and we would not be able to communicate with them, nor they with us. The problem is not so much an issue of size or life span (think viruses, humans, and black holes) so much as a complete lack of any way of transmitting or receiving messages, combined with development within a truly alien (each its own) environment (Faulconer 2003; Gato-Rivera 2003). As Faulconer notes in relation to transcendence "How is it possible for there to be a relation to that which is, and remains, truly other?" (Faulconer 2003: 2). Whether such life forms live at the speed of quantum decay on a neutron star or consist of the accretion of processes by floral ecologies (Chang 2010), they are as inaccessible to us as we are to them.

11.10 Summary

In this chapter, we discussed various frameworks under which communication between ILF civilizations and cultures, could take place. These frameworks vary from simple contact (i.e. identifying correctly one another's existence) to the intricacies of intimate forms of communication such as law and warfare. Much of what has been said here is speculative. Even more relies on our own human experience. Yet, alien or not, our relations with other ILFs in the universe will take place within frameworks that make sense *to us*, not necessarily to 'them'. That is, we will always try, being human, to make sense of what is going *in our terms*.

ETI cultures may not have concepts such as law, war or economics. This does not mean they do not engage in practices that we would define as such. While I do not share the fear of invasion, warfare, or cultural domination that some, including Hawking and Brin express, we are nonetheless enjoined to be careful in trying to understand and preempt potentially troubling possibilities.

If there is any threat at all, I would suggest is more likely to be within a solar system, in which case the barbarians we might run into may be ourselves. As the human race emerges into space, new human cultures and technologies will also emerge. These, in turn, may become predatory. Certainly there will be conflicts. And *these* will require some universal set of laws of war, far more than any galactic agreed-upon set.

Once contact is established, the human race as a whole, with its multiple, intertwined, and quarreling social forms, as well as its nascent civilization, will be faced with a plethora of decisions. What this chapter has tried to do is point out some—by no means all—of the possible frameworks.

Civilization may be the only response to the problem. Inherent in the term civilization is the idea that there exists a set of behavioral standards and a set of rules to which people are expected to adhere. A globalized civilization on Earth is still struggling to be born, and, for all we know, may be stillborn and never come to maturity. A *Solar* civilization could also emerge, either after a long period of settlement of the Solar System, from an existing world civilization, or on its own, as result of interaction between different and multiple space settlements. In either case, room will likely be found for behaviors during times of conflict, as well as with potential 'others' that are not human.

References

Adams, D. et al. (Eds.). (1984) *Undermining rural development with cheap credit*. London: Westview Special Studies in Social Policies and Development.

Agnew, J. A. (1982). Technology transfer and theories of development: Conceptual issues in the South Asian context. *Journal of Asian and African Studies, 17*, 1–2.

Anderson, B. (2011). Facing the future enemy. *Theory Culture Society, 28*(7–8), 216–240.

Araki, H. (1978). The wandering infant-noble theme in Japanese legends and mass media. *World Anthropology Folklore in the Modern World,* 147–54.

Arbib, M. A. (1979). Minds and millenia: The psychology of interstellar communication. *Cosmic Search, 1*(3), 21–25.

Ashkenazi, M. (1991). Local Japanese responses to missionary activities. *Social Compass, 38*(2), 141–154.

Ashkenazi, M. (2012). Development is destruction, and other things you weren't told at school. In V. Franke & R. Dorff (Eds.), *Conflict management and 'whole of government': Useful tools for U.S. national security strategy?* (pp. 91–125). Carlisle, PA: Strategic Studies Institute.

Augsburger, D. W. *(1992). Conflict mediation across cultures: Pathways and patterns*. Louisville, KY: Westminster/John Knox.

Avenhaus, R., & Zartman, W. I. (Eds.). (2007). *Diplomacy games: Formal models and international negotiations*. Berlin: Springer.

Avruch, K. (1998). *Culture and conflict resolution*. Washington, DC: United States Institute for Peace.

Avruch, K. (2001). Notes toward ethnographies of conflict and violence. *Journal of Contemporary Ethnography, 30*(5), 637–648.

Bailey, F. G. (1968). Parapolitical systems. In M. J. Swartz (Ed.), *Local-level politics* (pp. 281–294). Chicago: Aldine Publishing Co.

Bailey, F. G. (1969). *Stratagems and spoils*. New York: Schoken.

Baldacchino, G. (2009). Thucydides or Kissinger? A critical review of smaller state diplomacy. In A. F. Cooper & T. M. Shaw (Eds.), *The diplomacies of small states: Between vulnerability and resilience* (pp. 21–40). New York: Palgrave/MacMillan.

Balikci, A. (1973). Conflict and society. In T. Weaver (Ed.), *To see ourselves* (pp. 370–379). Glenview, IL: Scott, Foresman and Co.

Barbrook, R. (2007). New York prophecies: The imaginary future of artificial intelligence. *Science as Culture, 16*(2), 151–167.

Barlett, P. F., & Brown, P. J. (1991). Agricultural development and the quality of life. In A. Podolefsky & P. Brown (Eds.), *Applying cultural anthropology: An introductory reader* (pp. 76–83). Mountain View, CA: Mayfield Publishing.

Barth, F. (1966). *Models of social organization*. Glasgow: The University Press.

Barth, F. (1981). *Process and form in social life*. London: Routledge & Kegan Paul.

Beattie, G. (1983). *Talk: An analysis of speech and non-verbal behaviour in conversation*. Maidenhead: Open University Press.

Befu, H. (1974). Gift-giving in a modernizing Japan. In T. S. Lebra, & W. P. Lebra (Eds.), *Japanese culture and behavior* (pp. 208–21). Honolulu, Hawaii: The University Press of Hawaii.

Bercovitch, J., & Jackson, R. (2009). *Conflict resolution in the twenty-first century: Principles, methods, and approaches*. Ann Arbor: University of Michigan Press.

Biddle, B. J. (1986). Recent development in role theory. *Annual review of sociology 1986*, 67–92.

Biddle, B. J., & Theory, R. (2013). *Expectations, identities, and behaviors*. New York: Academic Press.

Bobrow, D. G., & Brady, J. Michael. (1998). Artificial intelligence 40 years later. *Artificial Intelligence, 103*(1–2), 1–4.

Bourdieu, P. (1986). The forms of capital. In J. Richardson (Ed.), *Handbook of theory and research for the sociology of education* (pp. 241–258). New York: Greenwood Press.

Brahm, E. (2003). Conflict stages. In G. Burgess & H. Burgess (Eds.), *Beyond intractability*. Boulder: Conflict Research Consortium, University of Colorado.

Burton, J. W. (1990). *Conflict: Resolution and prevention*. New York: St. Martin's Press.

Bushell, D., Jr., & Burgess, R. L. (1969). Some basic principles of behavior. In R. L. Burgess Jr. & D. Bushell (Eds.), *Behavioral sociology* (pp. 27–48). NY: Columbia University Press.

Bylinsky, G. (1981). *Life in Darwin's universe: Evolution and the cosmos*. Garden City, New York: Doubleday.

Campany, R. F. (2002). *To live as long as heaven and earth: A translation and study of Ge Hong's traditions of divine transcendents*. Berkeley: University of California Press.

Carrigan, R. A. (2006). Do potential SETI signals need to be decontaminated? *Acta Astronautica, 58*(2), 112–117.

Carrigan, R. A. (2010). Starry messages: Searching for signatures of interstellar archaeology. arXiv:1001.5455.

Censor, D. (2005). Relativistic electrodynamics: Various postulate and ratiocination frameworks. *Progress in Electromagnetics Research, 52*, 301–320.
Chadefaux, T. (2014). Early warning signals for war in the news. *Journal of Peace Research, 51* (1), 5–18.
Chang, S.-K. (2010). Editorial: A general framework for slow intelligence systems. *International Journal of Software Engineering and Knowledge Engineering, 20*(01), 1–15.
Ćirković, M. M. (2004). Forecast for the next Eon: Applied cosmology and the long-term fate of intelligent beings. *Foundations of Physics, 34*, 239–261.
Ćirković, M. M., & Bradbury, R. J. (2006). Galactic gradients, postbiological evolution and the apparent failure of SETI. *New Astronomy, 11*(8), 628–639.
Ćirković, M. M., & Vukotic, B. (2008). Astrobiological phase transition: Towards resolution of Fermi's paradox. *Origins of Life and Evolution of Biospheres, 38*(6), 535–547.
Cohen, A. (1977). The social organization of credit in a West African Cattle Market. In W. Steffen, S. Schmidt, L. Guasti, et al. (Eds.), *Friends followers & factions* (pp. 223–242). Berkeley CA: University of California Press.
Cooper, A. F. (2007). Beyond Hollywood and the boardroom: Celebrity diplomacy. *Geographic Journal of International Affairs, 8*, 125–129.
Cooper, A. F., & Shaw, T. M. (Eds.). (2009). *The Diplomacies of small states: Between vulnerability and resilience*. New York: Palgrave/MacMillan.
Crapanzano, R., & Mitchell, M. S. (2005). Social exchange theory: An interdisciplinary review. *Journal of Management, 31*(6), 874–900.
Curtis, P., & Nichols, D. (1993). *MUDs grow up: Social virtual reality in the real world*. Palo Alto: Xerox PARC.
Dahrendorf, R. (1968). *Essays in the theory of society*. Palo Alto: Stanford University Press.
Dawkins, R. (2006). *The selfish gene*. Oxford: Oxford University Press.
Dawkins, R. (2009). *The god delusion*. New York: Random House.
De Nys, M. J. (2009). *Considering transcendence: Elements of a philosophical theology*. Bloomington: Indiana University Press.
Dick, S. J. (2010). The postbiological universe. *Acta Astronautica, 62*(8-9), 499–504.
Dror, I., & Stevan H. (2008). Offloading cognition onto cognitive technology. arXiv:0808.3569.
Epstein, A. L. (1973). The reasonable man revisited: Some problems in the anthropology of law. *Law and Society Review, 7*(4), 643–666.
Fagan, B. M. (1984). *The Aztecs*. New York: W.H. Freeman & Co.
Fasan, E. (1970). *Relations with alien intelligences*. Berlin: Berlin Verlag.
Faulconer, J. E. (Ed.). (2003). *Transcendence in philosophy and religion*. Bloomington: Indiana University Press.
Finn, H. K. (2003). The case for cultural diplomacy—Engaging foreign audiences. *Foreign Affairs 82*(6).
Firestone, R. B. (2011). The case for the younger Dryas extraterrestrial impact event: Mammoth, megafauna and Clovis extinction. http://www.escholarship.org/uc/item/8fj3d8mc. Accessed February 21, 2016.
Firth, R. (1977). Exchange on a Polynesian Island. In W. Goldschmidt (Ed.), *Exploring the ways of mankind* (pp. 265–273). New York: Holt, Rinehart, and Winston.
Forde, C. D. (1963). *Habitat, economy and society*. New York: E.P. Dutton & Co., Inc.
Forgan, D. H., & Elvis, M. (2011). Extrasolar asteroid mining as forensic evidence for extraterrestrial intelligence. *International Journal of Astrobiology, 10*(4), 307–313.
Fridman, P. A. (2011). SETI: The transmission rate of radio communication and the signal's detection. *Acta Astronautica, 69*, 777–787.
Gamst, F. C. (1974). *Peasants in complex society*. New York: Holt, Rinehart and Winston Inc.
Gardner, R., & Heider, K. G. (1968). *Gardens of war: Life and death in the New Guinea stone age*. New York: Random House.
Gato-Rivera, B. (2003). Brane worlds, the subanthropic principle and the undetectability conjecture. arXiv:physics/0308078.

Gilboa, E. (2008). Searching for a theory of public diplomacy. *Annals of the American Academy of Political and Social Science, 616*, 55–77.

Glozman, E., Barak-Corren, N., & Yaniv, I. (2015). False negotiations: The art and science of not reaching an agreement. *Journal of Conflict Resolution, 59*(4), 671–697.

Gluckman, M. (1954). *Rituals of rebellion in South-East Africa*. Manchester: Manchester University Press.

Gluckman, M. (1972). *The ideas in Barotse jurisprudence*. Manchester: Manchester University Press.

Goffman, E. (1956). The nature of deference and demeanor. *American Anthropologist, 58*, 473–503.

Goffman, E. (1959). *The presentation of self in everyday life*. New York: Anchor Books.

Goffman, E. (1967). *Interaction ritual*. New York: Anchor/Doubleday.

Goffman, E., & Best, J. (2005). *Interaction ritual: Essays in face to face behavior*. Rochester NY: Aldine Transaction.

Goldsmith, D. (1990). Who will speak for Earth? Possible structures for shaping a response to a signal detected from an extraterrestrial civilization. *Acta Astronautica, 21*(2), 149–151.

Goodall, J. (1986). *The Chimpanzees of Gombe: Patterns of behavior*. Cambridge MA: Belknap Press of Harvard University Press.

Grady, S. M., & Reality, V. (2003). *Simulating and enhancing the world with computers*. New York: Facts on File.

Grandi, F. (2013). Why do the victors kill the vanquished? Explaining political violence in post-world war II Italy. *Journal of Peace Research, 50*(5), 577–593.

Griffith, S. B. (Trans.) (1963). *Sun Tzu—The art of war*. New York: Oxford University Press.

Guthrie, S. (1980). A cognitive theory of religion. *Current Anthropology, 21*(2), 181–204.

Hall, E. T. (1959). *The silent language*. Garden City, NY: Doubleday.

Hicks, F. (1979). Flowery war in Aztec history. *American Ethnologist, 6*, 87–92.

Hoebel, E. A. (1967). *The law of primitive man*. Cambridge, Mass.: Harvard University Press.

Homans, G. (1969). Elements of behavior. In D. W. Miner & S. Greer (Eds.), *The concept of community* (pp. 5–19). Chicago: Aldine Publishing Co.

Jaekson, J. A. (Ed.). (1972). *Role*. Cambridge: Cambridge University Press.

Jönnson, C., & Hall, M. (2003). Communication: An essential aspect of diplomacy. *International Studies Perspectives, 4*, 195–210.

Kahn, H., et al. (1978). Kahn, Mead, and Thompson: A three-way debate on the future. *Futurist, 12*(4), 229–232.

Kingsley, S. A. (2001). *Optical SETI observatories in the New Millennium: A review*. http://www.coseti.org/pdf/4273–06.pdf

Kohl, J. V., & Francoeur, R. T. (1995). *The scent of eros: Mysteries of odor in human sexuality*. New York: Continuum Publishing Company.

Kurzweil, R. (2010). Merging with the machines: Information technology, artificial intelligence, and the law of exponential growth. *World Future Review, 2*(2), 57–61.

Kuttner, R. E. (1981). The nature of extraterrestrial communication. *Cosmic Search*, 20–21.

Landes, D. S. (1998). *The wealth and poverty of nations: Why are some so rich and others so poor*. New York: W. W. Norton & Company.

Leach, J. W., & Edmund L. (Eds.). (1983). The Kula: New perspectives on Massim exchange. In *Proceedings of the Kula and Massim Exchange Conference 1978*. King's College, University of Cambridge. Cambridge: Cambridge University Press.

Learned, J. G., & Kudritzki, R., et al. (2008). The Cepheid galactic internet. arXiv:0809.0339.

Lefebvre, V. A., & Yuri, E. N. (2000). Cosmic intelligence and black holes. arXiv:astro-ph/0005546.

LeGuin, U. K. (1969). *Rocannon's world* (1966), *The left hand of darkness (1969).*

Liang, Q., & Wang, X. (1999). *Unrestricted warfare*. Beijing: PLA Literature and Arts Publishing House.

Litwak, R. (2007). *Regime change: US strategy through the prism of 9/11*. Baltimore: Johns Hopkins University Press.

Lyng, S. (1988). Theoretical observations on applied behavioral science. *Journal of Applied Behavioral Science, 24*(1), 101–117.

Maine, Sir H. S. (1906). *Ancient law, its connection with the early history of society and its relation to modern ideas, with an introduction and notes by Sir Frederick Pollock* (Vol. 4th American from the 10th London edition). New York: Henry Holt and Co.

Malinowski, B. (1961). *Argonauts of the Western Pacific*. New York: E.P. Dutton & Co.

Malinowski, B. (1968). In Tewara and Sanaroa—Mythology of the Kula. In R. A. Georges (Ed.), *Studies on mythology* (pp. 72–101). Homewood, IL: The Dorsey Press.

Marshall, C. (1999). *Warfare in the medieval world*. Austin: Raintree Steck-Vaughn Publishers.

Maruyama, M. (1976). Designing a space community. *Futurist, 10*(5), 273–281.

Mauss, M. (1966). *The gift* (I. Cunnison, Trans.). London: Cohen & West.

McGraw, A. P., & Tetlock, P. E. (2005). Taboo Trade-offs, relational framing, and the acceptability of exchanges. *Journal of Consumer Psychology, 15*(1), 2–15.

McLemore, S. D. (1970). Simmel's 'Stranger': A critique of the concept. *Pacific Sociological Review, 13*(2), 86–94.

Melin, M. M., Gartner, S. S., & Bercovitch, J. (2013). Fear of rejection: The puzzle of unaccepted mediation offers in international conflict. *Conflict Management and Peace Science, 30*(4), 354–368.

Merkle, R. C. (1992). Self replicating systems and molecular manufacturing. *Journal of the British Interplanetary Society, 45*(10), 407–413.

Michaud, M. A. G. (1979). Extraterrestrial politics. *Cosmic Search,* 11–14.

Michaud, M. A. G. (1998). Policy issues in communicating with ETI. *Space Policy, 14*(3), 173–178.

Molton, P. (1978). Exotic life and exobiology. *Journal of the British Interplanetary Society, 31*, 156–160.

Molton, P. (1981). The discovery, care and feeding of intelligent aliens. *Spaceflight, 23*, 176–180.

Moore, S. F. (1978). *Law as process*. London: Routledge & Kegan Paul.

Moriarty, J. R. (1969). Ritual combat: A comparison of the Aztec "War of Flowers" and the medieval "melée", presented at a meeting of the Southwestern Monuments Association in 1968.

Morris, D. (1982). *The pocket guide to manwatching*. London: Triad/Granada.

Morrison, P., Billingham, J., & Wolfe, J. (Eds.). (1977). *The search for extraterrestrial intelligence*. Moffet Field CA/New York: NASA/Dover Publications.

Narveson, J. (1985). Martians and morals: How to treat an alien. In E. Regis (Ed.), *Extraterrestrials: Science and alien intelligence* (pp. 245–265). Cambridge: Cambridge University Press.

Niven, L.1970. Ringworld. New York: Ballantine.

Niven, L. 1973. Protector. New York: Ballantine.

Niven, L. 1975. Tales of Known Space. New York: Ballantine.

Niven, L. 2006. The Draco Tavern. New York: TOR Books.

Nudler, O. (1990). On conflicts and metaphors: Toward an extended rationality. In J. Burton (Ed.), *Human needs theory* (pp. 177–201). London: MacMillan Press Ltd.

Ouwehand, C. (1964). *Namazu-E and their themes: An interpretative approach to some aspects of Japanese folk religion*. Leiden: E. J. Brill.

Palisi, B. J. (1970). Some suggestions about the transitory-permanence dimension of organizations. *British Journal of Sociology,* 200–206.

Polanyi, K. (1968). *Primitive, archaic and modern economies*. (G. Dalton, Trans. & Ed.). Garden City, NY: Doubleday & Co.

Powers, R. B., & Kirkpatrick, K. (2013). Playing with conflict teaching conflict resolution through simulations and games. *Simulation & Gaming, 44*(1), 51–72.

Rampelotto, P. H. (2010). The search for life on other planets: Sulfur-based, silicon-based, ammonia-based life. *Journal of Cosmology, 5*, 818–827.

Raup, D. M. (1985). ETI without intelligence. In E. Regis (Ed.), *Extraterrestrials: Science and alien intelligence* (pp. 36–43). Cambridge: Cambridge University Press.

Regis, E. (Ed.). (1985). *Extraterrestrials: Science and alien intelligence*. Cambridge: Cambridge University Press.

Reinhard, W., & Sturge, K. (2011). *A short history of colonialism*. Manchester: Manchester University Press.
Rice, C. (2008). Rethinking the national interest. *Foreign Affairs, 87*(4), 2–26.
Ross, M. H. (1986). A cross-cultural theory of political conflict and violence. *Political Psychology, 7*(3), 427–469.
Sahlins, M. D. (1972). *Stone age economics*. Chicago: Aldine Publishing Co.
Sakurada, K. (1980). The Ebisu-Gami in a fishing village. In D. Dorson (Ed.), *Studies in Japanese folklore* (pp. 122–132). New York: Arno Press.
Sanduloviciu, M., & Lozneanu, E. (2011). Ball lightning as a self-organization phenomenon. *Journal of Geophysical Research, 102*(D4), 41719.
Schank, R., & Birnbaum, L. (1994). Enhancing intelligence. In J. Khalfa (Ed.), *What is intelligence* (pp. 72–106). Cambridge: Cambridge University Press.
Schneider, J. (1977). A model for a non-chemical form of life: Crystalline physiology. *Origins of Life, 8*, 33.
Schneider, B. R., & Lawrence, G. (Eds.) (2001). *Battlefield of the future: 21st century warfare issues*. Maxwell Air Base: Air War College.
Scott, J. C. (1976). *The moral economy of the peasant*. New Haven: Yale University Press.
Shostak, S. (2010). Efficiency in SETI. *Acta Astronautica, 68*(3-4), 347–350.
Shuch, P., & Almar, I. (2007). Shouting in the jungle: The SETI tranmission debate. *Journal of the British Interplanetary Society, 60*, 142–146.
Simmel, G. (1950). *The stranger. The Sociology of Georg Simmel*. In K. Wolff (Trans, Ed.). Glencoe, IL: The Free Press.
Skinner, B. F. (1953). *Science and human behavior*. New York: Simon and Schuster.
Skinner, B. F. (1957). *Verbal behavior*. New York: Appleton-Century Crofts.
Skinner, B. F. (1969). Contingencies of reinforcement in the design of a culture. In Jr. R. L. Burgess & D. Bushell *Behavioral Sociology*. New York: Columbia University Press.
Skinner, B. F., & Ferster, C. B. (2011). *About behaviorism*. New York: Vintage.
Skinner, B. F., & Ferster, C. B. (2014). *Contingencies of reinforcement: A theoretical analysis*. MA: BF Skinner Foundation.
Skinner, B. F., & Ferster, C. B. (2015). *Schedules of reinforcement*. MA: BF Skinner Foundation.
Smart, J. M. (2012). The transcension hypothesis: Sufficiently advanced civilizations invariably leave our universe, and implications for METI and SETI. *Acta Astronautica, 78*, 55–68.
Solomon, M. R., Surprenant, C., et al. (1985). A role theory perspective on dyadic interactions: The service encounter. *The Journal of Marketing*, 99–111.
Sullivan, W. (1964). *We are not alone: The search for intelligent life on other worlds* (Rev ed.). New York: Signet.
Swartz, M. J. (Ed.). (1968). *Local-level politics: Social and cultural perspectives*. Chicago: Aldine Publishing Co.
Symons, D. (1974). Aggressive play and communication in rhesus monkeys (Macaca mulatta). *American Zoologist, 14*(1), 317–322.
Tan, K. (2006). The duty to protect. *Nomos*, 84–116.
Traphagan, J. (2014). *Extraterrestrial intelligence and human imagination: SETI at the intersection of science, religion, and culture*. New York: Springer.
Turnbull, C. M. (1972). *The mountain people*. New York: Simon & Schuster.
Venkatesan, P. (2010). "Nanoselves": NBIC and the culture of convergence. *Bulletin of Science, Technology & Society, 30*(2), 119–129.
Watson, O. M. (1970). *Proxemic behavior: A cross-cultural study*. The Hague: Mouton.
Watson, R. A. (1979). Self-consciousness and the rights of nonhuman animals and nature. *Environmental Ethics, 1*(2), 99–129.
Wood, M. (1934). *The stranger: A study in social relationships*. New York: Columbia University Press.
Yoshida, T. (1977). The Japanese concept of 'stranger'. Annual meeting. Houston: American Anthropological Association (Unpublished conference paper).

Chapter 12
Interaction and Intellectual Exchange

If the previous chapter painted a potentially unpleasant (at least, if you are a humanist) vision of interaction with ETI, it is necessary to balance that picture. The need for securing benefits (AKA economics) may well be at the back of every communicative activity, but it need not be in the forefront (as the SETI endeavor and the *kula* ring prove). Moreover, the simple question of 'benefit' does not necessarily imply that such benefits are zero-sum, and that one party's success comes at the expense of the other.

It is therefore worthwhile to consider some aspects of xenology that are not, inherently, based on the exchange of advantage. Such exchanges may be termed 'pollinating', in that they are transmitted without any benefit, and the results may well provide unexpected benefits from cross-fertilization. Not all of these can be identified or discussed here. The consequences of all such exchanges are very complex. In order to predict them, it would be necessary to examine factors which, at the present, and in absence of data, cannot even be foreseen. Here I will try only to address some of these realms: those that appear, at least, to be the most amenable to some sort of analysis. The basic questions to be asked are how, and to what degree is the particular realm transposable to the sensibilities and understanding of an ETI?

Given the vast distances involved, and the most likely mode of contact—electromagnetic messaging—it is highly unlikely that exchanges with ETI would be physical. ETI with great patience, a much longer lifetime, or some other physiological quirk, might be able and willing to wait centuries or millennia for their order to arrive. Not so humans. So the most likely exchanges are going to be exchanges of ideas. The market for 'useful' ideas—engineering, science and other information of potential or actual economic import—is obvious, since both sides would be striving to maximize gains. It is also likely that various languages developed for communication with ETI, such as Lincos (Freudenthal 1960) and other similar efforts (DeVito and Oehrle 1990; Musso 2010) are superfluous: ETI will have been there, done that.

M. Ashkenazi, *What We Know About Extraterrestrial Intelligence*,
Space and Society, DOI 10.1007/978-3-319-44456-7_12

The real problem will be in the communication of fundamental emotional and intangible issues ranging from music (Vakoch 2010a) to emotional-physiological states (Vakoch 2010b). One need only remember that many areas of human existence—sex, evacuation, family relations, and religious taboos—cannot be discussed in many human cultures (try discussing women's libwith a Saudi, or axillary hair with an American to test the point).

12.1 The Realm of Unreason

We do a lot of things for expressive and emotive reasons that have little if anything to do (insofar as scientific knowledge goes today) with survival or maximizing our economic benefits. It is true that some of these activities are beneficial economically, as artists from Michelangelo to Higashiyama will attest, but saying these are primarily economic activities puts the cart before the horse. We have 'fun', we crack jokes—some, such as puns, universally condemned as 'low humor', others considered, for whatever reason, of greater merit—we paint, dance, and juggle. From a purely instrumental point of view, these may seem like a waste of time. Yet we know that our ancestors—even those early ancestors not of our species (Hall 2007) created art, which leads one to suspect that they told jokes and made puns as well.

All of this non-purposive behavior may very well be a realm of complete unreason to any sober external observer. Yet these are fundamental to whom we are as humans: non-reasoned, non-purposive, wasteful behaviors are part of how we might define ourselves as human. How then do we explain ourselves to a correspondent 100ly away? Do we need to? And might ETI have some of the same quirks?

In practice, non-purposive behavior exhibits itself among humans in three endeavors: art, ritual, and play (we might add daydreaming to the list). As a general rule, animals do not engage in the first two to any degree. Insofar as play goes, as a rule of thumb, the greater the intelligence, the more an animal spends on play. Small mammals with less intelligence engage in play only during youth. Some larger mammals with more sophisticated brains engage in play even in adulthood. Otters build mud slides for the pleasure of playing on them (Beckel 1991). Right whales 'sail' using their uplifted flukes to catch the wind. Dolphins play in ships' wakes (Paulos et al. 2010).

Whether all, or any ETI 'play' or engage in other non-purposive behavior is something we shall have to find out. It may well be that play is a sign of great maturity/immaturity or merely a human peculiarity, take your pick. I suspect, however, that we would be able to engage in greater and better intercourse with another species that does indulge in non-purposive behaviors.

12.1.1 Aesthetics and Non-purposive Behavior

The aesthetic impulse we value so much may be merely a human peculiarity, and even if shared may be insufficient to foster ET communication (Kuttner 1981). Nonetheless, it is an important factor in defining ourselves vis-à-vis any ETI. One puerile reason has to do with economics. Species that engage in non-purposive behavior indicate that they are more intelligent and wealthier. There are two reasons for this. If you have more excess resources, you can afford to play. If your non-purposive behavior is complex enough, it indicates the degree to which your thinking is sophisticated, rather than simple (an argument advanced first by Huizinga 1955, in another form). This may very well reflect on your social credit, since it indicates your ability to grasp sophisticated issues. It definitely, however, indicates that you have vast resources available, societally, to engage in that sort of nonsense. To illustrate, one can easily compare the sums—absolute or per-capita—spent in the US on "play" such as sport, television, games, films, and so on, to the same consumption items in Zambia. High rates of play—of non-serious activity (ignoring, for a moment, the economic element of entertainment in the world)—imply large resources, *and* the willingness to expend them. A playing species may be, at the very least, more adaptive than a non-playing species all other things being equal, since it has opportunities for more experiences in a non-threatening environment. At the societal level this does not matter: the US is not more "intelligent" than Zambia, merely luckier, more powerful, and richer. But when comparing species, I would argue that the ability to dedicate more resources to play is indicative of greater success and, either a longer history allowing you to accumulate more resources, or greater ability in handling what you have. Parenthetically it must be noted that the opposite is not true: that a species does *not* dedicate much resources to play may be a species quirk, and is not indicative of the *lack* of resources.

12.1.2 Aesthetics and Art

We tend to conflate *art* and *aesthetics*, possibly because, certainly in Western education, the two are conflated or at least associated. However, since both terms are deeply contested, we need to note several things. First, art is clearly artifactual. It is something that humans *do*. Aesthetics are a sense of evaluation of the juxtaposition of certain elements—material, artifactual, natural, and emotional, perhaps even social—in some socially constructed, meaningful, pleasurable way. Art is judged according to a set of cultural criteria in force at a particular time. That is, one can consider a peasant's rice bowl made a millennium before to be a piece of art, judging by criteria of the present. The *peasant* might not have considered his hand-made bowl art, but in a different era, in a different place, using different criteria, it becomes art (cf. Novitz 1997; Peterson 1974).

Art—whatever its essence is—is a socially imbued practice carried out by humans. This means that it may differ in its essence between different times and places. Novitz (1997: 53) argues that "… there were no 'works of art' in our sense until about the seventeenth century, so that what we now see and understand as art was not so understood by Europeans much before that time." Perhaps so, but that consigns entire historical art schools—Iranian, Indian, Chinese, Japanese, Korean—to nothing. So Art can emerge from a variety of social impulses. Like cuisine, it can be argued that art is the product of complex societies with well-distributed surpluses, and consequently the leisure and ability to afford non-essential pleasing activities.

It is also clear that art is embedded in a society. To cite Novitz once again "… there are no art-making universals; no natural features shared by all works of art that make them art" (Novitz 1997: 77). This would manifestly be true of any art produced by an ETI: humans might decide that the production is artistic based on *human* perceptions that the ETI might not share, and of course vice versa. This argument has a number of complements. First, the concept of 'fine art' is a relatively modern invention (in Europe, a derivative of the eighteenth century and later, in Sinitic societies, probably since the eighth century, perhaps earlier). Second, art is some form of artifact that generates a variety of emotional responses, from the sublime (see for example Confucius's opinions on music. DeWoskin 1982; Yue 2008) to simple pleasure, suspense, or amusement (Carroll 2000). Third, the *appreciation* of art is a function not only of the producer's intention, but more so, of the standards and style practiced by the observer. To return to our peasant's rice bowl, for the peasant it is an artifact that is pleasing to hold, observe, and perhaps even make, but it's primary value is based on the fact that it holds a generous measure of hot cooked rice conveniently, is comfortable to grip, and has no jagged edges. For the connoisseur a millennium later, the sensory evaluation has much to do with feel, color, shape, and embeddedness in a particular culture, which are more important than that the bowl keep rice warm a few minutes longer.

Finally, and not the least important, art is *artifactual*. We can appreciate nature or a natural phenomenon using aesthetics as varied as Zen 'balance' and the Navajo concept of *hozô* (beauty), but for it to be *art* there must be some artifactual dimension, however limited or disguised. Thus the most significant single thing about art is that it is an artifact. True, it may be designed to emulate nature to a very close degree, as the great Noh actor and playwright Zeami tried to make his very artificiality natural (Raz 1976), or it even may try to immerse itself into nature, as does "environmental art". Nonetheless, it is, and remains, an artifact.

As for aesthetics, we must accept the fact that human and non-human aesthetics will be markedly different if only because of different physiological limitations and advantages. The best generalization is one based on Zangwill's (2002: 111) argument that aesthetic properties can be delineated with reference to beauty and ugliness. Concepts that relate to beauty (that is, desirable and positive sensible qualities) *from any actor's point of view* and those that relate to ugliness (those that are undesirable and negative from that point of view) are concepts used to evaluate natural and artifactual/social phenomena. This dichotomy, one would think, is essentially universal *even if we cannot agree on what is beautiful or ugly*. My own personal preference is that art be pleasing as well, though that is manifestly not the case in much modern music, dance, painting, or sculpture, nor is it intellectually necessary. Art is the creation of objects *for* an aesthetic. Now, we might not be able to share the aesthetic of an ETI, but we might be able to identify alien art using two basic parameters. One, is the object/behavior fashioned in some specified manner (artifactuality). Two, is aesthetic value assigned to it beyond utilitarian value. From that point it might well be possible to examine in greater detail the aesthetics that dominate that particular art.

As Zangwill also notes, artistic production lies between two other poles: the aesthetic and the narrative (Zangwill, 2002: 115). Some productions lean more towards one or another of the poles, some balance them. We can assume that, *at least from a human point of view* the same is true of ETI art, *even if ETI have no theory of aesthetics*. We must also accept that there may be additional dimensions we are unaware of, by which ETI with different physiologies, phylogenies, culture, and history, will characterize art, and which we ought to learn about.

Art is also related to *patterning*. That is, art often incorporates within it a simple or complex pattern, or, perhaps, it merely implies that pattern (Hofstadter 1979). Weaving is an example of the former. Metal cans of soup examples of the latter: the pattern of empty and full soup cans is all around us, the artistic object brings the heretofore ignored pattern into focus. It is possible that some of the patterns are in effect multidimensional (in the mathematical sense) in their repetitiousness. For example, let us consider a complicated three-dimensional curve. Seen in its totality, such a curve will have repetitiousness and a clear identifiable pattern. Cut that pattern by a plane and examine the resultant points. They will exhibit, to the untutored eye, either a random collection of dots, or, perhaps, may hint at the underlying curve. A mathematician might be able, analytically, to reconstruct that curve, even though we could not see it. An artist may see this patterning, intuitively, as something worth pursuing, and may make “art” of that pattern. The analogy to ETI art stands out. We might not be able to see the pattern of ETI art because we lack the aesthetic. We might, however, given the right tools, be able to describe it.

Whether or not an ETI society has something that is analogous to art, or perhaps art itself, depends on one’s initial assumptions. Stephan (1990) relates art to human physiology. He claims that art derives from physiological differences between the brain hemispheres of early *homo*. Galaburda et al. (1978) have presented evidence that *Homo neanderthalis* brains exhibit asymmetries, particularly in the auditory regions: the areas in left hemispheres are far larger than those on the right. Neanderthal tools exhibit patterning, and what we would call ‘art’ has been found among Neanderthal remains. Marshack (1976) argues that right handedness evolved partially because of the enlargement of the speech (and thus logic) faculty in early *homo* brains. Right hands were used for actions requiring precision and control, helping to refine that faculty in the left hemisphere. In contrast, the emotive and impressionistic processes in the right side of the brain—necessary for any animal as part of the need to quickly and holistically evaluate surroundings in order to prepare for fight or flight—retained less logic and more emotive content.

The process described in the previous paragraph is part of human biological history. If that is indeed the case, then we are drawn to the conclusion that art is a specifically *human* activity, and that unless an ETI has an analogous phylogenetic history to H*omo sapiens*, then it is unlikely to have art (Kuttner 1981: 20). We have already seen how the odds against repeating phylogenetic history are stacked very high indeed.

A more generalized position is suggested by Diamond's views of the genesis of art (Diamond 1992). Art is also a form of display, and many animals participate in extravagant displays, whether as a means of showing predators that they are in prime health (and thus, saving energy for both predator and potential prey) and as a means of attracting mates, or both. Art can therefore be seen as a derivative of a universal principle at least in a species in which intra-sexual competition is the norm. This supposes that ETI, who, for some phylogenetic reason need mating (or perhaps other physiologically required) displays will develop art as well.

In human societies, the ability to appreciate and understand the art form of a different culture reinforces and strengthens inter-cultural contact. By accepting another's criteria of aesthetics or at least by demonstrating some knowledge of the rules, we gain social credit. Besides the appreciation that alter has for ego if ego bothers to learn alter's rules, there is an additional, positive benefit: knowledge of all and any aspects of a strange culture makes other aspects of the culture more comprehensible. To give an example, all human societies assign values to certain colors. Some basic colors: red, black, white, may even have universal values as Turner (1966) argues. The juxtaposition of certain colors may symbolize certain effects in a culture. Red and gold implies a wedding to a Japanese, black and silver, a funeral. Other color combinations apply for other things. Knowledge of these effects, whether we choose to follow and use them or not, may well help explain (as more than one Western company in Japan has discovered, often too late) why one exchange strategy works better than another.

12.2 Mathematics and Art

Mathematics and art share a number of things in common. Human art, of whatever form, is an attention-getting, purposeful arrangement of a medium or media. These often express themselves in the form of repetitive, or semi-repetitive patterns. Mathematics, whatever algorithms are in practice, also represents patterns at varying depths of complexity. While the patterns in mathematics are innate and unrelated to whether or not an intelligence discovers or makes sense of them, they exist in both nature and artifice.

12.2.1 Mathematics

Mathematics is a purposeful arrangement of abstract concepts that are universal in their 'reality'. One can create odd mathematics, but once the intention is clarified, good mathematicians can follow one another's footsteps. For example, we have a simple rule in most human mathematics: the product of zero and any other number is zero. Minsky (1985) has argued convincingly that all mathematics are the same, that is, there exists a rule that one and only one mathematical way is appropriate. That may well be the case but it is immaterial. What is important in Minsky's argument is that he set out to explore alternative possibilities, regardless of whether they would be fruitful. In other words, mathematics exhibits some of the qualities of art in its regular patterning, and possibilities of play. Moreover, Minsky is to some extent wrong in his claim that there is only one way to 'do' mathematics. Though the relationship between the numbers is apparently a universal, there are many ways of using those universals. Many people are familiar with different algorithms for performing the same mathematical operations. Reverse Polish Notation, a system evolved by East European peasants for calculations, is the basis of a computer language, LISP. One can multiply the nines table by remembering each number, or by making up a rule to fit, or by adding nines a sufficient number of times.

Significantly, many of those ways of "doing" mathematics have, to the human mind, an aesthetic to them. Indeed, mathematical patterns when translated into auditory signals are the essentials of music, which has a deep resonance in human emotions, as observers from Aristotle to Confucius have noted early in human history. Translated into visual signals, they have impact as well (Hofstadter 1979; Silva 2002). Given that mathematical relationships are likely to be universal, recognition of 'good' (that is, properly articulated principles) or mathematical behavior *translated into a visual, aural, or other medium* may well be a first indicator of an aesthetic or artistic impulse among ETI.

12.2.2 Alien Art

Human art and mathematics *differ* in that art is *intended* to affect the participant in some emotional manner. While mathematical arrangements *may* do so, that is not their major purpose. Suppose, at some level, for some ETI, that were not the case: the aesthetics of mathematics, and the mathematical expression of art were identical. Would that change anything in either the cultural or the communicative aspect of dealing with them?

It is unquestionable that the conventions of ETI art (if there is any) will be radically different, in form, content, and interpretation, from human art. Partly this is likely to be true because ETI physiology will be different, even if we do not accept the premise raised earlier that art is derived from human physiology. They will, in other words, perceive things at the physical level, quite differently from humans. To a great degree ETI art will differ because art and its appreciation are the derivatives of particular historical realities and processes.

However, the meta-rules of patterning and representation that apply to both human art and mathematics will still most likely apply. At this point we must reconsider some of the physiological parameters presented earlier in the book. Aesthetics, in the final analysis, depends on sensory factors. Therefore, as a basic rule, any ETI aesthetic will have to be understood from the framework of that ETI's sensory experience. To add to that, sensory experience cannot be detached from *interpretation* of that sensory experience. Attentional blindness is one such example (Bressan and Pizzighello 2008; Simons 2000; Simons and Chabris 1999). A frog's eye responds to certain stimuli, such as a small flying shape (insect: eat) or a large dark shape (predator: escape). Other shapes may well be indistinguishable to the frog: the animal literally cannot see them, because the wiring to the brain does not allow that shape to be attended to. Humans will ignore a large gorilla posing next to people passing a basketball between them. This sort of hardwiring may also be the case among ETIs, where it may place limits on the ability of an ETI to comprehend some sets of relationships, colors, shapes, or other factors. This means that there might well be limitations on *our* ability to understand strange art forms *not* because we cannot understand the aesthetic, but because we cannot perceive the phenomenon.

Examples of phenomena that can or cannot be seen abound in the human experience. Numerous experiments have been carried out in human perception What some of these demonstrate is that different human experiences enable individuals to perceive different things within the same pattern. Now in the human case, these can be learned, because we are all human. But, if such phenomena do occur, then it is possible at least (and we would not know, unless we questioned an ETI) that such particular phenomena we are blind to even exist.

This aspect points to an important value to be gained simply by communicating with ETI. It is at least *possible* that 'blind spot' phenomena do exist. Because they are 'blind spot', we do not even know they are there, let alone study them. Once it is pointed out to us that these phenomena *exist* they may be open for study. This means that one major benefit of contact with ETI may well derive not from any material, or even cultural exchange, but from parallax: from the fact that the same world is studied by different eyes (or other organs) which see the same things but in different ways. This has indeed been the case in human experiences of contact, albeit not often. In such cases, too, the willingness of one culture to accept and deal with the insights derived from a different point of view, have not been adopted easily. In some cases they have been more successful, as for example the gradual introduction—slow and tortuous to be sure, but there—of ideas of holistic medicine from the medical experiences of simple societies into Western medical practice.

12.2.3 Music and Its Effects

Humans have known two things about music from early history: music appeals, somehow, to deep emotional feelings among humans (Aristotle POLITICS Book VIII, Chap. 5–7; Van Gulik 1969). Music also has definite social aspects: it promotes togetherness for many purposes [mating (Denora 1997), warfare (van Creveld 2008), collectiveness (Paddison 1997; Dueck 2006), economics (Hul et al. 1997), etc.]. Now, of course all of these might be simply human peculiarities: a reflection of our internal wiring. However, as we know, music *also* presents us with a series of mathematically related sounds, and the complexity and uniqueness of musical compositions as mathematical phenomena, are attractive and interesting. This is to say that music may well not have the same emotional and social effect on ETI, but that the patterned formalism that music exhibits would be attractive to an enquiring mind.

The question of whether music would have an effect on ETI has been discussed several times in the literature (Kaiser 2004; Roederer 2002, 2009; Vakoch 2010a, b; Zaitsev 2008). As a general idea, it would seem that inasmuch as music lends itself to both patterning and iconicity (which in turn can be create more patterns), the transmission of music to and the potential receipt from ETI, would provide a good sign of intelligence, and the potential for communication. Of course, given the wide ranging effects of human music, what we transmit and/or receive, may mean anything from "We want to mate with you" through "Buy Vuzd, the drink the Galaxy has been waiting for!" to "We challenge you to a duel".

So we are still left with the problem not of contact, but of communication and meaning. In other words, under what conditions are we likely to find music a useful tool (and an interesting one) to communicate with ETI, in the first instance, and would ETI have music? Does biology or mathematical art win the day? Music among humans may well be the product of the need for language learning among infant humans. Human foetuses are subject since before birth to acoustical signals which they receive passively, and, once born, become an active process of seeking acoustic signals from the mother, combined with emotional reinforcement, which also yields musical discrimination (Roederer 2002). This is a reasonable explanation for musical sense among *humans*. It may even operate among ETI *provided* that ETI communicates audibly (which we cannot assume is the case), and there is a lengthy gestation process. As a *universal* principle that argues for musical universality, it does not have validity. It *would* appear, that the recognition of patterning in music, whether in the form of some golden ratio (Kaiser 2004), or the recognition of artificial fractals, which are present in many human musical compositions, would provide a broader base for ETI music, or rather, for translation of music between humans and ETI. Moreover, simple musical signaling has another weakness. We know of several terrestrial animals—humpback whales, some birds —which produce musical compositions of great complexity. It would not be out of the question to find natural phenomena which ETI are familiar with, that would do the same, obviating music as an inherently intelligent signal.

So the usefulness of music as a contact signal seems to be moot. What about music as an art and (we might as well be practical) a form of exchange? While many musical forms such as e.g. Trance music have very repetitive patterns, the complexity of other music, ranging from classical Western symphonies (cf. Hofstadter 1979) through Javanese gamelan (Lindsay 1992) means that the complexity alone could make the phenomenon of music attractive to non-humans, if (a) they recognize that there is an aesthetic framing the music, (b) they have a concept of art, and notably (c) if they are members of a civilization, and sufficient human music is present to indicate civilizational processes, that is, standards of performance, a praxis, and canonical forms, all of which are present in major musical traditions, different though they may be to human ears.

12.3 Games and the Mind: Play

To return once again to play. Many human activities are not designed towards a concrete end. They may be purposeful *within* their own realm, but otherwise, do not necessarily provide realizable benefits. Play is observable in all human societies, and involves many distinct forms (Huizinga 1955). Play serves unquestionably as a training ground for social and productive activities, but that does not seem to be its main effect in human society (Sutton-Smith 2009). It is a complex activity involving display, patterning and ritual, relaxation, creativity, varied stimuli and many other factors. Within the realm of play come varied things such as games (artificial rule-bounded activities) and creativity That play is particularly prominent among children reflects partly an ethnocentric bias (the interest in childhood as a concept, and play as a major focus of childhood only emerged in the late nineteenth century). In simple societies, play is common to adults (women less than men) as much as children.

Whether or not play will occur among ETI is a matter for speculation and a search for empirical data. The consequences of having play may be quite important. The ability to suspend the normal rules of life and to assume artificial and temporary rules for behavior can be analogized. A species that is capable of play, particularly of the sort of play we call 'games'—highly structured, rule-finite activities—is probably capable of admitting the validity, if only *pro-tem*, of alternate rules. Such a species is probably better equipped to deal with ETI, since the ability to abide by non-normal rules is analogizable. A race that is incapable of relaxing the rules *pro-tem* or of adopting other, bounded, perhaps counter-factual rules for a bounded period of time, is someone we ought to worry about as a difficult customer for everything from trade to communication.

12.3.1 Play and Imagination

Among games are also those played *intellectually* by humans, which do not appear to be "functional". Play requires the use of imagination that is the consideration of multiple options for activity, thinking, and expression, which will form an attractive whole. Clearly, play is also determined by the cultural surroundings, as well as potential participants, etc. (De Oliveira and Valsiner 1997; Thomas and Brown 2009).

Humans apparently have an affinity for elaborate, rule burdened, purposeless activity. Such activities are reinforcing in their own right: they relate, presumably, to the aesthetic bonds described in the previous section. We find people counting cars, collecting leaves, playing word games, or even idly dreaming (Singer 1966). Not all individuals do this, and no one does this all the time. In practice, of course, it is very difficult to separate the pure intellectual pleasure of thinking games, from the benefits that accrue from it.

Challenging Groups: Making Teams Up Play allows humans to experiment with social alliances and enmities without too much risk. Admittedly, some forms of play such as traditional football (from which both American and world football emerged through different paths) can be injurious to individuals, but the play itself usually bounds the damage to the participants (including supporting onlookers, cf. Elias and Dunning 1971). At any age, the construction of play teams is a socially constructive act, where careful negotiations over rules, membership, and boundaries are conducted. Thus games are, at all times, bound by artificial rules which the players are expected to abide by (even if, in some games, lying and cheating are an element of game rules). The importance of choosing sides, and managing the blend of players is so great, that several IT game simulations have emerged whose focus is on choosing teams. The ability to play in teams may also be related to participation in civil life in general, at least among humans (Schulzke 2011).

While the choosing of teams for games might seem to be a puerile issue (except for the fans and the industry) the ability to choose a team properly, and manage it for success goes well beyond the issue of games. There is significant evidence that the right blend of personnel in a team is one major pathway to success, and that *failure* in choosing the right team is an almost certain recipe for failure (Djalali et al. 2014; Rico et al. 2009; Zornoza et al. 2009). The same is likely to be true for two aspects of communicating with ETI. On the one hand, who has selected, and who is a member of the *human* team that is responsible for the communication? As humans, we all bring our biases along with us. The result could be disastrous in a number of scenarios: if the team is not representative, if they bring strong ideological components to the discussion (whether reinforcing or clashing ideologies), or if great personality and/or professional rivalries emerge. What this means is that a great deal of thought needs to be given, in the case of a unified response from Earth, to whom the team will be composed of. The other side is also true: can our ETI correspondents make teams up? On the fly or only after deliberation? Who do they represent? Is the team a collection of individuals, a corporate unit, or a monolith?

Crucially, how would these variables affect the communication? *If* we knew that the ETI we are communicating with are able to assemble teams—from street corner ball games to mass display activities—we would be in better shape to understand how 'serious' matters are handled by teams we are in contact with.

Challenging Individuals: Dominance and Participation Games Within a play team, individuals might play against one another as well. In Western culture, individuals within teams are often competitive, and many games are competitions between individuals. This is not the case in other cultures. Burmese, for instance, play a game called *kyat*, whose objective is to maintain a ball in the air *collectively*. Games (including sports) allow individuals opportunities to compete in dominating the team, displaying prowess, and otherwise exhibiting fitness. Notably, this is less prominent among women players (Weisfeld et al. 1982), and there is some evidence that females engage in cooperative rather than competitive games more than males.

The degree to which a society glories in, and promotes individual versus collective dominance in games may provide a key towards their relationships with one another, as well as with strangers. The correlation between individual dominance games and the need for social dominance among humans is probably rather weak. Among non-humans, however, dominance games might be a very prominent and significant feature of society: something that we might need to account for in dealing with ETI.

Challenging Oneself: Made-up Worlds A subset of games are intellectual exercises —philosophical, fantastical—that may, eventually be shared with others, or even propagated with greater or lesser force. A regular section in a very respectable scientific magazine is dedicated to nothing else than play of this sort. Where such games are shared, two processes occur. First, the originator of the idea is forced to put her musings into a form that is accessible to others (one the one hand making it lose some of its force and beauty, on the other, perhaps clarifying it even to herself) and second, the idea comes up against alternative ideas offered by others.

Crucially, play is something that one can engage in oneself. Crossword puzzles, sudoku, computer games, and juggling are all examples of games that individuals play in order to please themselves, with no, or little input from others. Such games may be intimately related to daydreaming and the development of the imagination (Singer 1975). Imagination, both the wild untrammeled kind that takes place within one's mind, and the formal kind as expressed in scientific writing, also has an intermediate step: literature. Much literature allows the human imagination full play, without the restrictions imposed by reality. A case in point is science fiction.

Science fiction and science have an interesting relationship. To develop an analogy, science fiction is the crowd-sourced sandbox version of science. It is a form of imaginative, orderly thinking that allows unbounded speculation, and by doing so, both poses questions and attempts to provide solutions for them. Because it is crowd sourced, not all, in fact very few, of its suppositions actually come into existence. Significantly, however, because it is *literature*, the fact that science fiction ideas fail to find real counterparts is not a negative incentive. HG Wells's

Martians do not exist, nor are they likely to exist on Mars, yet that does not detract from the novel. Unsurprisingly, a great many scientists—physicists in particular (e.g. David Brin), but also astronomers (Carl Sagan), chemists (Hal Clement) and even political scientist-psychologists ('Cordwainer Smith'/Paul Linebarger)—are or have been successful science fiction writers. From a sociology of science perspective this should be obvious: writing science fiction allows the development of ideas without the highly critical and conservative reactions most unorthodox, unbounded and blue-sky thinking tends to evoke in the scientific community. The other direction is of course the bedrock of science fiction: as science produces more and more answers and more and more questions, science fiction writers and readers are out there in the forefront muttering to themselves and others "What does *this* development mean for the human experience?"

In summary, we ought to be aware of the potential of imagination and games to bring us closer to understanding ETI. Where games occur, we might be better prepared to understand some analogies to our own existence, and thus be able to communicate better. Whether this will also bring about more opportunities for cooperation remains to be seen: on the one hand, using human practice as a norm, over indulgence in collective games could make ETI dangerous correspondents if they insist on playing games throughout communicative processes. On the other hand, an ETI race addicted to pure imagining would be extremely difficult to communicate with, since they might view humans as just another form of their imagination.

12.4 The Body Beautiful

Styles and modes in behavior have been discussed earlier, as has the issue of physical presentation. Here we enlarge on that topic and its implications. The most commonly used locus of the term 'style' in modern speech, is the process of body adornment. This encompasses a wide array of universally engaged-in human practices and artifacts. With very few exceptions, all human cultures include some form of body adornment, whether on the body itself (tattooing, cicatricion, skull and neck modifications, piercing) or using external artifacts such as clothing, makeup, and jewelry.

All of these forms are subject to mode and style, which in turn are partially the effect of technological changes and social contacts. In the modern networked world, more successful/rich societies have been plundering the cultural ideals of other cultures to create, sustain, and change styles for personal gratification as well as commercial advantage.

Body modifications among humans take place in societies ranging from the simplest and most primitive, to the most complex. In the modern human world, bodily modification has developed the potential of helping an individual to provide greater survival fitness. This has come about by the use of gene engineering

(Fincham and Ravetz 1991), melding human organs and IT devices (Warwick 2010), and the potential development of nano machinery, including 'medichine': nano machines providing autonomous targeted medical repairs to the human body (Abeer 2012; Kroeker 2009; Nerlich 2008). Though all these technologies are still in their infancy, there is sufficient vision and some evidence to indicate that they could become major factors in the cultural evolution of the human race.

The development of all three areas of 'modern' body modification technologies are technical extrapolations of sciences: molecular biology, computerization, and nano-science. These sciences, perhaps with different perspectives, seem so obvious that we should assume that they are known to any culture that is scientifically and technically more advanced than current humanity. *If* this is the case, then we need to keep in mind that our first contact with ETI may be contact with a species in which the very idea of 'species' is defunct. That is, within the ETI race will be found as many somatotypes, genetypes, and degrees of automation as individuals can conceive of. Moreover, the somatic nature of this collective (one hesitates to call it a 'species' since genetic variation would be so high) could be constantly in flux, as individuals strive to maximize their fitness for particular contexts and environments: different natural (space, water, arctic, etc.), and social and cultural environments (from individualistic hermetical associations to hive-style living).

This intra-racial variability has a number of implications for humans. First and foremost, we shall have to modify our scientific understanding of terms such as 'species' and 'race' (see also Roughgarden 2004 for a well-reasoned critique of the 'species' concept). Second, we are more likely to be dealing with collectives of individuals, perhaps with radically different somatotypes, who have, somehow, reached accommodations for living in proximity. This would likely be the greatest gift we could receive from an advanced ETI: the erasure of intra-human human frontiers, and the 'trick' of nonetheless accommodating difference. And, unless we posit a social system of endless conflict, á la Brin, accommodation would be a principle, and the subjects would be infinitely mutable as individuals, within the framework of a prevailing fashion, modify their bodies and perhaps their minds to make themselves more fit for their significant groups and relationships.

For beginning-of-the-twenty-first century humans, who are still arguing about the many manifestations of 'race' and sexual identity, dealing with such an ETI race might be a shock. We can postulate that inasmuch as genetic engineering et al. will have occurred to any advanced ETI society more technologically advanced than us, these advances will be visible among the commonality. That is to say, it is highly likely that any ETI race we contact will have:

- Very fluid notions of sexuality and gender, with boundaries between the male and female poles of reproduction being fluid, permeable, and amorphous, and possibly a matter of choice.
- Varied, perhaps radically different, bodily forms (ignoring for a moment differences brought upon by living in different environments such as space stations) chosen for their aesthetic.

- Sexual, cuisinal, artistic, sporting, and other entertainment forms that mutate rapidly, are subject to fashion, and appeal to networked but physically separated small communities, which are themselves ephemeral.

In other words, rather than meeting a race of more-or-less single size, shape, coloration, and practices, we would be dazzled by a kaleidoscopic mass of individuals, practices, and artifacts somehow held together by civilizational practices *which we might not be equipped to apprehend*: somewhat like patterns in four and five dimensions.

12.5 Conclusion: Intelligent Life Forms We Would like to Meet

The "non-rational" and preoccupation with non-rational activities are a major element of intelligence. Such expressions of the non-rational and non-instrumental as art, philosophy, mathematics, and ritual may well be the expressions of a physiological necessity to dream, to exercise imagination. They are an element in play, but they are more than play, because they force the actor(s) to engage in activities which require personification and objectification. Animals enjoy play, and many of them engage in it even in adult form. Intelligent beings not only enjoy play, but they create elaborate artificial and artifactual devices dedicated to little *but* play.

In practical terms, identifying a species that expends significant resources on non-rational and non-instrumental activities, we are probably looking at an intelligent species. This does not prejudice the form such activities may take. In the human species there are some such activities that are universal. All human groups share a love of ritual, and expend great effort on performing elaborate and expensive ceremonies. All individual humans expend much energy on bodily modification: tattooing, jewelry, elaboration of dress are present *in all human groups*, primitive or technologically advanced. What expressions will be found among other intelligent species depends presumably, on factors we cannot guess at. We can, however, be certain they will be there.

Finally, the essence of play as humans do it—the artificial creation and manipulation of rules—also reinforces something else we know about the nature of advanced ETI. If we extrapolate from the science base we have today, we see that certain branches of science such as genetic engineering (which may be necessary for survival in space, as well as for developing energy sources), nano-engineering, and computerization will create a situation of flux. That is, our inherent tendency to look for a 'race' of ETI—where individuals look somatically alike—and an ETI society/culture—where the mass of individuals behave in predictable manner, and social institutions are identifiable and more-or-less fixed—will likely be falsified very quickly, and if we are to catch up, we shall need to adjust our parochial ideas.

At another level, ETI that are able to play, that is, to suspend belief and their own daily roles, would seem to be preferable as our opposite numbers across space, than many alternatives. One would expect them to incorporate and accept our own little (and not so little) foibles with a degree of understanding. On the other hand, they would also likely point the way, with all the best intentions, towards the revolutionary destruction of our own culture(s) and their replacement by something else, perhaps even better suited to an advance life-form.

We tend to assume that our cultural data set is fixed. That is, we have, as a humans., a tendency to assume the existence of "real" categories in our lives. In reality, all our perceptions are far more fluid than is thought. Sex and gender are examples of this. We have the category "woman" and we have the category "man" and tend to assume that these equate with divergent sets of paired differentiations: female-male, …. In practice, things are far more ambiguous than that (Roughgarden 2004): there are male females and female males, neuters, and other imprecise sexual categories. The same is true about other qualities that are associated with people: 'race', nationality, class, etc. While all may have *some* basis in 'reality' they are social constructs that have hardened into stereotypes. In practice, even among humans, such categories are fluid, and many people assume different categorical identities when possible and convenient. In an early paper, Max Gluckman, an anthropologist, argued that a significant difference between simpler and modern societies is that in simpler societies, people could only distinguish between different roles in their society—say if the same person were both 'father', 'chief', 'friend', 'war leader', member of and age group—by rituals which allowed personal fluidity (Gluckman 1962). To take that argument a couple of steps further, in contrast, *modern* societies allow for enormous role fluidity, as well as, if sufficiently advanced, great physical, and even genetic fluidity as a consequence of modernization and beyond.

Culture, as we have seen, is the product of the three ecologies—natural, technical, and social—to which individuals-in-groups adapt themselves. The technical environment of all greatly advanced ETI is likely to be very fluid. Even if there is some reachable technological horizon, the presence of styles, and the vast array of technical and role choices available, will mean that change and ferment, rather than stasis, are likely to be the norm. Hard-and-fast social realities may well not exist, and the ability to deal with subtle and not so subtle ambiguities, may well be what determines 'class' just as wealth and access to luxuries and technical choices does for early 21st century humans.

The ambiguity, inherent in culture may well be extremely important when we deal with an ETI. First, because we will try to force our categorizations on the ETI. This is an obvious danger, and much of this book and specifically this chapter have been addressing the issue. There is also, however, a far more subtle issue here, which has to do with our understanding of our own culture. The assumption that there exist hard-and-fast basic categories when those do not, in fact, exist except as social constructs, means that we run the risk of communicating, perhaps even unwittingly, false information about ourselves. When done intentionally, this might be of benefit, say during trade, when done unconsciously, it can be extremely dangerous, since, for the ETI we are in contact with, unless other subsidiary data are available *all datum are of equivalent value*. By arguing strenuously for an interpretation of ourselves that we manifestly believe in, but which for the ETI concerned is manifestly false, is an automatic game loser.

References

Abeer, S. (2012). Future medicine: Nanomedicine. *JIMSA, 25*, 187–192.
Beckel, A. L. (1991). Wrestling play in adult river otters, Lutra canadensis. *Journal of Mammalogy, 72*(2), 386–390.
Bressan, P., & Pizzighello, S. (2008). The attentional cost of inattentional blindness. *Cognition, 106*(1), 370–383.
Carroll, N. (Ed.). (2000). *Theories of art today*. London: The University of Wisconsin Press.
De Oliveira, Z., & Valsiner, J. (1997). Play and imagination: The psychological construction of novelty. In A. L. B. Smolka, M. C. R. Góes, et al. (Eds.), *Dynamics and indeterminism in developmental and social processes* (pp. 119–133). New York: Psychology Press.
Denora, T. (1997). Music and erotic agency—Sonic resources and social-sexual action. *Body & Society, 3*(2), 43–65.
DeVito, C. L., & Oehrle, R. T. (1990). A language based on the fundamental facts of science. *Journal of the British Interplanetary Society, 43*(12), 561–568.
DeWoskin, K. J. (1982). *A song for one or two: Music and the concept of art in early China*. Ann Arbor: Center for Chinese Studies, University of Michigan.
Diamond, J. (1992). *The third chimpanzee: The evolution and future of the human animal*. New York: Harper and Collins.
Djalali, A., Ingrassia, P. L., et al. (2014). Identifying deficiencies in national and foreign medical team responses through expert opinion surveys: Implications for education and training. *Prehospital Disaster Medicine, 29*(4), 364–368.
Dueck, B. (2006). Suddenly a sense of being a community: Aboriginal square dancing and the experience of collectivity. *Musiké, 1*(1), 41–58.
Elias, N., and Dunning, E. (1971). Folk Football in Medieval and Early Modern Britain. *The Sociology of Sport*, 116–32.
Fincham, J. R., & Ravetz, S. (1991). *Genetically engineered organisms: Benefits and risks*. Toronto: University Of Toronto Press.
Freudenthal, H. (1960). *Lincos: Design of a language for cosmic intercourse*. Amsterdam: North Holland Publishing Co.
Galaburda, M., LeMay, M., et al. (1978). Right-left assymetries in the brain. *Science, 199*, 852–856.
Gluckman, M. (1962). Les Rites De Passage. In M. Gluckman (Ed.), *Essays on the ritual of social relations* (pp. 51–67). Manchester: Manchester University Press.
Hall, R. L. (2007). Evolution of intelligence and consciousness. *Reviews in Anthropology, 36*(3), 215–244.
Hofstadter, D. R. (1979). *Gödel, Escher, Bach: An eternal golden braid*. New York: Basic Books.
Huizinga, J. (1955). *Homo Ludens*. Boston: Beacon Press.
Hul, M. K., Dube, L., & Chebat, J.-C. (1997). The impact of music on consumers' reactions to waiting for services. *Journal of Retailing, 73*(1), 87–104.
Kaiser, A. (2004). Sound as intercultural communication: A meta-analysis of music with implications for SETI. *Leonardo, 37*(1), 36–37.
Kroeker, K. L. (2009). Medical nanobots. *Communications of the ACM, 52*(9), 18–19.
Kuttner, R. E. (1981). The nature of extraterrestrial communication. *Cosmic Search* Spring, 20–21.

Lindsay, J. (1992). *Javanese Gamelan: Traditional Orchestra of Indonesia*. Singapore; Toronto: Oxford University Press.
Marshack, A. (1976). Some implications of the Paleolithic symbol evidence for the origins of language. *Current Anthropology, 13*, 455–477.
Minsky, M. (1985). Why intelligent aliens will be intelligible. In E. Regis (Ed.), *Science and alien intelligence* (pp. 117–128). Cambridge: Cambridge University Press.
Musso, P. (2010). A language based on analogy to communicate cultural concepts in SETI. *Acta Astronautica, 68*(3-4), 489–499.
Nerlich, B. (2008). Powered by imagination: Nanobots at the science photo library. *Science as Culture, 17*(3), 269–292.
Novitz, D. (1997). Art by another name: The identification of art across cultures. Transcultural Aesthetics. In E. Benitez (Ed.), *Proceedings of the Pacific Rim Conference in Transcultural Aesthetics University of Sydney* (pp. 67–80). Sydney: University of Sydney. June 18–20.
Paddison, M. (1997). *Adorno's aesthetics of music*. Cambridge: Cambridge University Press.
Paulos, R. D., Trone, M., & Kujcak, S. (2010). Play in wild and captive cetaceans. *International Journal of Comparative Psychology, 23*(4), 701–722.
Peterson, S. (1974). *Shoji Hamada: A potter's way and work*. Tokyo & New York: Kodansha International.
Raz, J. (1976). The actor and his audience: Zeami's views on the audience of the Noh. *Monumenta Nipponica, 31*(3), 251–274.
Rico, R., Alcover, C.-M., et al. (2009). The joint relationships of communication behaviors and task interdependence on trust building and change in virtual project teams. *Social Science Information, 48*(2), 229–255.
Roederer, J. G. (2002). Biological conditions for the emergence of musical arts in a civilization of intelligent beings. *International Atomic Energy Agency (IAEA) Bulletin, 34*(1).
Roederer, J. G. (2009). *The physics and psychophysics of music: An introduction*. Berlin: Springer.
Roughgarden, J. (2004). *Evolution's Rainbow: Diversity, Gender, and Sexuality in Nature and People*. Berkeley: University of California Press.
Schulzke, M. (2011). How games support associational life: Using Tocqueville to understand the connection. *Games and Culture, 6*(4), 354–372.
Silva, F. (2002). *Secrets in the fields: The science and mysticism of crop circles*. London: Hampton Roads Publications.
Simons, D. J. (2000). Attentional capture and inattentional blindness. *Trends in Cognitive Sciences, 4*(4), 147–155.
Simons, D. J., & Chabris, C. F. (1999). Gorillas in our midst: Sustained inattentional blindness for dynamic events. *Perception, 28*(1059), 1074.
Singer, J. (1966). *Daydreaming: An introduction to the experimental study of inner experience*. New York: Random House.
Singer, J. (1975). Navigating the stream of consciousness: Research in daydreaming and related inner experience. *American Psychologist, 30*, 727–738.
Stephan, M. (1990). *A transformational theory of aesthetics*. London and NY: Routledge.
Sutton-Smith, B. (2009). *The ambiguity of play*. Cambridge MA: Harvard University Press.
Thomas, D., & Brown, J. S. (2009). The play of imagination: Extending the literary mind. In K. Leidlmair (Ed.), *After cognitivism* (pp. 99–120). New York: Springer.
Turner, V. (1966). Colour classification in Ndembu ritual. In M. Banton (Ed.), *Anthropological approaches to the study of religion* (pp. 47–84). London: Tavistock Publications Ltd.,.
Vakoch, D. A. (2010a). An iconic approach to communicating musical concepts in interstellar messages. *Acta Astronautica, 67*(11–12), 1406–1409.
Vakoch, D. A. (2010b). What does it mean to be human?: Reflections on the portrayal of pain in interstellar messages. *Acta Astronautica, 68*(3–4), 445–450.
Van Creveld, M. (2008). *The culture of war*. New York: Presidio Press.
Van Gulik, R. (1969). *Lore of the Chinese Lute*. Tokyo: Tuttle.
Warwick, K. (2010). Implications and consequences of robots with biological brains. *Ethics and Information Technology, 12*(3), 223–234.

Weisfeld, C. C., Weisfeld, G. E., & Callaghan, J. W. (1982). Female inhibition in mixed-sex competition among young adolescents. *Ethology and Sociobiology, 3*(1), 29–42.
Yue, J. (2008). Confucius on music education. *Nebula, 5*(1-2), 128–134.
Zaitsev, A. L. (2008). The first musical interstellar radio message. *Journal of Communications Technology and Electronics, 53*(9), 1107–1113.
Zangwill, N. (2002). Are there counterexamples to aesthetic theories of art? *The Journal of Aesthetics and Art Criticism, 60*(2), 111–118.
Zornoza, A., Orengo, V., & Penarroja, V. (2009). Relational capital in virtual teams: The role played by trust. *Social Science Information, 48*(2), 257–281.

Chapter 13
The Consequences of Contact

By now, many readers will be, or have been thinking a very human question: What's in it for me/us? This is perfectly reasonable and even fruitful. Though the issue of the consequences of possible contact stretches the objectives of this book somewhat, it is, has been, and will be a cardinal question in SETI. Basically, we need some approach to the question of how humanity will respond to the proven existence of extraterrestrial intelligence. Alongside other issues, we need to carefully parse the concept 'contact'. We also need to break the egotistic question down into the components that it hides: "What will happen to me/us, once we know/contact/communicate with ETI?" Even without the chance of bi-directional enduring communication, we face a problem in contacting ETI. This is a set of questions which has been often discussed among those engaged in SETI. Much of what I have to say in this chapter is drawn from a seminar on the consequences of contact held by NASA in the early 1990s (Billingham et al. 1994). However, I emphasize that the views here are my own spin and interpretation of the proceedings.

What have we learned from our extant example of long-line communication (see the Tairô and the Oranda king, above)? First and foremost, the idea that communication with an ETI will absolutely and immediately upset our social order is possible, but not inevitable. Compared to the overall roster of knowledge any civilized species has (absolutely revolutionary scientific observations aside), the small amount that could be transmitted by radio communication (due to the long time factor) might be minuscule.

Second, to quote Larry Niven TANSTAAFL: There Ain't No Such Thing As A Free Lunch. One can assume that if they/we want an encyclopedia, it shall have to be paid for. Of course there may be highly advanced, utterly benevolent ETI out there, but we should probably also prepare ourselves for Others who also ask themselves "What's in it for me?"

Third, while unlikely, we also need to keep ourselves aware of the possibility of potential antagonistic action for reasons we know nothing about. Fears of Trojan

M. Ashkenazi, *What We Know About Extraterrestrial Intelligence*,
Space and Society, DOI 10.1007/978-3-319-44456-7_13

horse infiltration of our social and political systems may be the stuff of horror films, but can one say confidently they completely unjustified?

Fourth, the idea that Earth could act with one voice to discuss things with an advanced or equivalent ETI civilization is ludicrous. Unless kept a deep dark secret, everyone and his/her relatives will be composing messages to the ETI, and some will actually be sent. Bureaucracies—national, scientific, international—will battle mightily for the privilege of sending the 'official' message, probably to no avail. In practice (as well as morally) they will and should fail. This position opposes both commonly accepted wisdom and official dogma, and will be discussed below.

13.1 A Return to 'Contact'

At some point or other, an event we can describe as 'contact' may occur. That stage is defined by the *general public awareness* of the human scientific establishment of *incontrovertible facts* that indicate there are other *intelligent beings* in the universe. We shall parse each of these terms.

General public awareness It is of course possible that some sinister hidden government agency, bent on extracting as much information as possible that will benefit itself, will come secretly upon this scientific evidence and hoard it to itself. If only given the poor record of secrecy by government agencies, we ought to be somewhat skeptical of such claims. And, if some government *does* come upon such evidence, it is more than likely that it would be unable to keep such claims to itself. The scientific community is fairly small, very communicative, and fairy tight-knit. Except in really dire cases, it seems unlikely that any government would be able to (a) collect sufficient scientists of sufficient stature, (b) sequester these from their normal quite public jobs as teachers and faculty members for a lengthy period of time, and (c) suppress their urge to tell all and sundry of the marvelous discoveries they have made, (d) keep the entire process completely secret. Possible, but highly unlikely.

Incontrovertible facts For SETI, facts (that is, data) could come in several flavors: coherent electromagnetic or other (e.g. the moot gravity waves) form of emissions, that are *undoubtedly* of intelligent, non-human source(s); evidence of ETI technological life such as planetary pollution (gas mixes, planetary light pollution) or massive space construction, archaeological remains in the space of the Solar System, or archaeological remains in Earth, in order of likelihood. Electromagnetic signal detection devices, from radio waves through lasers, have been scanning the skies for decades. The possible construction of large radio and orbital telescopes, now in its infancy, will become commonplace as humanity ventures more easily into space. What characterizes all of these sensors is that (a) most of them are collective efforts, with scientists of various backgrounds and nationalities having access to the data, and (b) there are a multitude of redundancies, ranging from several large telescopes to very savvy and adept backyard amateur astronomers. Any random sample signal is thus likely to be collected by *several* instruments, and not one. To add to that, there is a formal agreement among most astronomers to share information on possible artificial signals (International Academy of Astronautics, ND), and, most signal interpretation needs a team effort, which is one of the hallmarks of modern science: virtually all major scientific discoveries require a team of interpreters to ensure that the data *are* incontrovertible. Once again, though it is possible that a secret government agency *could* intercept, analyze, and demonstrate the facts and analysis are unfalsifiable (and thus to be followed-up by budget increases, if nothing else), *and* sufficiently patterned that they are caused by ETI, it stretches belief somewhat that that could be done in secret, without anyone else being aware of it.

13.1.1 Radiation Contact

In a very real sense, radiation contact is the 'orthodox' view or 'strong' proposition in SETI. After all, the universal velocity limit cannot be ignored. Light travels at the velocity C. Period. This means the fastest form of communication at stellar distances is going to be electromagnetic, whether by laser, maser, or radio. So the first contact, if it occurs, will be by some form of radiation contact. Two types of signals might be detected: active and passive.

Passive signals Passive signals would be radiation evidence that is a side product of some ETI activity that is not intended for communicative purposes. This could be evidence of space structures, or violations of what we assume are the natural order, such as e.g. evidence of Kardashev cultures (Carrigan 2012) or tampering with stars (Learned et al. 2008). It could also be changes in the spectra of planets that indicate industrial or population processes over time (Carrigan 2012).

Active signaling detection Active signals are signals that are produced *for the purpose of communication*. This may not be aimed at us humans, but the intentionality is there. This means that everything will be done by the emitters to ensure the signal is as clear and perceptible as possible. Of course the information content may vary from 0 to 100—completely non-patterned but somehow indicating it is of intelligent origin, versus full of comprehensible information—but in this scenario we would have the possibility of incontrovertible evidence.

13.1.2 Physical Contact

The less likely, and weaker proposition is physical contact. That means that the existence of ETI comes in the form of physical, material evidence. Here there are three possibilities, all of them remote.

Archaeology on Earth On the assumption that somehow, ETI have visited Earth in the past, some so-called evidence has been produced in the past few decades for claims that ETI have visited Earth. So far, all the material collected and claimed for has proven to be the product of human societies, and can be characterized, with the greatest amount of courtesy and open-mindedness, as mistakes. There is no public evidence of any visitation at the macro or micro (that is, nanobots or von Neumann machines) scales. The Earth is still not completely explored, so *perhaps* some such material remains will be found in the future. Once again, it is unlikely some government agency will hide that in Area 51 or 51⊠.

There is *some* evidence of biological contamination of Earth by early cellular organisms (Wickramasinghe 2010) though the evidence is disputed. In any case, biological contamination does not intelligent contact make. Evidence of the products of ETI on Earth are so far nonexistent, notwithstanding a number of inflated claims to the contrary.

Solar System archaeology There have been persistent suggestions about, as well as some effort made to find archaeological traces of ETI visitations in the asteroids or around other planets (Forgan and Elvis 2011). The search is known as SETA. Inasmuch as less than one thousandth of one percent of the volume of the Solar System has been studied at the appropriate resolution for discovery of archaeological remains, it is impossible to say with certainty that there *are no* archaeological remains in space (Freitas 1980, 1983). But there is no evidence that any exist either (Carrigan 2010).

Visitors at the door The orthodox view does not take into account something we discussed in Chap. 2: The ILF perspective. A physical voyage of a thousand years is prohibitive for *humanity*. This does not mean it is prohibitive for a species with a longer lifespan, greater persistence, the ability to aestivate, excellent von Neumann machines, or more need. Crossing interstellar space at STL velocity is certainly possible *in theory*. It would require great expense, persistence, and unless we have in mind micro or nanobots it would literally cost the Earth (or an earth). And well before the fleet of giant spaceships loaded with aestivating ETI arrives, we would know about it from radiation sources. Simply put, Newton's laws would dictate braking emissions aimed more-or-less directly at the Solar System (and thus, our telescopes) would provide incontrovertible *radiation* evidence of a spaceship slowing down to Solar velocities.

13.1.3 Contact Effects

At the initial contact stage, we therefore have five alternatives, each one of which fans out into a wide range of sub-scenarios. Going more or less from the most likely and conservative/orthodox, to the least likely and most radical possibilities, we can examine the implications of each form of contact. As a general rule, these alternatives also account for signal *content*, ranging from 0 (no comprehensible content beyond existence of a signal from ETI), to 100 (fully comprehensible, information rich content).

Passive signal detection For the scientific community of astronomers, bioastronomers, and SETI, the discovery of valid and reliable evidence of emissions from an ETI culture would be the cause for a great deal of jubilation. It would demonstrate the validity of advances in astronomical techniques at the least (not incidentally, increasing astronomy budgets) and of bioastronomical and SETI assumptions, validating decades of dedicated research. What it would mean for the general public is something else. Most likely the discovery will first be contested vigorously (and properly, one might add) by other scientists. This scientific argument would likely dilute the message for most literate people, since most laypeople do not live the it-is/it-isn't nature of scientific argument. After all facts are facts, aren't they? Well, no, they aren't until they have undergone some careful assessment, including falsifying all alternative hypotheses. This seemingly pointless scientific wrangling will likely mean that after a week or two, the information will move over from page one, to inner pages, to the pages of scientific journals. The net result, unless someone can point out evidence of significant advantages to be gotten from the information, the public reaction is likely to be minimal.

Active signal detection The effects of an active signal will likely depend on the content, which as said, might vary between 0 (evidence of a beacon) and 100 (Galactowiki). The type of signal detected, that is, it's communicative content, are an essential variable which will affect human reaction to it (Tarter 1991). Moreover, assuming the signal is clearly a *communicative* signal, that in itself will likely ensure some reaction from the public. Clearly, if it is *also* address to us humans, interest will increase.

Archaeology on Earth Like any kind of contact, the reaction to physical evidence will depend on the content. Discovery of a rusted, unidentifiable metal object would likely do little more than excite fringe publics, whereas the verified finding of the corpse of a little green man or woman, dressed in a spacesuit, would bring about a festival of speculation about origins, importance, and significance. As is always the case, for the rational-minded and the scientific community, 'verified' is the key. Others will no-doubt celebrate or not according to their dispositions.

Solar System archaeology If, in the future, human-designed robots or, further in the future, human explorers find evidence of ETI—machinery, fragments, bodies—the reaction of the public is likely to be stronger for two reasons. First, space preserves remains better than the corrosive oxygen atmosphere of Earth. Second, it is likely that more of the human race will have access to the media than currently. As a result, the human reactions would likely be far more extensive. As with all contact, the greatest issue will be the content. At the higher values, more interest, at the lower values, the news will eventually be subsumed by issues of greater concern. Another factor might enter as well: a greater likelihood that the finders (assuming it is not by telescope) will prefer to keep the information to themselves, since there is a greater likelihood of technological discoveries being preserved in space. The ability to hide such information has been discussed. In a context in which space is a freewheeling arena for private, institutional, corporate, and state interests, it might be possible to sequester such a find. In the tightly controlled exploration environment that currently, and for the foreseeable future characterizes space exploration, the news will likely leak.

Visitors at the door again We have to assume that an actual (preferably alive and moving) sample of ETI will have a great deal of effect on how people react to contact. Physical contact implies, as a public event, a dynamic situation, one that can evolve, potentially produce surprises, and possibly yield either benefits or threats. As a developing story—from the detection of an incoming spacecraft, through attempts to communicate, to actual communication—touches that intellectual element of most humans that is aroused as well by dynamic arts such as music, stories and plays. Disregarding the scientific fallout of such an event, it would be, most crucially, a *story*, and thus something with which people in the street can identify. Whatever the reality of physical contact will be, the story will have multiple texts and interpretations. In a world in which stories—aural, visual, written—constantly bombard us from many sources, and in which virtually everyone is exposed to such stories, the story of contact will generate multiple reactions.

Dynamic and static contact events What has been said about different types of contact, and most prominent in the (least likely) possibility of visitors at the door, also raises another dimension of the effects of contact: dynamism. As a general rule, content would have varied value specifically (to scientists, the military, or capitalists) or generally, as part of the human corpus of narratives, ranging, as said from 0 to 100. For the average individual (however defined or characterized) a critical value of interest would be the dynamics of the story. Most people know little of what happens over the hill. All of us receive most of our information second and third hand, whether by word of mouth written media, or even immediate internet video feed. The vast mass of this information is a form of narrative. The more dynamic the narrative—action shots, people and things moving, events happening—the more likely it is to draw and hold our attention. Yet all of these narratives come to us through intermediaries. These intermediaries—photographers, reporters (professional and amateur), politicians and rabble rousers, uploaders—affect, manipulate and direct our attention in ways they want. In other words, 'we' whomever we are, are getting a biased picture. All things being equal, we are more likely to pay attention to a dynamic narrative (in whatever medium) than to a static one. And if the mills of the gods grind slowly, the mills of science are even slower, which means that the dynamic story—religious, cultural, political, cultural, social interpretations and stories—is the one that will attract and hold the attention of the public.

13.2 Can We Keep Them Away? No Professor Hawking, We Cannot

In 1976, Sir Martin Ryle, the British Astronomer Royal opposed active efforts to communicate with ETI due to concern that they might not be friendly. George Wald, a Nobel laureate, was similarly concerned that ETI would undermine our cultural confidence (Sullivan 1976). The concern was echoed in 2010 by Stephen Hawking, eminent physicist and writer. Unfortunately, we're already exposed. Everyone within about–119 light years (in 2016) is a convenient number—possessing a decent radio telescope and some skill in electronics, presumably knows we are here. Let alone if they have better state-of-the-art electronics than we do. And yes, they know we engage in war, make love, eat, and probably, if they have expended any resources on trying to examine our broadcasts, what we look like, and that some segment of our population expends enormous amounts of energy on the question of body odors.

In other words, if they want to, they will come. Whether that coming will be electronic or physical, we have no way of either hiding ourselves (though greater interstellar distance and interference from stars and dust will attenuate and eventually hide the signals we are emitting) or stopping others from reacting to us. Sending the odd active SETI signal out to space is neither here nor there, nor, for that matter, are the Pioneer 10/11 gold plaques, which, in any case will take centuries to reach any stars, if ever.

There are three major fears expressed about contact with ETI: War between humans and other races, who, being more advanced than us, would likely prevail, subversion of our up-and-coming digital technology, and cultural disruption. It is legitimate to ask for each of these potential scenarios "Why would an ETI do any of these things?" The ethical issue has been dealt with above, but to briefly summarize, there is no reason to suppose that ethics of ETI will *a priori* be humanitarian ones, or that ETI will be benevolent. But between that and active hostilities, there is a large gap. This means that whether we can, or cannot, assess the likely *motivations* for threat is irrelevant. As any military planner knows (or should), the question is what the potential opponents capabilities are. No less, and perhaps more important, to cite Sun Tzu, is knowing oneself: can we assess how humans would react to any of these potential threats?

War of the Worlds War, we are reminded, is a case where two state actors have incompatible objectives that (one of the actors at least, believes) can only be achieved by forcing the other to accept against its perceived self-interests. Achieving this state is done by the organized use of armed force, AKA war. As we noted earlier, war is only the end state in the mutation of disagreements, with more force and less persuasion brought to bear. We also need to be careful about defining all armed conflict as 'war': piracy, brigandage, economic and legal sanctions or impositions backed or protected by strong force may or may not be defined as war.

So the critical question we need to ask about war of the worlds is this: could human interaction with ETI include the use of force, and if so, how could this occur?

The sheer economics of distance between star systems would seem to make the exercise of force between different political entities greatly unlikely. This is not to say that ETI will have states, which are the product of a particular historical development on Earth, but that *some* form of political entity may exist. The problem is that human states are currently defined as entities which have the sole legitimate monopoly on force within their territories. This is not universal. So-called 'pirate states' such as the historical Minangkabau and Sulu maritime states (Antony 2013) and the Norse states in Europe had very little, if any, claim to legitimate monopoly of force. This means that force entrepreneurship was very common: a wealthy chieftain might mount a seaborne expedition to collect 'taxes' from other collectives within a claimed jurisdiction, which the victims would consider 'piracy' (Daxecker and Prins 2013; Jablonski and Oliver 2013).

This means that we cannot assume that forceful action will necessarily be 'war' as we define it currently under the Laws of War, nor that an entire political association (AKA 'state' or some analogy) is involved. In other words, we may be faced with a situation in which force is projected against humanity at large, or certain categories of human collectives specifically by individuals, or associations of different political and economic flavors. To examine this, we shall use the standard (more-or-less) military assessment paradigm: what are the opponent's capabilities and intentions?

Capabilities As noted before, the possibility of mounting a full-scale invasion, whether of conquest or domination would seem to be insuperable. As a general rule of thumb, armed assault against an equally determined and well-established enemy requires personnel odds of 3:1 (or closer to 6:1 if the opponent is well dug in for defense). Though great technological superiority is a modifier, modern human warfare is able to adjust and improvise with great rapidity.

Total human armed forces currently stand at 0.35 % of the total population, that is, some 25–27 million soldiers (World Bank 2016). These of course vary in effectiveness from parade soldiers to fully equipped and trained troops. The potential for mobilization is even greater, even if, given the political situation on Earth today and past experience, many troops will fight against their potential human allies first, before turning their attention to some ETI foe. It does imply that any external invader should include, in their planning, the transportation of about 75 *million* troops to subdue Earth. That number would shrink if the technological efficiency they display were higher (Allenby 2014; Van Creveld 1991), or if they use some form of threat, or actual, total annihilation technology. The number would grow if they intended occupation, which we know from recent history is a complicated, time-consuming, and expensive process. In other words, invasion would likely be so expensive, and so time consuming, that it is doubtful any military power would consider it in any STL regime. At the other end of the spectrum of military possibilities are pinprick attacks, or irregular warfare, perhaps by an ETI that considers all others to be prey, and in which warfare is a socially accepted process, part of a rite of passage, an attempt to establish economic dominance or political hegemony, an economic practice AKA piracy, or for some other reason. Here too, humans are familiar with this type of game, and indeed, have played it in various forms and permutations for centuries (Ferguson 2003; Ferreira 2010; McGinnis 1990; Omeje 2008; Ortiz 2007; Warren 1978; Wadley 2003). There are of course many hybrid varieties of these two extremes, a mix of subversion, force and positive rewards, divide-and-rule, and other forms *all of which humans are familiar with*, many of which have one or several counters.

There are two other possibilities which emerge from the physics of the universe. At one extreme, it is *relatively* easy to commit total genocide if the target race is confined to a single planet. Select a few orbiting rocks of reasonable (1 km < diameter); calculate orbits; speed the rock up to a reasonable fraction of C; Sit back and watch the fireworks. The Earth has been there, seen that (Hildebrand et al. 1991). Here too, we have the advantage of the universal constant. Given Newton's and Einstein's laws it is extremely difficult to sneak up across space on a technological civilization that is reasonably alert: whatever energy one uses to boost one's spacecraft up to cruise speed will also need to be expended to brake, preceding the spacecraft in the form of visible particles, and thus be visible to the target. And by the time the craft has arrived, even if only to the Kuiper belt to establish a construction base, the locals will likely be there already, even if they need to get together and create the necessary technology in a hurry.

A second possibility is a 'slow sneak' attack by use of self-replicating probes (Brin 1983; Bjoerk 2007; von Neumann and Burks 1966), which, if 'properly' programmed, would eliminate the objections to invasion produced above. Such probes would establish a base somewhere outside the reach of humanity, build manufacturing bases for aggressive drones, and finally invade in force.

Of course, 'invasion' might be a misnomer. Hostile action could be in the form of electronic subversion, to which we are extremely vulnerable. Such Trojan horse attacks could be hidden in supposedly friendly messages, or even in a Galactowiki, and then followed, even years later, by physical invasion.

To summarize, there are number of potential capabilities that could be employed for interstellar war, at least in theory. Even assuming great technological advantages, any ETI race would be as restricted as humans by the nature of the universe, unless we are ready to concede exceptions to Einsteinian physics.

Intentions Intentions are, alas, far more difficult to assess than capabilities. Intentions for aggressive action may range from forms of play (Moriarty 1969), through proofs of individual skill (Rosaldo 1980; McGinnis 1990), economics (Abbot and Renwick 1999; Hobsbawm 2000) to national or ethnic domination and genocide.

In the context of interstellar relations, any or all of these might be operative intentions. However, given the universal constants, it seems that the major motivation one needs to be aware of is an attempt to impose an ideology, secular or religious. Humans of course have plentiful experience with that as well (Juergensmeyer 1993; Kippenberg 2003; Konnert 2008; McKinley 2007). The advantage ideological war of any stripe has is that ideas can be imposed (or attempted to be imposed) by a mix of means, including selective force, subversion, argument, and exploitation of the opponents' own beliefs (Humphrey and Davidson 2003; Steger 2008). Whether or not such intentions *succeed* depends on too many factors to tell. However, in the context of humanity, in which we cheerfully engage in ideological debate, sometimes in the form of conflict, one supposes that any ideological invasion would have its work cut out for itself. And if there is one thing we know about war (and humans do know quite a lot), it is that outcomes are unpredictable. An ETI trying to invade might have plenty of practice and knowledge about the process, but there is always something for them to learn.

The Trojan spacehorse We live in a world that is increasingly globalized. Much of that globalization is heavily dependent on the invention and implementation of IT, upon which almost the entire human population depends to some degree or another. In wealthy, well-developed societies, virtually everyone is extremely dependent on IT to produce, sometimes to reproduce, to exchange, and to consume. Virtually every institution is infiltrated by IT and its attendant hardware and software. Even in less-developed societies, IT becomes a part of life, if only in that it determines the prices of products worldwide, the nature of processes that even people in underdeveloped societies are subject to, and often local communications (Agar 2003; de Sola Pool 1977; Nye 1997; Yonazi et al. 2012).

The disruptive potential of an IT attack (by either humans or ETI) is both more likely (since the capability is demonstrable) and more feasible than a physical war of any kind. It has already been noted that the most likely thing to be traded/in demand, are ideas. A sophisticated virus, as we know, is capable of overcoming and transforming a computer system, and providing it with instructions to forward information to a recipient determined by the virus. In effect, interstellar piracy by IT subversion answers both the capacity question (yes, it is possible, and the global web makes it even attractive) and the intentionality question (yes, robbing a race of its arts, ideas, and other intellectual property may well be very attractive for a culture or civilization, or for collective or individual actors). Discussions in the SETI community about electronic warfare (Carrigan 2003) should therefore be taken seriously, even if we have no clue about, nor should we assume, hostile intent from ETI.

Natives and outsiders: Corrupting a culture For many less-developed cultures, the emergence of a world civilization, in fact, any kind of contact with technologically powerful outsiders, has been a losing proposition. By its very nature, the introduction of development to less-developed cultures disrupts the social fabric, creates changes in economic realities, and, intentional or not, is a destructive force

(Ashkenazi 2012; Bordonaro 2009; Epstein 1962; Harrell-Bond 1986; Harris 2008; Kelly 1993). Historically, many simple societies, even where direct physical means were not employed against them, simply disintegrated in the face of new ideas, technologies, and forms of organization and economics.[1]

The question is whether this can happen to complex societies and civilizations as well? One answer has to do with the degree to which the social group is open to, or embraces new ideas. The other has to do with the *variety* of cultural practices that are acceptable. Cultures which are rigid, which are unable to accept heterogeneous sub-cultures encapsulated within them, are more likely to resist the intrusion of new ideas and cultural practices, and thus, paradoxically, are more vulnerable to cultural disruption. Cultures that are able to accept without much ado the injection of new ideas, and the incorporation of new cultural forms, are probably much more difficult to disrupt. Japanese culture, with a general acceptance of many subcultures, which are formalized and have a place within the general culture is an example of the latter (Kinsella 1998). More rigid cultures, where conformity is the norm and there are no frames that allow divergence, would probably be far more vulnerable.

Another way of looking at the issue brings us back to fashion and fads. Where fashions direct towards newness, and fads are acceptable social practices and cultural desiderata, the injection or even disruption of a particular cultural facet or fashion is little more than an addition, or change: something the fashionable society engages in as a matter of course. Societies in which fashion is suppressed by cultural concerns or political fiat, are not likely to be able to resist intentional cultural corruption from the outside, making the work of an ETI cultural attack (intentional or otherwise) much easier.

Can civilization save our cultures? The question is whether there is an antidote to the possibility of cultural corruption? Inasmuch as cultural expression is dynamic, and in particular modern cultures tend to change with fads and fashion, could a technologically and scientifically disadvantaged race defend its culture against interest, power, and dynamics of a more advanced scientific and technological power?

Perhaps civilization is the answer. As noted, a civilization is an immaterial ideological glue that can hold individuals and cultures together in an agreed-upon framework consisting of standards, praxis, and texts. Absent *physical* disruption, which has been discussed above, civilizations as a whole would generally be able to absorb and contextualize new ideas, technologies and techniques, judging them on the basis of established standards. *Of course* civilization standards change with time. They will always be judged on the basis of a selection of earlier standards, and are mutable over the long range. Over the short range, new things—ideas, technologies, social practices—can be judged against extant standards, which serves as a filter and a brake on immediate change.

[1]Not because they 'gave up' as some authors would have it, but because they were overwhelmed by force, ideologies, and economics.

In summary, yes, war is possible under a number of different scenarios. This is not to say that is inevitable, nor that it is likely, given the huge investment required by the aggressor. Certain forms of war are more easily practiced than others. The case of corrupting a culture, which for the current discussion may be considered a hostile act, blends into the issue of cultural merging and even development by well-intentioned powerful ETI.

As a matter of course, the ability to resist any form of intrusion—physical war through cultural disruption—depends heavily on the ability of the receiver to be flexible about the relationship. The existence of a world civilization, while it does not *guarantee* success against cultural intrusion, does ensure that injection of powerful new ideas and technologies can be blended, perhaps encapsulated within the general frame of the civilization.

13.3 A Line with Two Ends: The Bidirectionality of Contact

A song from my youth went "… every fishing line has two ends: at one end, the fisher, at the other, the fish; fishing is about determining which is which" (Hagashash Hakhiver. My translation) In other words, in a social version of Heisenberg's principle, communicating with ETI will change us as well as them; the question is, who's change will be greater?

13.3.1 Will CETI also Change Our ETI Interlocutors?

One of the most crucial things about intercultural contact, including CETI, is the degree to which the 'other' side, whichever it is, understands, and is able to change in respond to communication and contact. We do have examples from human experience of change. The opening of communication to East Asia brought about an entire phenomenon in the arts of *chinoiserie*, in which Chinese, and later Japanese art idioms affected European culture: tea, ceramics, and later painting changed the face of European high culture as the new idioms were incorporated into local culture. African artistic idioms influence modern art. Chinese civilization resisted the intrusion of foreign culture for several centuries, but eventually was forced to incorporate ideas, organizations, and technologies from Europe. In the modern world, the cultural search for new faddish experiences has brought about the incorporation (often without context, uncertain, and vulgar) of new cultural, linguistic, and technical idioms. The search for economic and political advantage inevitably drags along cultural, artistic, and social change. Several parameters will determine the degree to which human culture could affect ETI who communicate with us.

History of contacts The number of previous contacts between the ETI and other ILFs would be a major parameter. If previous exchanges with other ILFs have been undertaken, notably if these exchanges have yielded positive results from the recipient's point of view, then it is likely that the addition of human ideas to the mix will be met with indifference at worst, and enthusiasm at best.

Distance Greater distance implies longer message times, which in turn implies that the receiving society would have more time to adjust to new ideas offered by the other side, and incorporate desirable new things into its own repertoire. We know from the human experience that a rush of new ideas precipitates a great deal of opposition, some of it substantive, much of it politically or economically motivated. The distance of the time filter allows a more balanced absorption of the new.

ETI environments and physiology We cannot ignore the possible issue of different physiologies and environments. Notwithstanding some arguments that ETI are likely to be physiologically similar to human beings (Morris 2011), we can entertain the opposite. And ETI who do not have the same sensorium, or whose environment is radically different from our own, are less likely to adopt our ideas, and thus also less likely to be wary of them, simply because in their experience, these ideas would be irrelevant.

The demands of politics and economics Economics plays a part, since *if* the ETI economy sees value in new things (as does modern humanity) there is likely to be a constant mining (plunder, if you will) of human culture, which in its turn, creates demand and leverage for economic benefits for humans. As a rule of thumb, *if* some communication from human beings—an idea, a technology, a mode of thought or perception—becomes a resource (something desirable in short supply), we can expect two effects. On the one hand, the content will become incorporated, possible 'nativized' in quick order (something that, on Earth, the Japanese excel at. Cf. Ashkenazi 1989). On the other, we can expect powerful political and economic interests to seek to control this resource, perhaps limit its distribution (to ensure it stays a resource) and thus act as a counterbalance to the incorporation and change of their society.

Civilization and culture Perhaps, in the broader scheme of things, the question that needs to be asked is whether we are dealing with representatives of a star-spanning civilization, or of a more volume-restricted local culture. Civilizations by their nature are likely to take change and the incorporation of new ideas in their stride: so long as the change does not catastrophically threaten the pillars of the civilization, and, notably, so long as the transmitters demonstrate the potential for becoming members of the civilization, no harm done. Cultures, on the other hand, are likely to be far more defensive. Whether the culture is coterminous with a political or economic order, cultures normally demand and expect a great degree of stability, and tend to be more rigid than civilizations. As a consequence, we can expect that alien ideas arriving from humanity *to a culture* could be seen (or be portrayed, by e.g. the political classes) as a threat, and thus be suppressed or limited.

13.3.2 What Should They Fear from Us?

We should also ask what, in our communications, could arouse suspicion and/or fear of humanity? Again, given the vast distances involved, the one thing they would be unlikely to fear is physical invasion. However, certain meta-phenomena in human messaging—notably in unintentional signaling such as our electromagnetic broadcasts—might be a cause for worry. Once again, the *primary* variable is likely to be the nature of the ETI society concerned, their environment, physiology, economics and so on. I would suggest however, that certain aspects of our communication (and the list presented here does not pretend to be either axiomatic or complete) could lead an ETI society to scratch its collective head at least, and to be actively suspicious at worst.

The single voice and Babel We are a talkative species. We have opinions, ideas, self-presentation and opposition, all displayed and broadcast at the top of our metaphorical (and sometimes actual) voices. All the time. From an external observers viewpoint, this is a good thing for two reasons. First, of course, a great deal of cross-cutting data about humans can be extracted from this Babel/babble. Big data analytics, only now coming into prominence as a scientific tool (Zakir et al. 2015) are a case in point. Given that we expect ETI to be technically in advance of us, it seems reasonable that they have better data analytics as well. The second point is to my mind more compelling. As a general rule, the fewer voices speaking for a given collective, the more likely it is that the collective is a totalitarian system. Totalitarian systems (e.g. Nazi Germany, Communist Russia and its successor the Federated States, the PRC [China] and the DPRK [AKA North Korea]) are all of this stripe: a small group of political elites are the only ones allowed to air their views. Of course, if the receiving ETI polity is of the same mind, that is, a totalitarian system, they would be overjoyed to receive a message from a fellow traveler. The rest of us would, and should, shudder at the implications. Overall, unless we assume that the well-advanced, wise, benevolent ETI are *also* a totalitarian system, the presentation of multiple voices from Earth is beneficial, rather than the reverse. The lesson to be learned is keep talking: eight billion people speaking in a single voice is cause for concern.

Is your deodorant good enough for ET? News topics aside, a great deal of the mass of broadcasts we produce has to do with presentation of self in its various forms (which is in turn linked directly to our reproductive processes). Body odor and axillary hair (the latter a peculiarly American fetish which puzzles most other cultures), skin whitening or darkening, and body adornment of various forms predominate. An external observer could be led to some very unfortunate conclusions based on the content of these advertisements, notably not being aware of the economic form they represent. In particular, since an active SETI message is *unlikely* to mention these issues (see for instance the anemic and bland message in the Pioneer 10/11 plaques), the shock of coming into contact with such advertisements would be, at the least, puzzling.

Bugs Bunny and the Tasmanian Devil: Parody or reality, who's to tell? Human beings have an individual and collective penchant for the production and expression of lies: counterfactual narratives. These—tales, stories, novels, jokes, plays, movies—occur in all human societies as a form of amusement, a teaching method, explanation for facts, opposition to the way things are, and so on. *For humans* this is an essential and very desirable form of narrative. It is among other things a way to explore ideas, to relieve stress, and to avoid repercussions by presenting dilemmas and problems in a non-threatening form. Talking animals such as Bugs Bunny and the Tasmanian Devil, if taken seriously (as they are, by people exposed to them for the first time, usually, but not always, children), can be threatening. For a race that does *not* fictionalize its experience (or not to the degree that humans do) the presence of such strange creatures among humans would require a radical rethink of human society and of the environment in which the human race lives.

The mere idea of telling lies for fun (for that is what storytelling amounts to) might shock and horrify a race that does not do so. Moreover, for such a race, notably if there is a certain amount of cultural stasis and a swell of demand for change, this innovation might be extremely disturbing, perhaps damaging, and conceivably a move in a conflict.

The crucial point here is not that we tell stories and they don't. What we need to keep thoroughly in mind is the potential of human cultural features—innocent and normal as they might appear to humans—to subvert other cultures, even unintentionally. Of course we are led to suspect that ETI will be more advanced than we humans are, and, by implication, that they will be more advanced in all fields of knowledge and ideas, and therefore inured to phenomena such as Bugs Bunny. Whether this is the case or not, we ought to remember "There are more things in heaven and earth, Horatio, Than are dreamt of in your philosophy." (Shakespeare, *Hamlet*, Act 1 scene 5). In our own history, less technologically developed cultures have had massive impact on their more developed neighbors and even conquerors: consider tobacco and trousers.

13.4 The Consequences of Contact

Between 1990 and 1993 when it was cut short by politicians, NASA engaged in a formal set of workshops called CASETI: Cultural Aspects of SETI (Billingham et al. 1994). The objective was to assess, using a panel composed of astronomers, engineers, social scientists, theologians and philosophers, what the consequence of a successful SETI would be. Of course, we need to say at the outside, the CASETI workshop focused on the 'orthodox' scenario: electromagnetic contact with a species interests in communication. Here, however, we need to have a finer, more nuanced approach.

The crucial question is simple: "How would contact with an ETI (individual or corporate/collective actor) affect humanity?" The issue has repeatedly been asked, and discussed in the literature (e.g. Almár and Race 2011; Baum et al. 2011; Billingham et al. 1994; Davies 1995; De la Torre 2014; Dominik and Zarnecki 2011; Fasan 1986; Finney and Jones 1986; Peters 2011; Sullivan 1979; Tough 1986).

Rather than trying to predict how humans will behave, it may be better to look at a framework for examining potential reactions in a way that is amenable to analysis. Of course, being human, some sort of attempt at prediction is almost inevitable. In a universe of unpredictability, humans *always* try to make generalizations that allow predictability, which implies safety and a comprehensible future.

It is also important to note, as Morrison (1977) has, that the study of any kind of message will continue for years, if not generations. A less comprehensible and unique event message will no doubt lose the public interest, but keep scholars of various disciplines going for a long time. A problem *might* arise with contact of a more comprehensible sort. In such a case, individuals with and without the qualifications they claim to bring to the table, scholars, religious experts, and above all political interests will make their own interpretations, using them, as all phenomena may be used, as an axe to build their particular hobby-horses. This in turn will have an effect on the scientific interpretation and the response of humanity to the message, both internally and in terms of the response to contact (see also, for a more detailed and nuanced discussion, Almár 1995).

The discussion of how humans will react to the evidence of ETI in the broadest sense has been ongoing for the past several decades. *Speculations* about the possibility of other intelligences outside the Earth have a long intellectual history going back centuries, perhaps millennia (Dick 2013). But the question of how humans will react to actual proof has become more prominent for two reasons: first, the scientific Search for Extraterrestrial Intelligence, now in its sixth decade, has raised the likelihood of finding ETI, if only because there now exists an active process. Second, the world is becoming globalized and media-driven, which means that *should* there be some contact, it will percolate throughout humanity with a speed and spread that did not exist before.

A number of formal attempts have been made to estimate what the effect of contact (however defined) would be. Baum et al. (2011) developed a comprehensive scenario analysis in which an attempt is made to create families of potential responses, depending on whether the contact is estimated to be beneficial, neutral, or harmful to humanity, whether there is one or multiple ETI voices from which contact emanates, the type of contact (physical or radiation) and the nature of the message and of the emitting ETI (Almár and Race 2011). The Rio Scale Index, proposed by Almár and Tarter (2010) accepted by the SETI Committee of the International Academy of Astronautics, quantifies the impact of a public announcement about evidence for the existence of ETI. The detailed scale is the product of several variables including credibility of the announcement the class of the reported phenomenon, type of discovery and distance to detected phenomenon. The London Scale Index similarly is an ordinal scale which attempts to assess the potential effects of different types of contact on humanity. LSI variables are life form, nature of the evidence, method of discovery, distance of phenomenon and assessed credibility (Almár and Race 2011). If anything at all can be said with absolute certainty about the effects of contact, it is that, notwithstanding much scratching of heads, and attempts to formalize predictions, there is simply no way to predict how humans will react to contact with ETI. What we *do* know is that there will likely be a vast spectrum of responses from individuals and groups ranging from an indifferent shrug to delighted ecstasy. This will depend on the individual concerned, the nature of the contact phenomenon, and intervening phenomena: media, human actual context (e.g. Is the world undergoing a crisis at the time?), political meddling, and so on.

Rather than reiterating the very detailed discussions and arguments on the topic (cf. e.g. Billingham et al. 1994; Connors 1987; Davies 1995; Denning 2013; Dick 2013; Dominik and Zarnecki 2011; Finney and Jones 1986; Vakoch 2013), I will focus here on four issues which seem to be cardinal, and certainly appear in almost every discussion of the topic: the effect of globalization on human reactions; the effect of contact on religion; the issue of 'moribund cultures' faced with an advanced culture; and the differences between near and long term effects.

The effects of globalization We are, as has been noted earlier, at the cusp of the emergence of a world civilization. This has been brought about by an unparalleled surplus of resources (even if a small fraction of humanity are hogging most of it) and by technologies in transportation and communication which allow individuals and groups unprecedented communication with one another (again, notwithstanding that many individuals have only minimal access to this network). Whether or not a world civilization (as defined above) will emerge is impossible to predict, certainly for the short term, notably as there are countervailing factors, such as e.g. nation state and global corporate attempts to sequester and control technologies for their own ends.

A critical mechanism in this word civilization is the emergence of unprecedented communication potential brought about by the internet and the collapse of the formal media such as newspapers. As Jones (2013) notes, one effect of the proliferation of non-professional journalists in the form of bloggers and film clip sites is a rise in sensationalism and a decline in journalistic standards. This in effect means that the search for sensationalist publication will take two forms. On the one hand, there is likely to be a wash of false, overblown, and unattributed reporting to *any* report of contact with ETI, whether valid or not. To add to that, there will also likely be a flood of pundits pontificating on the discovery, whether qualified or not.

In essence, what this means is that the actual details of the discovery are likely to be swamped by false reporting and pronouncements. Le Bon (1897) was one of the first to suggest that actions of crowds show regularities and patterns. When we come to popular reactions to information about possible contact we need to be aware of three factors that will affect individual and collective reactions over and above the variables presented in the Rio and London scales: *context*, *status*, and *opinion formers*.

The context in which individuals are living—personal, economic, political, international—does much to affect people's reactions to strange news. That is, during times of stress—the Cold War, the Depression, civil wars throughout the world—individuals and groups tend to react aggressively or in panic to added information, which is almost always interpreted in negative ways. This is not surprising: the need to be able to anticipate the immediate future is inherent in much human behavior (Malinowski 1954; Moore and Myerhoff 1977) and humans tend to get upset when unknowns pop up on the horizon, and they then demand 'something be done about that'. It would be surprising if all people who heard of proof of the existence of ETI would react with equanimity, even if the news would have no immediate effect on their lives. That means that we should expect a spectrum of *individual* responses to such information.

Status is crucial to every person. The value that others place on one's social status, as well as one's own understanding of the position are constantly negotiated and renegotiated in practice. Nonetheless, there are social norms which demand responses, and accept them, from particular statuses: a parent to a child, a leader to followers, and so on (Biddle 2013; Lenski 1972; Linton 1973). In many societies, the opinions of high-status individuals are expected to be echoed by their followers, whether they be parents, religious figures, managers, or political figures. And, in most cases, these individuals are understandably conservative, which means that potential disruptions, such as those posed by new and unknown features of the social landscape, are treated with suspicion and even denial. This feature of human societies has two consequences for the discovery of ETI. First, any worldwide communication about the discovery must keep high-status change resistors in mind, and must be tailored in such a way that their fears are accounted for. Second, high-status, conservative individuals are not automatically going to oppose the idea that ETI are "here" (whatever that means to them), but that they can be seeds of reaction (positive or negative) which could precipitate social movements with particular, and not necessarily scientific views, of the implication of the discovery of ETI.

Within any society there are individuals who exert great influence on many other people, what Weber called 'charismatics' (Weber 1964). Such individuals emerge in all societies and in all areas of endeavor (Breuer 1989; Giustozzi 2009; Lebra 1991; Turner 1977). Notably in times of social stress, charismatic individuals emerge who purport to be able to provide solutions to the stressful situations. And though, as Weber notes, the ineffable nature of charisma mutates into formal and bureaucratic forms soon enough, charismatics with a great deal of popular support can emerge urging, sometimes quite powerfully, certain courses of action that will affect the reaction of governments. Subsequent actions in e.g. pursuing more information about the contact, or even forbidding it, remain an issue that needs careful handling, notably since once a charismatic declares a social or political stance, his/her followers are likely to adopt that stance without too much opposition. Religious sects are of that form, but political and military charismatics occur as well. Once established, they remain a feature of the social landscape and are likely to exert a great deal of influence on the political policy-making that follows contact, and even potentially (through funding, if nothing else) on the scientific process.

We ought to keep in mind that all surveys of reactions to the proven discovery of ETI were conducted in developed countries (e.g. Bainbridge 1983; Pettinico 2011), and often among educated elites (university students, people interested in the topic, etc.). The majority of people in the world—residents of less-developed countries—have not been asked (Weigel and Coe 2013). The issue is more far reaching than described by Weigel and Coe, since their example is limited to the very small population of traditional hunting-gathering societies. Much of the population of less-developed countries consists of poor urban residents and peri-urban peasants, where social millenarian movements flourish (Meyer and Staggenborg 1996; Mutabazi 2012; Pinard 1969; Piven 1979). In the presence of stress from poverty and possibly oppression, given an impetus such as confirmed and publicized findings about ETI could create strong millenarian movements, raising expectations, and potentially disrupting 'normal' order, with potentially major consequences.

Will ETI "break" religion? I am not a particularly religious person, but I have a lively interest, and a great deal of patience for elaborate rigmarole and intellectual athletics, which, viewed from the outside, are a major phenomenological feature of religion.[2] Not all people see religion this way. For both opponents of religion (Dawkins 2009; Hitchens 2007, 2008) and its supporters (too many to mention) religion is a form of serious answer, rightly or wrongly, of fundamental questions. These two positions—religion as a behavior, a form of entertaining process and a way of using surpluses, and religion as an intellectual pursuit of fundamental questions—are not necessarily contradictory. However, *from an external perspective*, that is, from the perspective of a potential ETI observer, they might as well be two different phenomena: one a set of behaviors in public and private, the other an elite occupation of specially trained 'cognoscenti'. The possibility that ETI have religion has been discussed previously. On the whole, given both the performative and the intellectual views of religion, it seems possible that ETI *do* have religion. That aside, however, we should ask what kind of reaction would human religions have to the discovery of ETI.

At the intellectual level—that of theologians and professors of religious studies—the evidence is that virtually all religions would have very little difficulty in accepting the presence of ETI. That is to say that the intellectual foundations of most religions are strong enough to either encompass the presence of ETI, or robust enough to withstand potential intellectual challenges they present (Ashkenazi 1992; Griffiths and Oliveira 2009; Peters and Julie 2008; Peters 2009, 2011, 2013; Wiker 2009; Zubek 1961).

[2]For the record I am a Jewish Daoist Shintôist Buddhist polytheist who believes (sometimes) that all things are alive in some fashion, and that computer printers are inherently malevolent.

The somewhat rosy picture portrayed by the elite, that religion will not be unduly affected by contact (to which I am a culpable contributor) needs some reassessment at the grassroots level. Crucially, the higher the social stress (whether from economic or political [including violence] causes) the more likely that religious movements will emerge, and the more valid, well-publicized, and near the contact is (that is, high values on the RSI and LSI scales) the more likely that ETI will be woven into the religious narrative and doctrine, whether as villains or saviors.

At the behavioral, popular level, notably in those communities where religion is deeply combined with populist sentiment, millenarianism, and heterodox practices and beliefs, contact may trigger mass popular movements. There are some precedents that can be considered such as millenarian movements throughout the world. Some religious systems (at a guess, Umbanda/Candomblé/Voudon/Santeria complex, Japanese New Religions, Western New Religions such as Scientology) may well incorporate the contacted ETI into their existing beliefs. The likelihood that the members of other religions will do so willy-nilly is unlikely, as the debate between the positions of Sahlins and Obeyesekere presented by Lowrie demonstrates (Lowrie 2013).

Poorer sections of the population in well-developed countries (as well as some wealthier individuals, cf. Urban 2011; Melton 2001) are also prone to participation in what are called New Religious Movements (Aoki 1972; Beckford 1985; Bird 1982; Kaplan1997; Lewis 1995) in which populism, millenarianism, and occasionally pseudo-scientific elements may predominate in either ideology or ritual. Religious millenarian movements such as Cargo Cults and the John Frum movement in Papua (Burridge 1960), New Religious Movements in Japan (Ouwehand 1964) and others elsewhere are primarily reactions to poverty, powerlessness, and want. They are often triggered, however, by external events, one of the most fervent of which is the presence or potential presence, of powerful strangers. What needs a great deal of thought and possibly preparation, is the likelihood of the emergence of new religious movements in which charismatic leaders utilize the 'proven' fact of the existence of ETI (innocently or knowingly) as a tool to leverage religious, economic, political, and possibly militant sentiment.

Moribund cultures Another theme that emerges frequently in early discussions of contact, is the fear that, faced with the glory and magnificence of ETI technology and culture, human culture and science will shrivel away in despair. Evidence is cited from cultures such as Native Americans, Tierra del Fuegans, Aztecs, Melanesian societies, and Tasmanian Islanders. This seems extremely unlikely and fanciful. As Denning (2013) notes, all the examples cited, where cultures and entire people disappeared were the consequence of often brutal suppression, diseases imported intentionally and not by the colonizers, alcohol, and displacement. While it is possible that contact will challenge some of humanity's fundamental beliefs and scientific findings (Dominik and Zarnecki 2011), there is no evidence that these will bring about a cultural collapse. Some, perhaps many cherished beliefs and practices will go to the wall, but it is more than likely that human cultures, complex as they are, will retain much of their vitality. If anything, we have actual examples of societies that not only accepted cultural change brought from the outside, but actually relished it, and continue to do so. Japan, Turkey, and China represent examples of this kind of proactive cultural absorption (Beckmann 1962; Tsai 1986; Ward and Rustow 1964).

In practice, what is likely to happen is 'mosaic development'. Different actors will absorb (and emphasize) different benefits and concepts from the contact. Inevitably, as Epstein (1962) discovered, those with better initial resources—financial and intellectual—are likely to benefit more from contact. Clumps of those who have benefited more—because their nation, their corporation, or their network are savvy enough to understand and exploit the advances offered—will become evident. Others, those whose cherished ideas, economic position, or political strengths will be challenged or upset by whatever derives from contact, will probably complain, perhaps violently. Unless humanity undergoes a radical change from a capitalist, individualistic ethic to one of greater inclusiveness and sharing (somewhat less likely than discovering the proven existence of ETI), the fundamentals of human society will remain unchanged, and news of its demise unlikely.

Short- and long-term effects The difference between short- and long-term effects of contact is important to keep in mind. Unlike Harrison (2011), I believe that the short-term effects will be more difficult to predict than the long-term, since, as said above, they depend on many factors that are essentially unpredictable—context, how different statuses will interpret the news, how the media will treat the information, and which leaders—established or charismatic—will exploit the news for what purposes. In general, the news might create social disruption at one extreme end of the possibilities, at the other end it will be virtually ignored in favor of the newest crisis, disruption, or fad.

In the long term, we cannot assume the positive ideal results suggested by Tough (1986). As Almár notes, there will be effects in the long term, but these are unlikely to shake the foundations: "…only a really exceptional discovery would have a lasting impact on society" (Almár 1995). Nonetheless, I agree with Dick (2013), that fundamental changes in some areas are inevitable. What is likely to happen, as has happened throughout history, is the mosaic adoption of newness described above: certain actors will absorb more, and become more influential, for good or for ill, other actors will not. It *is* highly likely that information-rich messages will become a resource. That will happen particularly if the information is hard to acquire (that is, if large radio-telescopes, or even space-based telescopes are necessary to acquire the information) in which case technologically advanced nations, as well as private enterprises will engage in a technological gold rush. Two results would be expected in such a case: those who are able to produce better and more sensitive technologies, better translation algorithms, and a better understanding of the cultural nuances of the communicating ETI, will benefit. The rest—underdeveloped countries and their populations—will have the crumbs. The second effect, notably if better reception is the key to more information, is that the human space effort will become a well-funded enterprise with a multitude of interests competing to develop and possibly colonize space.

At the scientific and political levels, it is possible that changes will occur in the long term, notably in scientific techniques and possibly in government focuses: possibly a consequence of the parallax provided by ETI, discussed previously. At the socio-cultural level, presumably some actors will experiment with new social forms suggested by information rich messages. However, given differences in physiology and phylogeny, and the braking effects of the web of relationships between different social forms among humans, it is unlikely that ETI social ideas will find much immediate purchase. Humans, notably in those societies and cultures which crave change and are subject to fashion (which is most of the developed world) will adopt superficial features as a matter of fashion, whether these are some art form, particular aesthetics, or other features that can be translated into a human idiom. This is likely to be accelerated if and when the ETI counterparts are representing a civilization, rather than a specific culture or group, since entry into the civilization is conditional upon accepting the civilization's standards.

Mentifacts, provided they could be translated into a human idiom, would also likely be picked up by at least three types of actors: scientists, religious and philosophical experts, and fashion leaders. New ideas provided by ETI would likely cause a scientific revolution in many disciplines, something scientists *should* welcome, even if it upsets their hard-earned truths. Which disciplines would benefit, it is impossible to predict, since it depends to a great deal on the content of the message. For the philosophers, religious and otherwise, the potential for new ideas, and the reification of old ones in light of information from ETI would be irresistible. However, religious ideas might take years to become common wisdom, since established religions are notoriously conservative. Fashion leaders, whose primacy is dependent on being ahead of the crowd, and whose status is a function of their ability to access and selectively transmit information (Bikhchandani et al. 1992) are likely to be drawn quickly to ideas and aesthetics that *could* be derived from ETI information, even if this does not match the reality of the information itself; in other words, 'pseudo' ETI ideas will be presented as fashion statements.

Finally, we can expect that any artifactual information will be heavily sought after, and that even the smallest items of information will be considered resources. Among other things, this will mean, notably if the information is hard to come by (coded, or reception-poor) that useful snippets of information, as well as their technological potential might be jealously guarded by governments and corporations.

13.4.1 At the Individual Level

Appropriately, most individuals are concerned with their own affairs. Great news and alarums from afar are a background noise, unless they impact directly on the individual concerned. It is useful therefore to consider the impact of contact on the individual. Obviously, this should be qualified but the nature and content of contact, for which the Rio and London scales provide an ample picture. Nonetheless, for most individuals, who are unlikely to come into direct contact with the communication directly (Jones 2013), the content of the message may not be crucially important.

Subjective feeling The average individual is likely to ask whether and how the discovery will affect her livelihood and family. Anxiety about unpredictable changes is a normal human response to potential change (Hallowell 1977; Korten and Korten 1972; Moore and Myerhoff 1977). We can assume that news about ETI, and, in particular, the interpretation put upon it by opinion leaders the individual respects, may well cause either anxiety, or a sense of euphoria, and anything in between. However, it is unlikely that these feelings will persist over the long term, as processes of routinization take hold. Of course certain ideas will appeal to fringe groups composed of those suffering from anomie (Durkheim 1961), but the degree that influences other individuals is again contextual: basically, given that immediate effects of most forms of contact are unlikely, the entire issue is likely to be routinized, unless seized upon as a political lever. Of course, scientific advances offered by the contact process will affect individual lives. On the positive side, there is the potential for medical advances, new transportation and communication opportunities, and many other changes. On the negative side, a wave of new technologies could make occupations redundant, and entire industries might be sidelined. However, we live in an age where change is the norm, rather than the exception, most individuals have been faced at some time or other in their lives with occupational problems, and during the lifetime of many readers, many technologies, ranging from donkey carts through the steel industry have become moribund in more than one location. In summary, yes, there will be unpleasant disruptions of life for many, but this is something the human race has experienced throughout the twentieth century, and shows no sign of abating.

13.4.2 Collective Forms

Human collective forms—groups both face to face and complex organizations and corporate groups (Smith 1974)—come in many and varied forms. Moreover, as discussed previously, they are social adaptations to conditions—ecological, psychological, physiological—that humans have experienced for millennia. Occasionally, but rarely, humans come up with some new wrinkle, but on the whole, we have tried most possible variations that suit us as a race. It is *conceivable* that ETI might provide examples or arguments for new forms of organization based on their own experiences. However, given the fact that human social forms have had such a long time to evolve to fit human circumstances, it is highly doubtful that social forms from ETI would be easily adopted: they are after all the product of another environment.

What humans *might* learn from ETI is how well (or ill) certain forms of organization have worked for them. This does not mean that humans will jettison those organizational forms that they hold dear simply because someone living near a star light years away says it doesn't work. Often forms of organization have a strong ideological component that is held to dearly by those who practice it, even when the effects of that form are pernicious and dangerous: should advanced ETI announce that capitalism/communism are doomed to failure, would partisans of either socio-economic ideology jettison them? Highly unlikely, in fact, completely improbable.

Nevertheless, as evidence accumulates that one or another dearly-held system has been tried, tested, and found wanting by ETI, so to will more questions be raised by humans about that system's use. Of course there is unlikely to be an immediate change, but as new forms become familiar, so too negative evidence will help with the business of running our own lives. It is well to keep in mind that the exchange of social forms is something humans do all the time, as any industrial management consultant, or business manager will tell you. So borrowing a new form of organization from an ETI sources, trying it out, and making it a staple of administration or management, or rejecting it entirely, is likely more a matter of time than otherwise. What we can expect, as with most ETI-derived social innovations, is mosaic acceptance: some people or organizations will try the innovation out, others will adapt and then try it, and others will ignore it. Darwinian processes will then decide whether this idea, social form, process, or whatever, flourishes or fails.

Nation-states Nation-states are the primary form of overarching political system today. As noted, virtually all nation-states we know of are effectively empires, where a smaller or larger dominant elite calls the shots. Since such elites are reluctant (to say the least) to relinquish their dominance (benevolent and generous as it may be), it seems that even if ETI provide us with templates for a different form of political organization, that will be rejected by virtually all nation-states. It is of course possible that certain elements from the putative proposed templates (or examples of such) will be adopted piecemeal. I'd suggest no one hold their breath waiting for political change to be brought about by ETI advice.

The world community The assumption that ETI will serve as social messiahs to the human race, which is often expressed in SETI literature, does not stand the test of reality. The ways humans have of organizing themselves are extremely robust for three reasons. First, they represent long adaptations to the human ecology: they have proven themselves, sometimes for thousands of years. We have tried discarded, and developed many social forms at many levels of social life, and the ones we live with often have lengthy periods of fitting to their human purposes. Second, organizational forms among humans are deeply interlocked. Change, for most humans can be pernicious, and there is no guarantee that new forms will actually work for humans as advertised. Third, membership, notably senior membership, in large and complex groups (which usually determine agendae, both political and economic) have huge investments in the organizational forms they belong to. Unsurprisingly, they are likely to resist organizational changes quite strongly, and it is unlikely that substantive changes in organizational form will have much support, or will be implemented at all (see, for example, the case of Afghanistan, Ashkenazi 2012).

Having worked for a decade alongside the UN, I am under no illusions as to its benevolence, effectiveness, or ability to lead the world community. Nonetheless, for better or worse, the UN is a major international voice. The same is no doubt true of other international bodies, ranging from the FAO, through the OECD, the AU and EU, and beyond. Simply put, such organizations, once put in place develop a bureaucratic culture of their own. They tend to be more concerned about budgets and their own bureaucratic survival, than about the objectives they are ostensibly in place for.

To assume that suggestions from an ETI about improved services and organization will be implemented (unless the bureaucrats in charge see in this change some benefit to themselves and their offices) is slightly ludicrous. International organizations serve interests—of nations and of those who run them—and rarely accept demands for change or any form of criticism gracefully. Again, Darwinism would raise its head. An organization X might adopt some of the principles suggested, and, if successful, it would be imitated by others. The better fit would of course not provide too much pressure, and change would occur slowly, unless it offered immediate benefits: for example, in military operations and warfare.

The SETI declaration: Should we all speak with a single voice? A major debate in SETI is the question of who should formulate a response to a potential signal from a confirmed ETI source. The general and official point of view, exemplified in agreement is that was formulated in 1989 by a committee of the International Academy of Astronautics (IAA) —the SETI post-detection protocol—was subsequently been endorsed by the International Institute of Space Law (IISL), the Committee on Space Research (COSPAR) of the International Council for Science (ICSU), the International Astronomical Union (IAU) and the International Union of Radio Science (URSI). It mandated a single response devised by a joint scientific and political committee. Allow me to differ. Vigorously.

I have already noted above that international organizations—the UN, AU, EU, OECD or UEFA—are clubs of *nations*. They represent the interests of their constituent members. What they do *not* do, is represent humanity as a whole, or the planet Earth, or individual freedoms and interests (which, given the ways in which humans behave in a crowd, is probably a good thing. See Le Bon 1897). There are three other potential collective/transnational fora. Commercial organizations, ranging from one-country based ones such as Korean *chaebol* or Japanese *keiretsu*, to international conglomerates such as Monsanto and GE. These presumably have the economic clout to develop contacts with ETI, but could they speak for humanity? Doubtful, since their interests and control are shareholders: the small percentage of those wealthy enough to own large blocks of shares. We do have a large number of international organizations who claim to be passionate advocates of *human* and world interests: International Non-Government Organizations, who act in support of many worthy causes. These range from Amnesty International and

Oxfam to Save the Tiger and Greenpeace. Could they represent humanity to ETI? Doubtful. Like all large organizations, they are prisoners of both the human tendency to warp reality to their own design, ignoring inconvenient facts in the process, and the usual internal squabbling over prestige, personal power, and access to funds. Finally, we should consider the world of research scientists: after all, they would be responsible for identifying and verifying contact in the first place. Scientists, for all the posturing and the ideology of disinterested curiosity and methodological verities that drive their activities are nonetheless human. This means that they have the normal share of nationalism, religious belief, intolerance and other personal features, making them unreliable representatives of all of humanity.

In summary, the idea that any organized group in the 21st century can, or should be mandated to, represent a one-voice policy to an ETI correspondent is dangerous. *Of course* some organizations will take that mandate upon themselves, whether the government of the US/EU/Federated States/PRC/India/…, or a scientific body. But then so should Joe Bloggs and his friends in any backyard they can find: humanity is a babble, not a concert, and we would find it difficult to fool any reasonably sensible ETI into believing otherwise.

13.5 Modernity, Culture, and CASETI: Nothing New About Newness

Peculiarly enough, there is nothing catastrophic about the impact of successful SETI, however it comes about. One thing that has characterized humanity since the tenth millennium BCE, is change. In my lifetime, we went from propeller driven planes to hypersonic jets, from computers the size of a room to ones we carry in our hands and embed in our bodies, from dying of polio and a host of other diseases, to relative freedom from those things. In the social sphere, even greater changes have occurred. Greater human equality in many places, a better understanding and accommodation of human differences, greater attention to the environment we live in, in all its aspects. Not that things are perfect, by any means, but change has been the order of the day. So, in the broadest sense, a successful SETI encounter will, almost certainly, be a seven day wonder, not necessarily a game changer. Of course, in the highly unlikely event of a far-advanced totally benevolent society of ETI communicating with us, major changes could occur (though, being human, we will not all like everything they have to offer). The same would be true in the case of a belligerent contact (and again, being human, some people will be for, some against, what they have to offer).

I believe (and emphasize the 'believe') that it is unlikely (extreme cases above excluded) that once the newsworthiness of the communication has died down, there will be an immediate major upheaval. What is more likely to happen is a gradual series of economic, technical, and scientific changes, the same sort of changes we are experiencing in the early 21st century. As we have seen, the effects on science, thus on technology, and therefore on industry and economics, may well be profound if we are able to access new scientific and technical data and theories. The influence of new ideas usually has two results: the passing of oxcart industries, and the beginning of automotive industries. In other words, some people will find themselves out of jobs, but new jobs are likely to open up.

13.6 Some Conclusions from Chapter 13

We need to remember a few major points.

1. The future is, by definition, unpredictable. We can make assumptions—based on analogy, logic, projection or any other mechanism—about what will happen. Predict we cannot.
2. Based on a reasoned assessment of the odds, most thinking since the halcyon days of the start of the SETI effort, is that the immediate human response is likely to be muted. This, I believe, ought to be qualified in two ways. One, which has been emphasized by among others Baum et al. (2011), Neal (2014), and Lowrie (2013) is that the content of contact will determine the human response. In other words, the more accessible and rich the content, the more immediate impact it will have. Two, the context in which the information will be received—social economic or other stress or its absence—will have a major effect on the immediate response.
3. There will be a difference between the short-term and long-term effects. Short-term effects are difficult to predict due to the volatile nature of responses from a multitude of different actors with greatly varied interests. Long-term effects, which would follow more-or-less known patterns of change in human societies which have been, and are being explored by social scientists may be more predictable.

We might add a few comforting slogans, for those of us who are uneasy about what contact implies (not excluding this author).

13.6.1 Keep Cool! We've Been Here Before

Change is in our very blood. Since we, the human species, decided to try out a new form of evolution—cultural evolution—change has become the norm. Slowly at first, growing in speed over the past century, our society is changing, our experiences are changing. We know how to live with it.

13.6.2 Be Happy! Newness Is Part of Our Heritage

Various idiosyncratic old geezers, religious fundamentalists, and other intellectual fogeys aside, change as said is a regular occurrence to humans. This is not to say it is painless. To the contrary, like all births, change is extremely painful. Continuous change even more so. So painful that many ideologies, ranging from rank survivalists to religious fundamentalists, are making efforts, sometimes extremely violent ones, to stop change in its tracks and even turn the clock back. One can understand, even if one does not sympathize nor agree. Communication with ETI could bring some of these ideologies crashing down, whereas others might be reinforced (humans are amazingly capable of rationalizing poor decisions). What is crucial is not *stopping* the effects of change brought about by successful CETI, but being able to ride the changes, perhaps manage them at different levels.

13.6.3 Stay Safe! Sometimes, Greeks Bear Gifts

Though it is highly unlikely (certainly in the case of radiation communication, less so in the case of physical contact improbable though that may be) that communication with ETI would be belligerent, this does not mean we need be certain of benevolence. Even benevolence, as we have seen earlier, can mean different things to different species. Thus it behooves us to examine all gifts, free or paid for, with extreme care.

13.6.4 Enjoy! 'Cause There's Little One Can Do to Stop the Train

If SETI/CETI are successful, we need to accept and enjoy this new thing the universe has bestowed upon us. If it is ultimately (whatever that means) unsuccessful, we have nonetheless benefitted from the ride, and hopefully enjoyed the scenery.

References

Abbot, J., & Renwick, N. (1999). Pirates? Maritime piracy and societal security in southeast asia. *Global Change, Peace & Security, 11*(1), 7–24.

Agar, J. (2003). *Constant touch: A global history of the mobile phone*. Duxford, Cambridge: Icon Books.

Allenby, B. R. (2014). Are new technologies undermining the laws of war? *Bulletin of the Atomic Scientists, 70*(1), 21–31.

Almár, I. (1995). The consequences of a discovery: Different scenarios. In S. Shostack (Ed.), *Progress in the search for extraterrestrial life. Astronomical society of the Pacific conference series 74*. San Francisco: Astronomical Society of the Pacific.

Almár, I., & Race, M. S. (2011). Discovery of extra-terrestrial life: Assessment by scales of its importance and associated risks. *Philosophical Transactions of the Royal Society of London A: Mathematical, Physical and Engineering Sciences, 369*(1936), 679–692.

Almár, I., & Tarter, J. (2010). The discovery of ETI as a high-consequence, low-probability event. *Acta Astronautica, 68*(3–4), 358–361.

Antony, R. J. (2013). Turbulent waters: Sea raiding in early modern south east Asia. *The Mariner's Mirror, 99*(1), 23–38.

Aoki, T. (1972). Some remarks on the new religion movement in contemporary Japan. *East Asian Cultural Studies, 11*(1–4), 101–112.

Ashkenazi, M. (1989). Japanization, internationalization and aesthetics in the Japanese meal. In. A. Boscaro et al. (Eds.), *Rethinking Japan*. London: Paul Norbury.

Ashkenazi, M. (1992). Not the sons of Adam: Religious response to ETI. *Space Policy, 8*(4), 341–350.

Ashkenazi, M. (2012). Development is destruction, and other things you weren't told at school. In V. Franke & R. Dorff (Eds.), *Conflict management and 'whole of government': Useful tools for U.S. national security strategy*? (pp. 91–125). Carlisle, PA: Strategic Studies Institute.

Bainbridge, W. S. (1983). Attitudes towards interstellar communication: An empirical study. *Journal of the British Interplanetary Society, 36*, 298–304.

Baum, S. D., Haqq-Misra, D. J., & Domagal-Goldman S. D. (2011). Would contact with extraterrestrials benefit or harm humanity? A scenario analysis. *Acta Astronautica, 68*, 2114–2129.

Beckford, J. A. (1985). *Cult controversies: The societal response to new religious movements*. London: Tavistock Publications.

Beckmann, G. M. (1962). *The modernization of China and Japan*. New York: Harper and Row.

Biddle, B. J. (2013). *Role theory: Expectations, identities, and behaviors*. New York: Academic Press.

Bikhchandani, S., Hirshleifer, D., & Welch, I. (1992). A theory of fads, fashion, custom, and cultural change as informational cascades. *The Journal of Political Economy, 100*(5), 992–1026.

Billingham, J., Roger, H. et al. (Eds.). (1994). *Social implications of the detection of an extraterrestrial civilization: A report of the workshops on the cultural aspects of SETI*. Mountain View, California: SETI Press.

Billingham, J., Heyns, R. et al. (Eds.). (1994). *Social Implications of the Detection of an ExtraterrestrialCivilization: A report of the workshops on the cultural aspects of SETI*. Mountain View, CA: SETI Press.

Bird, F., & Reimer, B. (1982). Participation rates in new religious and para-religious movements. *Journal for the Scientific Study of Religion, 21*(1), 1–14.

Bjoerk, R. (2007). Exploring the galaxy using space probes. *International Journal of Astrobiology, 6*(02), 89–93.

Bordonaro, L. I. (2009). 'Culture stops development!': Bijago youth and the appropriation of developmentalist discourse in Guinea-Bissau. *African Studies Review, 52*(2), 69–92.

Breuer, S. (1989). Magisches Und Religioses Charisma: Entwicklungsgeschichtliche Perspektiven [Magical and Religious Charisma: Outlines of a Developmental History]. *Kolner Zeitschrift fur Soziologie und Sozialpsychologie, 41*(2), 215–240.

Brin, G. D. (1983). The 'great silence': The controversy concerning extraterrestrial life. *Quarterly Journal of the Royal Astronomical Society, 24*, 283–309.

Burridge, K. (1960). *Mambu: A Melanesian millenium*. London: Methuen & Co.

Carrigan, R. A. (2003). The ultimate hacker: SETI signals may need to be decontaminated. In R. Norris & F. Stootman (Eds.), *Bioastronomy 2002: Life among the stars. Proceedings of IAU symposium*. San Francisco: Astronomical Society of the Pacific.

Carrigan, R. A. (2010). Starry messages: Searching for signatures of interstellar archaeology. F—arXiv:1001.5455.
Carrigan, R. A. (2012). Is interstellar archeology possible. *Acta Astronautica, 78*, 121–26.
Connors, M. M. (1987). *The consequences of detecting extraterrestrial intelligence for telecommunications policy*. Unpublished paper.
Davies, P. C. W. (1995). *Are we alone? Implications of the discovery of extraterrestrial life*. Harmondsworth: Penguin.
Dawkins, R. (2009). *The god delusion*. New York: Random House.
Daxecker, U., & Prins, B. (2013). Insurgents of the sea institutional and economic opportunities for maritime piracy. *Journal of Conflict Resolution, 57*(6), 940–965.
De la Torre, G. G. (2014). Toward a new cosmic consciousness: Psychoeducational aspects of contact with extraterrestrial civilizations. *Acta Astronautica, 94*(2), 577–583.
de Sola Pool, I. (1997). *The social impact of the telephone*. Cambridge, MA: MIT Press.
Denning, K. (2013). Impossible predictions of the unprecedented: Analogy, history, and the work of prognostication. In D. A. Vakoch (Ed.), *Advances in astrobiology and biogeophysics* (pp. 301–312). Berlin, Heidelberg: Springer.
Dick, S. J. (2013). The societal impact of extraterrestrial life: The relevance of history and the social sciences. In D. A. Vakoch (Ed.), *Advances in astrobiology and biogeophysics* (pp. 227–257). Berlin, Heidelberg: Springer.
Dominik, M., & Zarnecki, J. C. (2011). The detection of extra-terrestrial life and the consequences for science and society. *Philosophical Transactions of the Royal Society of London A: Mathematical, Physical and Engineering Sciences, 369*(1936), 499–507.
Durkheim, E. (1961). Anomie. *Comparative Politics*, 504–512.
Epstein, T. S. (1962). *Economic development and social change in south India*. Manchester: Manchester University Press.
Fasan, E. (1986). Discovery of ETI: Terrestrial and extraterrestrial legal implications. *International Astronautical Federation*. IAA-86-476. Oxford: Pergamon Press.
Ferguson, B. R. (2003). The birth of war. *Natural history, 112*(6), 28–35.
Ferreira, R. (2010) Asymmetric warfare in African conflicts. *Scientia Militaria: South African Journal of Military Studies*. http://ajol.info/index.php/smsajms/article/view/55855. Accessed June 2, 2016.
Finney, B. R., and Jones, E. M. (1986). "The Impact of Contact." IAF Congress paper. IAA 86–471.
Forgan, D. H., & Elvis, M. (2011). Extrasolar asteroid mining as forensic evidence for extraterrestrial intelligence. *International Journal of Astrobiology, 10*(4), 307–313.
Freitas, R. A. (1980). A search for natural or artificial objects located at the Earth-Moon libration points. *Icarus, 42*, 442–447.
Freitas, R. A. (1983). If they are here, where are they? Observational and search considerations. *Icarus, 55*, 337–343.
Giustozzi, A. (2009). *Empires of mud: War and warlords in Afghanistan*. New York: Columbia Uiniversity Press.
Griffiths, M., & Oliveira, C. (2009). On god and aliens. http://www.lablit.com/article/487.
Hallowell, A. I. (1977). The social function of anxiety in a primitive society. *Culture, Disease, and Healing*, 132–137.
Harrell-Bond, B. (1986). *Imposing aid*. Oxford: Oxford University Press.
Harris, V. (2008). Development workers as agents of cultural change. In A. Kent & D. Chandler (Eds.), *People of virtue: Reconfiguring religion, power and moral order in Cambodia today* (pp. 293–309). Copenhagen: NIAS Press.
Harrison, A. A. (2011). The search for extraterrestrial intelligence: Astrosociology and cultural aspects. *Astropolitics, 9*(1), 63–83.
Hildebrand, A. R., Penfield, G. T., et al. (1991). Chicxulub crater; a possible cretaceous/tertiary boundary impact crater on the Yucatan Peninsula, Mexico. *Geology, 19*(9), 867–871.
Hitchens, C. (2007). *The portable atheist: Essential readings for the nonbeliever*. Boston: Da Capo Press.

Hitchens, C. (2008). *God is not great: How religion poisons everything*. Toronto: McClelland & Stewart.

Hobsbawm, E. J. (2000). *Bandits*. London: Weidenfeld & Nicolson.

Humphrey, M., & Davidson, A. (2003). Political violence and terrorism. *Social Analysis, 47*(1), 152–169.

Jablonski, R. S., & Oliver, S. (2013). The political economy of plunder economic opportunity and modern piracy. *Journal of Conflict Resolution, 57*(4), 682–708.

Jones, M. (2013). Mainstream media and social media reactions to the discovery of extraterrestrial life. In D. A. Vakoch (Ed.). *Advances in astrobiology and biogeophysics* (pp. 313–328). Berlin, Heidelberg: Springer.

Juergensmeyer, M. (1993). *The new cold war?: Religious nationalism confronts the secular state*. Berkeley: University of California Press.

Kaplan, J. (1997). *Radical religion in America: Millenarian movements from the far right to the children of Noah*. Syracuse: Syracuse University Press.

Kelly, P. F. (1993). *Development as degradation: Aquaculture, mangrove deforestation and entitlements in Bataan, Philippines*. McGill University.

Kinsella, S. (1998). Japanese subculture in the 1990s: *Otaku* and the amateur manga movement. *The Journal of Japanese Studies, 24*(2), 289–316.

Kippenberg, H. G. (2003). Aktuelle Religiose Gewalt Aus Handlungstheoretischer Perspektive [Current religious violence from an action theory perspective]. *Paideuma, 49*, 31–59.

Konnert, M. (2008). *Early modern Europe: The age of religious war, 1559–1715*. Toronto: University of Toronto Press.

Korten, D., & Korten, F. (1972). *Planned change in a traditional society: Psychological problems of modernization in Ethiopia*. New York: Praeger.

Learned, J. G, Kudritzki R.-P., et al. (2008). The Cepheid galactic internet. F—arXiv:0809.0339.

Le Bon, G. (1897). *The crowd: A study of the popular mind*. London: Fischer.

Lebra, T. S. (1991). Resurrecting ancestral charisma: Aristocratic descendants in contemporary Japan. *Journal of Japanese Studies, 17*(1), 59–78.

Lenski, G. E. (1972). Status crystallization: A nonvertical dimension of social status. In P. F. Lazarsfeld, A. K. Pasanella, & M. Rosenberg (Eds.), *Continuities in the language of social research* (pp. 73–82). Princeton, NJ: Princeton University Press.

Lewis, J. R. (1995). *The gods have landed: New religions from other worlds*. New York: SUNY Press.

Linton, R. (1973). Status and role. In L. Bohannan & H. Glazer (Eds.), *High points in anthropology* (pp. 187–199). New York: Alfred A. Knopf, Inc.

Lowrie, I. (2013). Cultural resources and cognitive frames: Keys to an anthropological approach to prediction. In D. A. Vakoch (Ed.), *Advances in astrobiology and biogeophysics* (pp. 259–269). Berlin, Heidelberg: Springer.

Malinowski, B. (1954). *Magic, science and religion*. New York: Doubleday Anchor Books.

McGinnis, A. J. (1990). *Counting coup and cutting horses: Intertribal warfare on the northern plains* (pp. 1738–1889). Cordillera, CO: Johnson Books.

McKinley, M. (2007). *Economic globalisation as religious war: Tragic convergence*. London: Routledge.

Melton, J. G. (2001). The church of Scientology. *Studies in contemporary religions* (Vol. 6(1)). Chicago: Signature Books/CESNUR (Center for Studies on New Religions).

Meyer, D. S., & Staggenborg, S. (1996). Movements, countermovements, and the structure of political opportunity. *American Journal of Sociology, 101*(6), 1628–1660.

Moore, S. F., & Myerhoff, B. (1977). Introduction: Secular ritual: Forms and meanings. In Moore, S. F., & Myerhoff, B. (Eds.) *Secular ritual*. Amsterdam: Van Gorcum.

Moriarty, J. R. (1969). Ritual combat: A comparison of the Aztec "war of flowers" and the medieval "melée". Presented at a meeting of the Southwestern Monuments Association in 1968.

Morris, S. C. (2011). Predicting what extra-terrestrials will be like: And preparing for the worst. *Philosophical Transactions of the Royal Society of London A: Mathematical, Physical and Engineering Sciences, 369*(1936), 555–71.

Morrison, P., Billingham, J., & Wolfe, J. (Eds.). (1977). *The search for extraterrestrial intelligence*. Moffet Field, CA/New York: NASA/Dover Publications.

Mutabazi, M. M. (2012). *Militia movements in Africa: A conceptual examination and reformulation*. SSRN 2041763 .

Neal, M. (2014). Preparing for extraterrestrial contact. *Risk Management, 16*(2), 63–87.

Nye, D. E. (1997). Shaping communication networks: Telegraph, telephone, computer. *Social Research, 64*(3), 1067–1091.

Omeje, K. C. (2008). *Extractive economies and conflicts in the global south*. London: Ashgate Publishing Ltd.

Ortiz, C. (2007) The private military company: An entity at the center of overlapping spheres of commercial activity and responsibility. *Private Military and Security Companies*, 55–68.

Ouwehand, C. (1964). *Namazu-E and their themes: An interpretative approach to some aspects of Japanese folk religion*. Leiden: E. J. Brill.

Peters, T. (2009). Astrotheology and the ETI myth. *Theology and Science, 7*(1), 3–29.

Peters, T. (2011). The implications of the discovery of extra-terrestrial life for religion. *Philosophical Transactions of the Royal Society of London A: Mathematical, Physical and Engineering Sciences, 369*(1936), 644–655.

Peters, T. (2013). Would the discovery of ETI provoke a religious crisis. In D. A. Vakoch (Ed.), *Advances in astrobiology and biogeophysics* (pp. 341–355). Berlin, Heidelberg: Springer.

Peters, T., & Julie, L. F. (2008). Peters ETI religious crisis survey. See http://www.counterbalance.org/search/search.php.

Pettinico, G. (2011). American attitudes about life beyond earth. In D. A. Vakoch & A. A. Harrison (Eds.), *Communication with extraterrestrial intelligence* (pp. 102–117). New York: Berghahn.

Pinard, M. (1969). Poverty and political movements. In J. L. Price (Ed.), *Social facts: Introductory readings* (pp. 470–477). London: Collier-Macmillan Ltd.

Piven, F., & Cloward, R. (1979). *Poor people's movements*. New York: Vintage.

Rosaldo, R. (1980). *Ilongot headhunting, 1883–1974: A study in society and history*. Palo Alto: Stanford University Press.

Smith, M. G. (1974). *Corporations and society*. Chicago: Transaction Publishers.

Steger, M. B. (2008). *The rise of the global imaginary: Political ideologies from the French revolution to the global war on terror*. Oxford & New York: Oxford University Press.

Sullivan, W. (1976). Astronomer fears hostile attack; would keep life on earth a secret. *New York Times, 4*.

Sullivan, W. (1979). What if we succeed? *Cosmic Search, 1979*, 37–39.

Tarter, D. E. (1991). Signal typology as a variable in interpretation and reporting of a SETI discovery. Paper presented to International Space University, Toulouse, France.

Tough, A. (1986). What role will extraterrestrials play in humanity's future? *Journal of the British Interplanetary Society, 39*, 491–498.

Tsai, W. H. (1986). The modernization of four Chinese societies: China, Taiwan, Hong Kong, and Singapore. *Kuo Li Tai wan Ta Hsueh She Hui Hsueh K'an/National Taiwan University Journal of Sociology, 18*, 163–190.

Turner, P. R. (1977). Witchcraft as negative charisma. *Culture, Disease, and Healing*, 226–230.

Urban, H. B. (2011). *The Church of Scientology: A history of a new religion*. Princeton: Princeton University Press.

Vakoch, D. A. (Ed.). (2013) Astrobiology, history, and society: Life beyond earth and the impact of discovery. *Advances in astrobiology and biogeophysics*. Berlin, Heidelberg: Springer.

Van Creveld, M. (1991). *Technology and war: From 2000 B.C. to the present*. New York: The Free Press.

von Neumann, J., & Burks A. W. (1966). *Theory of self-reproducing automata*. Urbana: University of Illinois Press.

Wadley, R. L. (2003). Lethal treachery and the imbalance of power in warfare and feuding. *Journal of Anthropological Research*, 531–554.

Ward, R. E., & Rustow D. A. (Eds.). (1964). *Political modernization in Japan and Turkey*. Princeton: Princeton University Press.

Warren, J. (1978). Who were the Balangingi Samal? Slave raiding and ethnogenesis in 19th-century sulu. *The Journal of Asian Studies, 37*(3), 479–490.

Weber, M. (1964). The routinization of charisma. In A. Etzioni, and E. Etzioni (Eds) *Social Change*, (pp. 53–63). New York: Basic Books.

Weigel, M. M., & Coe, K. (2013). Impact of extraterrestrial life discovery for third world societies: Anthropological and public health considerations. In D. A. Vakoch (Ed.), *Advances in astrobiology and biogeophysics* (pp. 283–299). Berlin, Heidelberg: Springer.

Wickramasinghe, C. (2010). The astrobiological case for our cosmic ancestry. *International Journal of Astrobiology, 9*(2), 119–129.

Wiker, B. D. (2009). "Alien Ideas: Christianity and the Search for Extraterrestrial Life." Inside Catholic: Spiritual and Other Discussions.

World Bank. (2016). Armed forces personnel, total. Retreived July 1, 2016 from http://data.worldbank.org/indicator/MS.MIL.TOTL.P1.

Yonazi, E., Kelly, T., et al. (2012) *The transformational use of information and communication technologies in Africa*. Washington DC: World Bank Publications.

Zakir, J., Tom S., & Kristi, B. (2015). Big data analytics. *Issues in Information Systems*. Feb 16, 2015.

Zubek, T. J. (1961). Theological questions on space creatures. *The American Ecclesiastical Review, 145*, 393–399.

Chapter 14
Final Words

This book revolves around two premises. The first, which is contentious enough, is that N > 1. That is, there are other intelligences somewhere in the Galaxy. And that we are at the first stages of contacting them (or they us). The second is somewhat more insidious: just as there are physical and chemical regularities which, insofar as is known by humans, are true throughout the universe, so too there are social and cultural ones, which will be expressed in ETI behaviors and productions. This is not to say that these regularities express themselves in the same way. Far from it. The regularities—of the consequences of reproductive processes, organization, and practice—are merely frames that allow for infinite variety. However, if we peer beneath surface appearances, this book argues we shall see similar underpinnings.

Is it possible to provide a protocol, a set of flexible steps to allow *homo sapiens* to communicate successfully with ETI? *On the whole* humanity has been notoriously unsuccessful in interacting with other cultures, not to mention species, in a congenial and mutually supportive manner. The results of such contacts as have occurred between humans have been rather depressing. "Do not do unto others what is distasteful to yourself" is a dictum coined by Hillel the Elder two millennia ago. But the first step in making it work is convincing "you" whomever you are, that "other" is sufficiently like yourself to merit consideration. The cetacea, all other primates, and almost all communities of humankind have been at the receiving end of not being recognized as sufficiently human to merit consideration.

14.1 Principles

The first question is therefore the question of recognition. At the danger of being considered a woolly-minded intellectual, I would suggest that in all cases, erring on the side of caution is the prudent thing to do. That is, if we, as a species, believe in any way that a particular other species outside Earth is an intelligent life form, we had better treat them as such *pro tem* at least. At a minimum, we ought to keep in

M. Ashkenazi, *What We Know About Extraterrestrial Intelligence*,
Space and Society, DOI 10.1007/978-3-319-44456-7_14

mind that even weak individuals, not to mention species, might have strong brothers. More important is the fact that alien insights, activities, knowledge, are of value in themselves, not to mention the benefits we might garner from them.

Some identifying criteria for intelligence have been suggested above. All of them (and others) would have to be tried before an extraterrestrial species could be considered non-intelligent, rather than the reverse. In other words, on practical grounds I am advocating that we assume sentience, and try to disprove it carefully, rather than the reverse.

Science Fiction on television to the contrary, it is highly unlikely that we would run across a single representative of any ETI (though this is different from recognizing that we will be in contact with *individual* members of a species). The *association* of members of a species (or of several species, if the speculation in Chap. 9 about civilizations is true) in some fashion is likely to be an important clue to the existence of an intelligent species. Technological intelligent species are a product of a certain population density, as well as being its cause. Unusually high concentrations of members of a species, with few natural enemies, and an un-ecological density, would be a sign of more than animal existence. Vast physical heterogeneity in a single collective another.

Artifacts are also signifiers of intelligence, though here, as the bees would teach us, there are a number of side-paths down which we must be careful not to stray. As a general definition, therefore, a species suspect of intelligent is one that exists with several members in concert in an artificial environment (begging the question of what is artificial, for the moment).

I would suggest that play—consistent evidence of non-purposive behavior—is important. This begs the question of identifying purposive behavior, and is made more complicated by the mix of purposive and non-purposive behavior in human play (consider "Who wants to be a millionaire?" and football games). Nonetheless, where a species is able to devote significant resources to play, notably such spectacles as sport and other festive activities that consume, but do not produce resources, we ought to assume sentience and even technology until proven otherwise.

A second, no less important, question is "Now what?" What do we do with an ETI species? Fortunately, in some hypothetical cases (more powerful/technological than us, and equivalent to us) the answer to the question is not solely up to us: it is interactive. We must come to some sort of accommodation in order to define our own goals. We need then to clarify our own goals first, and the nature of the other end second.

Well then, what *are* humanity's goals? In other words, in what way do we define ourselves as human? This of course presupposes that we have many goals that we share with other species, intelligent as well as not: the biological imperatives built into our genes are incontestable, though it is possible to overcome them. The broader the definition of "human" the more likely we are to share it with other species.

At the "animal" (non-sophont) level, we can presume that any species (let alone intelligent) will be survival oriented. This orientation is a matter of aggregation of individual actions. *Individuals* are survival-oriented, and therefore the species is too, whether successful or not. Intelligent life-forms, one may assume, are *conscious* about their survival as a species. Survival of individuals is not sufficient. Perhaps, at a higher level, as Michaud (1982) states, all intelligent species are bound together in an unconscious recognition of shared interest—their interest in countering, surviving, or surmounting entropy. Conscious recognition of shared fate may well be a higher level of intelligence. Certainly, however, beyond survival there are other goals.

14.1.1 Defining Ourselves and Our Objectives

The idea of defining "human" is one which has raised passions and created more trouble than most other issues. For an observant Christian, for example, humanity is defined by the presence and existence of a non-material, meta-physical element peculiar to humanity, called a soul. Possession of a soul is *ipso facto* a definition of humanity, and indeed, the absence of a soul, or the presence of an improper soul has historically been sufficient cause for regarding humans as objects. Thus, soul possession defines the nature of human goals: refining and improving one's soul. Other related religions do not have the same problem. Islam would most probably see no problem in accepting ETI converts (Ashkenazi 1992), whereas Judaism would probably not accept them any more than it accepts other converts: the first is less concerned with souls and more with ritual and obedience to God, the latter more concerned with group membership defined by ritual.

Being a harsh rationalist, and in the absence of any substantial proof of souls (at least scientifically), I would be forced to define *human* goals, rather than theological ones. At a basic level the goal of humanity, as a species, is to flourish. That implies a delicate balance between numbers (at a basic biological level there is safety for a species in numbers, avoiding catastrophe at the demise of individuals) and quality. Quality, in itself, demands diversity, demands the ability of individuals to choose and experiment with various ecological niches. Now, here we are not speaking solely about a biological species, but about something far more complex: ILFs. "Choice", "experimentation", "ecological niche" have far broader implications for ILFs than they do for other biological species. *Because* ILFs manufacture (at least partly, e.g. in space, and possibly completely, if gene engineering technology and AI are available) their own ecologies, their ecological relationship is far more complex, far more synergistic than that of other biological species.

There is a synergistic effect in human culture: its whole is larger than its parts. Partly this appears to be so because we really do not yet know, how human culture is constructed. Partly this is because of our isolation as an ILF, and the consequent arrogance that derives from it. Partly, however, we recognize something—spirituality? humanity?—that defines humans as something distinct.

14.1.2 Knowing ETI

Just as we need to know ourselves and our objectives, so too do we need to know our opposites.

1. The potential variability of ETI is too great to be encompassed by any simple model, even if we rely only on our own experience of contact with other cultures.
2. It is possible to make some branching assumptions about ETI, but *only* if we are ready to abandon them at a moment's notice: they are merely platforms for our ignorance until additional data turn up.
3. Whatever ETI do, it will be logical, but within *their* frame of reference. It would be a mistake, perhaps a deadly one, to view their actions or possible actions from our point of view.

Superficial knowledge: What they look like, what they say To most humans, the first question to be asked will be "What do they look like?" We are a visual species, and, no less, we create stereotypes on the basis of visual cues. This poses obvious risks we need to keep in mind. First, for humans, there is a wide emotional difference between seeing Little Green Men (or women) and seeing Bug-Eyed Monsters. The first are cute, or wise, or inscrutable. The latter are icky, slavering, and dangerous. The physical shape of our correspondents will affect the way humans view them. More importantly, as we have seen, ETI may not even look like a single race. This will force the average human to reassess—positively and negatively—their view of ETI, as well, hopefully, their view of their fellow humans.

The second question most will ask will be "What do they say?" which expresses both hope and fear. We fear change, we respect the word, we want better, and we are universally predisposed to want to hear good things. ETI may not be benevolent, may not be truthful, and may well maneuver to get things from us we are reluctant to give. So assumptions of benevolent super-beings need to be treated with great caution, along, hopefully, with an acceptance of the first rule of Zen: "Shit happens". If it does, we must be ready to fix things, and we must recognize that accommodating ourselves to more powerful ETI agencies may be the order of the day.

Deep Knowledge: How does their culture work We don't *really* know how our own culture(s) work, and how the puzzles of human societies and cultures fit together. Nonetheless, it is essential that we try to peer beyond the veil of self-presentation that ETI will no doubt wear, since they are thinking social beings. Developing both a willingness, and the tools, to understand ourselves better, is the first step towards using and applying those tools to ETI as well. Certainly, the more we know about our correspondents, the more successful we shall be in dealing with them.

If they're civilized, can we join the game? For ETI that are not 'transcendent' (and thus beyond human analytical powers), I've suggested that life as a multi-stellar medium will evolve into a civilization. Can we fit into that particular form? No less importantly, can we give up our cherished somatic and physiological realities—our shape, skin colors, size, the 'purity' of our bodies—and our cherished social institutions—religion, political states, heterosexual families—in favor of 'polysomatism' where neither sex, gender, race, or social membership are fixed and immutable, and where constant change and ferment are the norm? I argue that in any case, that is the fate that awaits humanity if scientific and technical changes we experience today continue. We may as well accept the fact that many of our most treasured institutions, habits and forms will, eventually, be changed by our association with an ETI civilization. Otherwise we risk being relegated to the role of an outlandish, ineffectual, peasant community. Like the Amish in Pennsylvania, a cute, nostalgia-triggering, relict of the past.

14.2 Final Recap

There are a number of fundamental points which, if nothing else, I would like the reader to go away with.

1. We *do* know some things about ETI even before we meet them, for the simple reason that certain problems—acquiring and distributing energy, organizing life, reproducing, social continuity—will be faced by any organism in the universe. This does not prejudge the myriads of solutions to these problems, but at least we have a hint as to where to look.
2. An intelligent species that develops technology in the broadest sense (biological, chemical, and other sciences included) will gradually, over a period of indeterminate time, improve these sciences, and be forced to develop other sciences to complement those.
3. There are a finite set of ways to respond to scientific/technological development, but all require *social* adaptations, changes, and development as well.
4. The benefits of certain technologies—genetic engineering, space sciences, computerization, nano-technology—are so great that any intelligent technological race will develop them at some point.
5. The development of these 'second wave' sciences will almost automatically create both greater connectivity, and greater heterogeneity in a culture.
6. From what we know of cultures, they are unlikely to remain unchanged and immutable under the pressures of new ecologies brought about by scientific and technological development.
7. Civilizations that might emerge as super-cultures will likely last far longer than the cultures that make them up, and probably will have niches for all variations from whatever norms established, humans included.
8. It would be unwise, however, to assume that any advanced civilization (in the sense used in this book) will be benevolent and nurturing: humans will have to find their place.

9. There are no free lunches.
10. When we do contact ETI, who are likely to be in advance of us technologically, we need to obey three rules: learn as much as possible about 'them'; know as much as possible about ourselves and our objectives; speak softly and listen carefully.

References

Ashkenazi, M. (1992). Not the sons of Adam: Religious response to ETI. *Space Policy, 8*(4), 341–350.

Michaud, M. A. G. (1982). The last question. *Cosmic Search,* First half, 6–7.

References

Ashkenazi, M. (1989). Japanization, internationalization and aesthetics in the Japanese Meal. In A. Boscaro et al. (Ed.), *Rethinking Japan*. London: Paul Norbury.
Ashkenazi, M. (1991). Traditional small group organization and cultural modelling in modern Japan. *Human Organization, 50*(4), 385–392.
Ashkenazi, M. (2000). Swords, collectors and kula exchanges. In M. Ashkenazi & J. Clammer (Eds.), *Consumption and material culture in contemporary Japan* (pp. 118–138). London: Kegan Paul International.
Astronautics, International Academy of. "Declaration of Principles Concerning Activities Following the Detection of Extraterrestrial Intelligence." (ND).
Barth, F. (1975). *Ritual and knowledge among the Baktaman of New Guinea*. Oslo: Universitetsforlaget.
Bloh, W., & Cuntz, M., et al. (2009). Habitability of super-earth planets around other suns: Models including red giant branch evolution. *Astrobiology, 9*(6), 593–602.
Carrigan, R. A. (2012). Is interstellar archeology possible. *Acta Astronautica, 78*, 121–126.
Crane, D. (2012). *Fashion and its social agendas: Class, gender, and identity in clothing*. Chicago: University of Chicago Press.
Dick, S. J. (2004). The new universe, destiny of life, and the cultural implications. In J. Seckbach, J. Chela-Flores, T. Owen & F. Raulin (Eds.), *Life in the universe: From the Miller experiment to the search for life on other worlds* (pp. 319–326). Berlin and Heidelberg: Springer.
Edmondson, W. (2012). The intelligence in ETI: What can we know? *Acta Astronautica, 78*, 37–42.
Finney, B. R., & Jones, E. M. *The impact of contact*. IAF Congress paper. IAA 86-4711986.
Finney, B. R., & Eric, M. J. (eds). (1985). *Interstellar migration and the human experience*. Berkeley: University of California Press.
Flores, M. C. L. (2015). Convergent evolution and the search for biosignatures within the solar system and beyond. *Acta Astronautica*. http://www.sciencedirect.com/science/article/pii/S0094576514004457. Accessed March 12, 2016
Flynn, J. R. (2006). Tethering the elephant: Capital cases, IQ, and the Flynn effect. *Psychology, Public Policy, and Law, 12*(2), 170–189.
Freitas, R. A. (1983). Terraforming Mars and Venus using machine self-replicating systems (SRS). *Journal of the British Interplanetary Society, 36*, 139–142.
Freitas, R. A. (2013). *Xenology: An introduction to the scientific study of extraterrestrial life, intelligence, and civilization*. Sacramento CA, ND: 1st Xenology Research Institute. (ND) http://www.xenology.info/Xeno.htm. Accessed June 1, 2013.
Gould, S. J. (1991). *Bully for brontosaurus*. Boston: W. W. Norton.
Ioirysh, A. I. (1979). The problem of extraterrestrial civilizations and metalaw. *Soviet Law and Government, 18*(1), 79–94.
Klass, P. J. (1975). *UFOs explained*. New York: Random House Inc.

M. Ashkenazi, *What We Know About Extraterrestrial Intelligence*,
Space and Society, DOI 10.1007/978-3-319-44456-7

Kroeber, A. L. (1919). On the principle of order in civilization as exemplified by changes of fashion. *American Anthropologist, New Series, 21*(3), 235–263.
Laumer, K. Retief series (1963–2005).
Lowe, J. G., & Elizabeth, L. (1982). Cultural pattern and process: A study of stylistic change in womens' dress. *American Anthropologist, 84*(3), 521–544.
MacDonald, L. W., & Vince, J. (Eds.). (1994). *Interacting with virtual environments*. Chichester, UK: Wiley.
Maruyama, M. (1976). Social and political interactions among extraterrestrial human communities —contrasting models. *Technological forecasting and social change, 9*(4), 349–360.
Matía, I., González-Camacho, F., Herranz, R., Kiss, J. Z., Gasset, G., van Loon, J. J., et al. (2010). Plant cell proliferation and growth are altered by microgravity conditions in spaceflight. *Journal of plant physiology, 167*(3), 184–193.
Matloff, G. L., Johnson, L., & Bangs, C. (2007). *Living off the land in space: Green roads to the cosmos*. New York: Springer.
Niven, L. *Protector; Tales of known space; The Draco Tavern; Ringworld.*
Pinker, S. (2002). *The blank slate: The modern denial of human nature*. New York: Viking.
Pratchett, T. *Guards! Guards!*.
Radcliffe-Brown, A. R. (1940). On social structure. *The Journal of the Royal Anthropological Institute of Great Britain and Ireland, 70*(1), 1–12.
Redmond, R. (2008). *The folk dances of Shotokan*. Holly Springs, GA: Rob Redmond.
Schönbeck, Y., et al. (2012). The World's tallest nation has stopped growing taller: The height of Dutch children From 1955 to 2009. *Pediatric Research, 73*(3), 371–377.
Seedhouse, E. (2009). *Martian outpost: The challenges of establishing a human settlement on Mars*. Berlin: Springer.
Shahar, M. (2008). *The Shaolin monastery: History, religion, and the Chinese martial arts*. Honolulu: University of Hawaii Press.
Sukumar, R. (2003). *The living elephants : Evolutionary ecology, behavior, and conservation*. Oxford: Oxford University Press.
Sullivan, W. (1976). *Astronomer fears hostile attack*, Would Keep Life on Earth a Secret. *New York Times*.
Sullivan, W. (1979). What if we succeed? *Cosmic Search, 1979*, 37–39.
Swartz, M., Turner, V., & Tuden, A. (Eds.). (1966). *Political anthropology*. Chicago: Aldine.
Vallentine, J. W. (1985). The origins of evolutionary novelty and galactic colonization. In B. R. Finney & E. M. Jones (Eds.), *Interstellar migration and the human experience* (pp. 266–276). Berkeley: University of California.
Van Creveld, M. (2008). *The culture of war*. New York: Presidio Press.
Wiker, B. D. (2009). Alien ideas: Christianity and the search for extraterrestrial life. *Inside Catholic: Spiritual and Other Discussions* 6/27/09 (2009). http://www.maryqueenchurch.org/files/theology-Series/alien_ideas.pdf. Retrieved June 15, 2016.

Index

M. Ashkenazi, *What We Know About Extraterrestrial Intelligence*,
Space and Society, DOI 10.1007/978-3-319-44456-7

MIX
Papier aus verantwortungsvollen Quellen
Paper from responsible sources
FSC® C105338

If you have any concerns about our products,
you can contact us on
ProductSafety@springernature.com

In case Publisher is established outside the EU,
the EU authorized representative is:
Springer Nature Customer Service Center GmbH
Europaplatz 3, 69115 Heidelberg, Germany

Printed by Libri Plureos GmbH
in Hamburg, Germany